Lecture Notes in Geoinformation and Cartography

Series Editors: William Cartwright, Georg Gartner, Liqiu Meng, Michael P. Peterson

Georg Gartner • Felix Ortag
(Editors)

Advances in Location-Based Services

8th International Symposium on Location-Based Services, Vienna 2011

8th Symposium on Location-Based Services
Vienna, 21–23 November 2011
LBS 2011

Springer

Editors
Georg Gartner
Felix Ortag
Institute of Geoinformation and Cartography
Vienna University of Technology
Erzherzog-Johann-Platz 1
1040 Vienna
Austria
georg.gartner@tuwien.ac.at
felix.ortag@tuwien.ac.at

ISSN 1863-2246 e-ISSN 1863-2351
ISBN 978-3-642-24197-0 e-ISBN 978-3-642-24198-7
DOI 10.1007/ 978-3-642-24198-7
Springer Heidelberg Dordrecht London New York

Library of Congress Control Number: 2011940421

Typesetting: Camera-ready by Felix Ortag

Printed on acid-free paper

Springer is part of Springer Science+Business Media (www.springer.com)

8th Symposium on Location-Based Services
Vienna, 21-23 November 2011
LBS 2011

Preface

This book provides a collection of contemporary research papers related to Location-based Services. These activities have emerged over the last years, especially around issues of positioning, sensor fusion, spatial modelling, cartographic communication as well as in the fields of ubiquitious cartog-raphy, geo-pervasive services, user-centered modelling and context-awareness. The innovative and contemporary character of these topics has lead to a great variety of interdisciplinary contributions, from academia to business, from computer science to geodesy. Topics cover an enormous range with heterogenous relationships to the main book issues. Whilst contemporary cartography aims at looking at new and efficient ways for communicating spatial information, the development and availability of technologies like mobile networking, mobile devices or short-range sensors lead to interesting new possibilities for achieving this aim. By trying to make use of available technologies, cartography and a variety of related disciplines look specifically at user-centered and context-aware system development, as well as new forms of supporting wayfinding and navigation systems.

The chapters of this book are a selection of full-text blind reviewed papers submitted to the 8th International Conference on Location-Based Services in Vienna in November 2011. We are grateful to all colleagues who helped with their critical reviews. Please find a list of their names after the table of contents.

The symposia on LBS and TeleCartography have been held at

- 2002 Vienna, Austria
- 2004 Vienna, Austria
- 2005 Vienna, Austria
- 2007 HongKong, China
- 2008 Salzburg, Austria
- 2009 Nottingham, United Kingdom
- 2010 Guangzhou, China
- 2011 Vienna, Austria
- 2012 Munich, Germany (planned)

The meetings themselves were a response to technological developments in miniaturizing devices for telecommunication, computing and display and an increased interest in both incorporating cartographic presentations on such mobile devices and developing services that are specific to a particular location. The broad variety of disciplines involved in this research and the differences in approaching the basic problems is probably typical of a developing field of interdisciplinary research. However, some main areas of research and development in the emerging area of LBS can now be identified. The contributions to this book reflect the main areas of interest including positioning, modelling and awareness, visualisation and cartographic communication and application development.

The production of this book would not have been possible without the professional and formidable work of Manuela Schmidt and Huang Haosheng. The editors are grateful for their help.

Georg Gartner and Felix Ortag
September 2011 in Vienna, Austria

Table of Contents

List of Reviewers

Rein Ahas, University of Tartu, Estionia
Gennady Andrienko, Fraunhofer Institute IAIS, Germany
Thierry Badard, Laval University, Canada
William Cartwright, RMIT University, Australia
Keith C. Clarke, UC Santa Barbara, USA
Peter Fröhlich, Telecommunications Research Center Vienna, Austria
Georg Gartner, Vienna University of Technology, Austria
Muki Haklay, UCL, UK
Haosheng Huang, Vienna University of Technology, Austria
Mike Jackson, University of Nottingham, UK
Hassan Karimi, University of Pittsburgh, USA
Matthias Kranz, Munich University of Technology, Germany
Jukka Krisp, Munich University of Technology, Germany
Yan Li, South China Normal University, China
Chun Liu, Tongji University, China
Xiaolin Meng, University of Nottingham, UK
Liqiu Meng, Munich University of Technology, Germany
Peter Mooney, National University of Ireland Maynooth, Ireland
Takashi Morita, Hosei University, Japan
Felix Ortag, Vienna University of Technology, Austria
Michael Peterson, University of Nebraska at Omaha, USA
Martin Raubal, Swiss Federal Institute of Technology Zurich, Switzerland
Karl Rehrl, Salzburg Research, Austria
Günther Retscher, Vienna University of Technology, Austria
Tapani Sarjakoski, Finnish Geodetic Institute, Finland
Manuela Schmidt, Vienna University of Technology, Austria
Johannes Schöning, German Research Center for Artificial Intelligence, Germany
Josef Strobl, University of Salzburg, Austria
Stefan Van Der Spek, TU Delft, Netherlands
Kirsi Virrantaus, Aalto University, Finland
Stephan Winter, University of Melbourne, Australia
Kefei Zhang, RMIT University, Australia
Sisi Zlatanova, Delft University of Technology, Netherlands

Section I

Visualization and Context Awareness

Chapter 1

Pedestrian Navigation with Augmented Reality, Voice and Digital Map: Results from a Field Study assessing Performance and User Experience

Karl Rehrl[1], Elisabeth Häusler[1], Renate Steinmann[1], Sven Leitinger[1], Daniel Bell[2] and Michael Weber[3]

[1] Salzburg Research, Salzburg, Austria
{karl.rehrl, elisabeth.haeusler, renate.steinmann, sven.leitinger}
@salzburgresearch.at
[2] Factum OHG, Vienna, Austria
daniel.bell@factum.at
[3] Traffic Consultants GmbH, Salzburg, Austria
weber@trafficon.eu

Abstract

This paper reports on a field study comparing the interface technologies voice, digital map, and augmented reality in the context of pedestrian navigation. The in situ study was conducted in the city of Salzburg using a self-implemented smartphone application running on Apple's iPhone 4. The study involved 24 participants between 22 and 66 years with different experiences in using smartphones and navigation systems. Results show that in the context of GPS-enhanced pedestrian navigation digital map and voice-only interfaces lead to significantly better navigation performance and user experience in comparison to an augmented reality interface. The study also reveals similar results for digital map and voice-only interfaces given that voice instructions are carefully composed.

Keywords: Pedestrian navigation, field study, GPS, smartphone application, augmented reality, voice, digital map

G. Gartner, F. Ortag (Eds.), *Advances in Location-Based Services*, Lecture Notes in Geoinformation and Cartography, DOI 10.1007/ 978-3-642-24198-7_1,

1.1 Introduction

Smartphone applications supporting people in everyday navigation tasks are becoming more and more popular. While early implementations primarily addressed the domain of car navigation, recent developments put the focus on pedestrians, too (Arikawa, Konomi & Ohnishi 2007). There are two main reasons for this trend: The permanent availability of the Smartphone as personal assistant and the technical equipment of Smartphone devices with high resolution touch displays, GPS sensors and magnetometers, being perfectly suited to support personal navigation tasks. Although Smartphones are, from a technical point of view, capable with electronic navigation assistance, the question of effective communication of route knowledge remains. Following on to previous findings, this work reports on a field study comparing the performance and user experience of electronic navigation support for pedestrians in a real-world navigation task with the interface technologies voice-only, digital map, and augmented reality. As far as we know, this is the first empirical study that compares the three technologies under real-world conditions.

1.2 Related Work

In the discussion of related work we focus on empirical field studies assessing the performance and user experience of electronic navigation support. Generally, route communication in the context of electronic navigation assistance can either be (1) map-based, (2) photograph-based, (3) text-based, (4) voice-based, (5) tactile, (6) augmented reality overlay and (7) arbitrary combinations of these basic kinds. Most of the previous work in the context of pedestrian navigation has focused on map-based or text-based guidance (Baus, Cheverst & Kray 2005). Electronic map displays have been criticized due to the small display size (Dillemuth 2005) as well as problems of users aligning 2D maps to the real world. 3D maps do not lead to considerable improvements (Coors et al. 2005). Text-based guidance has been found to work basically well, but suffers from the drawback that people have to read the texts during their walk. Tactile guidance, as minimal attention interface, solves this problem and has been successfully proven in previous studies (Heuten et al. 2008). Augmenting navigation directions with street scene photographs has been tested by several authors. Chittaro & Burigat 2005 augmented audio directions with three different kinds of visual directions: (1) map, (2) combination of a map and photographs and (3) combination of large arrows and photographs and revealed that users find their way more quickly if the map is augmented with photographs or completely replaced with photographs and arrows. However, in a subjective rating, users prefer a combination of map and photographs. A recent study by Ishikawa & Yamazaki 2009 compares the effectiveness of maps and street scene photographs for providing orientation at underground exits. In this study, participants show higher performance with photographs compared to maps. Both studies do not take into account voice-only navigation or augmented reality interfaces, which have recently gained attention due to capable Smartphones equipped with enabling technologies (e.g. GPS, magnetometer, tilt sensor). To our knowledge, there are currently no field studies proving the effectiveness of augmented reality overlays for pedestrian navigation.

Stark, Riebeck & Kawalek 2007 tested four navigational concepts: (1) auditory instructions plus digital, dynamic route, (2) digital, dynamic route, (3) map with position and direction and (4) textual description by street name. The authors conclude that none of the concepts could be identified as the optimal one. Depending on the usage scenario, each of the concepts has its benefits. Nevertheless, the authors give hints on best practices for interface design. Goodman, Brewster & Gray 2005 investigated electronic navigation support for elderly people with the navigation concepts text, speech, text and speech and photographic instructions and came to the conclusion that elderly people benefit from using an electronic navigation aid with spoken and/or written instructions. However, the authors conclude that pedestrians with voice-only guidance (without GPS support) perform significantly worse than with other kinds of navigation support. Ishikawa et al. 2008 confirmed the poor performance of GPS-based navigation support in comparison to maps and direct experience. Although the authors tackled the aspect of GPS positioning, they did not separate voice from map-guidance. Additionally, Goodman, Brewster & Gray 2005 indicate that more research is necessary to compose high-quality voice instructions. This gap was closed with a series of studies addressing voice-only electronic navigation support (Rehrl et al. 2009, Rehrl, Häusler & Leitinger 2010). The authors addressed several shortcomings of voice instructions in the context of electronic navigation support and propose a method for composing cognitively ergonomic turn instructions (Klippel, Richter & Hansen 2009). A field study revealed that with a carefully composed instruction set voice-only navigation at minimal error rate and at standard walking times is feasible.

Summarizing previous work on navigation performance, we conclude that there is no satisfying answer to the question which impact each of the technologies has on navigation performance and user experience. While some of the studies reveal better results for photographs, the subjective rating for maps is typically better. However, most studies end up in mixing more than one technology so that no conclusions can be drawn for a single technology. Navigation performance with voice-only guidance is generally considered poor, although voice instructions are mainly used in addition to map display or textual instructions and not tested in voice-only settings. One of the first studies investigating voice-only guidance has been recently published, but does not compare the results with other interface technologies (Rehrl, Häusler & Leitinger 2010). Pedestrian navigation systems using augmented reality interfaces have not undergone empirical field studies so far. Findings concerning display layouts are rather preliminary (e.g. Shi, Mandel & Ross 2007) and do not provide best practice guidelines. Thus, the review of related work motivates a field study comparing the three technologies map display, voice-only guidance and augmented reality display without mixing the different modes. The study should contribute to the question, whether significant differences in navigation performance and user experiences can be attributed to one of the interface technologies.

1.3 Implementation

For studying navigation performance and user experience with different interface technologies we used a self-implemented mobile application running on Apple's iPhone 4 (due to the outstanding user experience and due to improved GPS positioning). The application is capable of switching between the three navigation modes (1) multi touch map interface, (2) voice-based interface and (3) augmented reality interface. The decision for self-implementing the navigation application is mainly based on the needs coming from the research method of user experience prototyping. It is envisaged to iteratively improve the application with the results from user studies, which can best be achieved if any design decision can be revised and changed at any time.

1.3.1 Digital Map

For designing the map user interface we considered related work of Agrawala & Stolte 2000 listing four essential design goals for effective route maps: *readability* (all essential components, especially the roads, should be visible and easily identifiable), *clarity* (the route should be clearly marked and readily apparent from a quick glance), *completeness* (the map must provide all necessary information for navigation), and *convenience* (the map should be easy to carry and manipulate). Based on these design goals, further literature (Gartner & Radoczky 2007, Gartner & Uhlirz 2005, Schmid et al. 2010) provides useful hints for route map design such as (1) providing an overview of the whole route at the beginning of the navigation task and also during route following, (2) automatic adaptation of the presented map section to the position of the user, (3) egocentric map view, (4) the route should be visible to the user at any time, and (5) the distinction between the past and the future path should be unambiguous. Finally, literature concerning generalization, scale, extend, colors was considered for map design (Darken & Peterson 2002, Dillemuth 2005). The aforementioned features of the map display were implemented as custom overlays.

Fig. 1.1. Map interface with route overview, egocentric view, distinction between past and future path, zooming and scrolling function, and turn direction arrow.

The route was visualized with a red line and a pattern of small white arrows pointing in the forward direction. To clearly separate the past and the future route at the current position, the two route sections are tinted differently. As suggested in related work, the already completed route part is displayed in the lower third of the screen, whereas the coming route part gets the two upper thirds of the screen. Next in the layer hierarchy is the visualization of the current position with an appropriate white/black arrow icon (egocentric view). The current position is determined by GPS signals, using a route matching algorithm to filter inaccuracies. On top of the layer stack a semi-transparent, blue-white directional arrow representing turn directions based on a 7-sector model (Klippel et al. 2004) fades in whenever a user is in range of a decision point. Finally, the user interface allows – by pressing the button in the middle of the header – to fade out the route for a better readability of street names on the map tiles (see *Figure 1.1*).

1.3.2 Voice-only Guidance

The audio-based interface for voice-only guidance is based on the wayfinding choreme theory (Klippel et al. 2005) as well as on the findings of previous studies (Klippel et al. 2005, Rehrl et al. 2009, Rehrl, Häusler & Leitinger 2010). Our proposed approach for generating cognitively ergonomic voice instructions is called activity-graph, which is mainly based on a topological route graph (Walther-Franks 2007) consisting of decision points and route segments connecting these decision points. For an activity-based modeling of routes, the route graph model has been extended with the concept of navigation activities, describing the course information as qualitative spatial actions (Shi, Mandel & Ross 2007). Following this approach, navigation activities are modeled with a sequence of one to four qualitative spatial actions. Each action consists of (1) an action verb (e.g. walk or turn), (2) an intrinsic (e.g. left, right) or extrinsic (e.g. across, in, to) spatial relation and (3) an optional spatial reference (e.g. a visible entity like a bridge or a prominent building). An example activity with a sequence of three actions is (1) "Walk half right" (2) "Walk in the direction of the hotel 'Crowne Plaza'" (3) "Walk to the roundabout".

For being processed by our navigation application, it was decided to add the activity-graph for the test routes to the OpenStreetMap database. Due to the open data model of OpenStreetMap, it was possible to flexibly cast activity-graphs into the data scheme of OpenStreetMap. We introduced a range of new tags with the namespace prefix *semnav*. When inserting new tags or extending existing ones in OpenStreetMap we left a comment in the appropriate field for informing the community about the purpose of the tags.

The user interface of the voice-only guidance mode consists of a single screen with a slider for controlling the sound volume and a button to offer the possibility to repeat the latest instruction. The voice-only navigation was realized in a way, that whenever a user comes close to a decision point, the application indicates the approaching decision point with a vibration alarm and plays the voice instruction describing the navigation actions from this decision point to the next. The location at which the alarm occurs is calculated by correlating the user's GPS position (respectively that of the mobile device) with the 'sensitive area' around the WGS84 coordinates of the

Fig. 1.2. Graphical indicators being displayed as augmented reality overlay over the camera view and changing their style upon the distance to a decision point.

decision point. A typical radius for the sensitive area is 5 meters, however, in order to compensate for GPS inaccuracies, e.g. in the near of high buildings or in narrow streets, and avoid untimely or delayed voice instructions for the users, the radii of the 'sensitive areas' for some decision points were increased to 10 or 20 meters.

Voice instructions were generated automatically by extracting the activity-graph data from the OpenStreetMap database and transforming the qualitative spatial actions to textual instructions following the rules and standards of a schema-based approach of natural language generation (Reiter & Dale 1997). Finally, the (german) text strings get transformed into voice commands by means of an appropriately configured text-to-speech engine.

1.3.3 Augmented Reality View

Concerning the augmented reality mode, route information is overlaid over the live picture of the iPhone camera. For calculating the correct position of the overlay information on the screen, the application uses the GPS position, the magnetometer and the tilt sensor. Since decision points coming from OpenStreetMap typically are not tagged with height-attributes, the application presets the height of the decision points to the current height calculated from the GPS signal. Concerning the design of the overlay information, we reviewed previous work (Hile & Borriello 2008, Kolbe 2004, Schmalstieg & Reitmayr 2007, Stark, Riebeck & Kawalek 2007) and decided to use implicit route instructions ("follow-me" approach). The basic idea of the "follow me"-approach is to annotate the camera view with a virtual route path. We decided for the "follow me"-approach, since it is an approach coming close to the basic idea of 'augmenting reality'. Our concept of realizing it is based on displaying 2D graphics without any perspective distortion. As shown in *Figure 1.2,* depending on the distance to a decision point, our guidance uses three different styles of graphical indicators: In case the targeted decision point is far away, the indicator is simply a semi-transparent circle, marking the position of the decision point (a). As the user approaches the next decision point, the circle additionally displays the remaining distance in meters (b). Finally, in the proximity of the targeted decision point, the indicator is ring-shaped, enclosing the waypoint to be entered (c). By enlarging the ring when approaching a decision point, users should get the feeling of crossing a portal. Indicators are changing from style one to style two when 40 % of the distance between two decision points is covered. Style three appears when reaching 70 % of the route segment. An additional vision/directional indicator is placed as a static interface element in the left top corner of the screen. It allows to control the overall orientation of the phone and to locate the next decision point at the beginning of a route segment. In addition to the graphical indicators a vibration alarm is raised when a decision point is reached and the subsequent decision point becomes the new target.

1.4 Study Design

Our study design is based on an empirical field test with 24 participants. Participants were asked to navigate along a pre-defined and for them unknown route. The aim of the study was to compare navigation performance as well as user experience in the context of a pedestrian navigation task with assistance of one of the three selected technologies.

The study area was in Salzburg, Austria, where we defined three routes in an inner city district (so called "Andräviertel"). Along the route we identified 27 decision points (a decision point is defined as a location, where a navigation instruction is necessary because of multiple choices to follow the route). The whole route is 1807 meters long and the surroundings of the route are characterized by residential and business areas. The route was split into three sub-routes (sub-route 1: 629 m, sub-route 2: 703 m, sub-route 3: 475 m) so that each person could navigate each of the sub-routes with a different navigation technology (see also *Figure 1.3)*.

1.4.1 Participants

We recruited a total of 24 participants, 12 female and 12 male. We classified our test persons in three age classes: younger than 30, 30–60, and older than 60. The youngest person was 22 years, the oldest 66 years (M = 39.55; SD = 14.91). Furthermore, we classified participants concerning their experience with smartphones in general and especially the experience with the devices in the context of car navigation and pedestrian navigation. None of the participants were familiar with the test route.

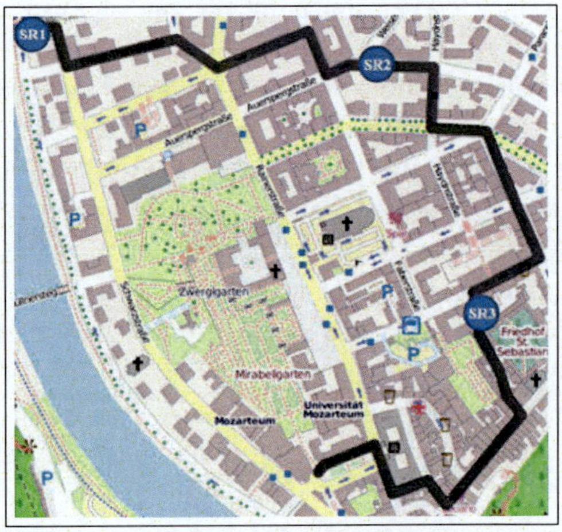

Fig. 1.3. Test route in Salzburg divided into 3 sub-routes.

1.4.2 Experimental Design

Our field test uses a within-participants design. The order in which the navigation technologies were tested along the test route was pre-defined across participants, considering an approximately balanced subdivision concerning mentioned navigation experience. However, we counterbalanced the order of each interface technology across the participants to ensure that each sub-route was completed by 8 participants with each of the three navigation technologies. Each participant was accompanied by two researchers (always the same), one observing the test run and guiding through the interviews and the other collecting quantitative as well as qualitative performance measures (e.g. number of stops or reasons for stops).

1.4.3 Measures

First of all, we collected detailed device logs allowing in-depth analysis of participants' behaviors. This allowed us to measure walking speed, task completion time, GPS accuracy and user interaction behavior. In addition, the observing researcher recorded the number of stops, the duration of stops, and the reasons for stops. At the end of each sub-route, the observing researcher assessed participants' task load as well as user experience and asked quantitative and qualitative questions. Furthermore, the standardized Santa Barbara Sense-Of-Direction Scale (SBSDS) (Hegarty et al. 2002) was used to get insights on the spatial abilities of the participants.

1.4.4 Procedure

At the beginning of the test session each participant had to complete the SBSDS. Next, we explained the basic usage of the pedestrian navigation system to the participant and gave a short demonstration of the prototype on the iPhone. After a brief training session, the test persons were led to the pre-defined starting point of the first sub-route. The navigation system was switched to one of the three interface modes, following a pre-defined plan. The task for the participants was to navigate to the end of the pre-defined route. If participants decided wrong at a decision point, the observing researcher used gestures to indicate the right choice. No other assistance was given during navigation. In order to avoid any influence on participants, the researcher walked several meters behind participants and recorded observations with a voice recorder. When reaching the end point of a sub-route, participants had to answer three questionnaires concerning spatial knowledge, usability and task load. After completing one sub-route, the interface was switched and the same procedure was repeated for the next sub-route. When finishing the experiment participants were asked final questions to gather further qualitative feedback and experiences. Finally, participants were rewarded with an amount of money as a token of appreciation.

1.5 Results and Analysis

All participants could successfully complete the navigation tasks. As indicated before among others we assessed sense of direction, task completion time, stops, duration of stops, task load and system usability.

1.5.1 Self-Report Sense of Direction

For each participant we calculated the mean values of their answers to the fifteen questions of the SBSDS. Following previous work (Ishikawa et al. 2008) we reversed positively stated questions to negatively stated ones so that a higher score means a better sense of direction. The results of the SBSDS ($M = 4.63$; $SD = 0.89$) revealed no significant differences between the 24 participants, which is an indication for a balanced group of participants. As revealed in previous studies, female participants estimated their sense-of-direction worse than male (Female: 4.36 (0.98), Male: 4.91 (0.72)).

1.5.2 Task Completion Time

Task completion time was analyzed according to the three test routes and the three different technologies. Since the sub-routes are not equally long, task completion times vary between the routes. However, as *Table 1.1* reveals, there is a difference in task completion time between the different technologies.

Participants navigating with augmented reality interface needed the longest time to complete one of the three sub-routes. Voice-only guidance was the fastest technology along sub-route 2 and sub-route 3. Digital map technology can be classified in the middle of all three technologies except with sub-route 1 where digital map was faster than voice guidance.

Table 1.1. Average task completion times by sub-route and navigation technology.

Sub-Route	Navigation Technology	Mean [s]	SD [s]
Sub-Route 1	Augmented Reality	524.8	74.3
	Voice	507.3	71.9
	Digital Map	461.1	77.3
Sub-Route 2	Augmented Reality	621.5	94.7
	Voice	541.5	56.4
	Digital Map	551.8	74.2
Sub-Route 3	Augmented Reality	385.0	26.6
	Voice	340.1	25.7
	Digital Map	343.4	67.5

A One Factor ANOVA was computed for all of the three test routes in order to assess the effect of the three interface technologies on walking times. There were no significant results found on either of the three test routes. However, the three different test routes had a significant effect on the walking speeds of the participants ($F(2,69) = 58.483$, $p < 0.001$). Completion time of the navigation task was the fastest on test route 3 ($M = 356.17$ sec.) and the slowest on test route 2 ($M = 571.58$). This is mainly due to the different lengths of the three sub-routes. Furthermore, there was no significant inter subject effect between the sub-routes and the three technologies ($F(4,63) = 0.640, p = 0.636$).

1.5.3 Stops

The number of stops indicates how often participants stopped their walk during navigation. *Table 1.2* compares the mean number of stops and the standard deviation for each sub-route and interface technology.

Again participants navigating with augmented reality had to stop their walk more often compared to participants with voice or digital map. But also with voice-only guidance participants had to stop their walk more often than with digital map guidance. This result is interesting since it was expected that the benefit of voice-guidance is to navigate a route without stopping the walk and should be further discussed.

Table 1.3 shows the mean duration of stops and the standard deviation for each sub-route and navigation technology. Again navigation with augmented reality results in the longest stops, whereas voice-only navigation leads to shorter stops in comparison to augmented reality, but to longer stops in comparison to the digital map. However, special care should be given to the rather high standard deviation, which indicates significant differences between participants.

Table 1.2. Mean number of stops for one person ordered by sub-route and navigation technology.

Sub-Route	Navigation Technology	Mean [# stops]	SD [# stops]
Sub-Route 1	Augmented Reality	7.25	4.464
	Voice	4.25	2.053
	Digital Map	2.50	2.507
Sub-Route 2	Augmented Reality	8.13	3.758
	Voice	4.63	2.326
	Digital Map	2.63	1.506
Sub-Route 3	Augmented Reality	5.50	2.878
	Voice	2.25	1.581
	Digital Map	1.50	2.390

Table 1.3. Mean duration of stops for one person ordered by sub-route and navigation technology.

Sub-Route	Navigation Technology	Mean [s]	SD [s]
Sub-Route 1	Augmented Reality	68.00	64.821
	Voice	44.75	39.978
	Digital Map	23.50	31.650
Sub-Route 2	Augmented Reality	101.25	47.680
	Speech	65.75	52.227
	Digital Map	44.75	35.796
Sub-Route 3	Augmented Reality	30.75	12.992
	Speech	21.50	42.275
	Digital Map	8.63	15.811

Again one Factor ANOVAs were computed in order to assess the effects of each individual sub-route and the three navigation technologies on the number of stops and the duration of the stops. There were no significant results found regarding the effect of the technologies on the duration of the stops, while the different interface technologies significantly affected the number of stops on all three sub-routes:

- Sub-Route 1: $F(2,21) = 4.552, p = 0.023$
- Sub-Route 2: $F(2,21) = 8.531, p = 0.002$
- Sub-Route 3: $F(2,21) = 6.576, p = 0.006$

The sub-routes showed a significant effect on the duration of the stops ($F(2,69) = 8.056, p = 0.001$), with the longest stops observed on sub-route 2 and the shortest on sub-route 3. No significant inter subject effect was found between the navigation technologies and the three sub-routes. Computation of a univariate ANOVA to assess the overall effect of the three technologies again on both number of stops and duration of stops shows significant results regarding both variables:

- Number of stops: $F(2,69) = 18.018, p < 0.001$
- Duration of stops: $F(2,69) = 4.998, p = 0.009$

Analyzing the reasons why participants stopped during navigation revealed interesting results *(Figure 1.4)*. While the usage of the application triggered only few stops (1%, N=3), a number of stops was triggered by various traffic related issues like red traffic lights or intense traffic avoiding the crossing of a street (28%, N=132). However, most of the stops were triggered by orientation (36%, N=170) and GPS positioning issues (36%, N=169) such as poor GPS signal reception or multipath effects.

1.5.4 NASA TLX

In order to assess the cognitive work load caused by the process of navigating with the different interface technologies the NASA Task Load Index (NASA TLX)

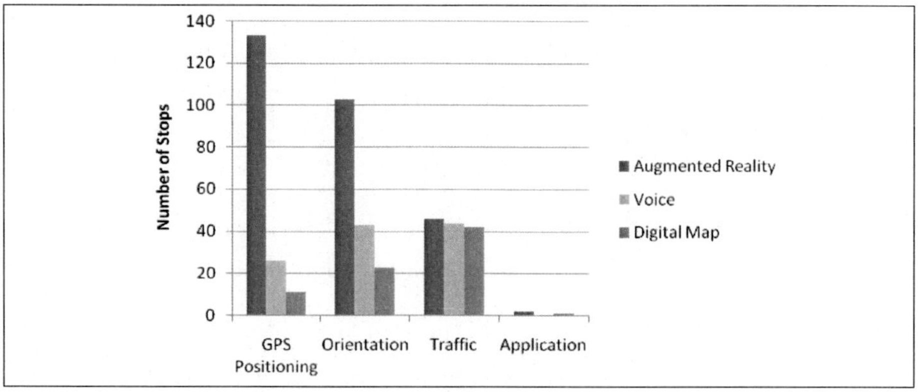

Fig. 1.4. Reasons for stops by navigation technology.

(Hart & Staveland 1988) was collected for all 24 participants. The NASA TLX is a subjective, multidimensional assessment tool that rates perceived workload on six different subscales: Mental Demand, Physical Demand, Temporal Demand, Performance, Effort, and Frustration. Participants filled in the questionnaire after successfully completing a sub-route to assess the effects of the different interface technologies on the cognitive load. As the RAW TLX version of the test was used, no weighing process of the different subscales was applied. Therefore the individual task load scores for each person were computed by summing the scores of each of the 6 scales and calculating an average value.

Due to the distribution of the data in regard to the requirements of multivariate analysis methods, a non-parametric Kruskal-Wallis-H test was performed in order to assess the effect of the interface technology on the six subscales of the NASA TLX and the overall NASA RTLX score. The test yielded one significant result regarding the subscale effort (*Chi-squared* = 6.264, $p = 0.044$). It shows that the augmented reality interface leads to a higher effort when navigating the sub-routes than both other interfaces. Again voice and digital map show no significant differences in regard to the perceived effort during the field test. In order to assess the effects of usage of the three interface technologies on the task load of the participants a One Factor ANOVA was computed.

The results as shown in *Figure 1.5* indicate a significant difference in usage between the three technologies (F (2, 69) = 4.921, $p = 0.010$). Navigation with augmented reality leads to a significantly higher overall task load ($M = 23.125$) than navigation by voice ($M = 14.549$) or digital map ($M = 14.875$). The two navigation interfaces voice and digital map show no significant differences regarding the overall NASA RTLX score.

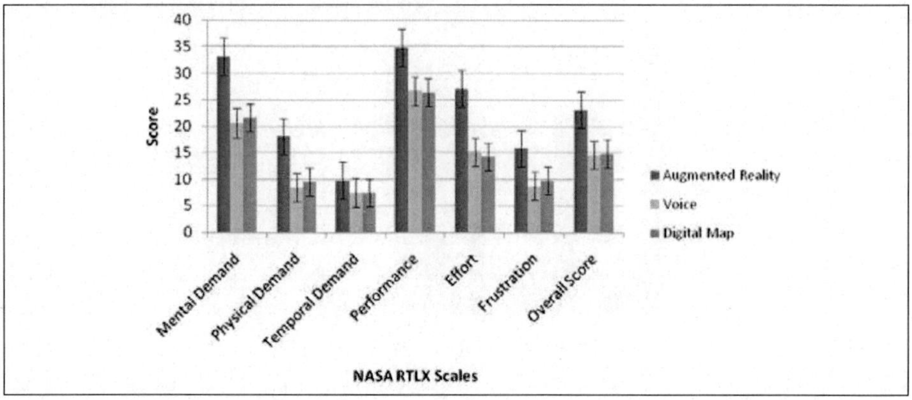

Fig. 1.5. Results of the NASA RTLX for the three tested navigation technologies.

1.5.5 System Usability

In order to assess the usability of the tested navigation technologies the SUS – System Usability Scale (Brooke 1996) – was used. The SUS is a proved and reliable test to assess the usability of a given device or software by determining the three dimensions: satisfaction, efficiency, effectiveness, by asking participants to rate ten Likert scaled items. The SUS test was handed out to the participants in a German version of the original English item battery. The tested technologies yielded significantly different SUS scores (F (2, 69) = 7.349, p = 0.001), with augmented reality yielding the lowest mean SUS Score of 65.83 and with digital map (M SUS Score = 81.25) and voice (M SUS Score = 82.40) showing no significant differences.

The results of the final questionnaire (Please rate the tested system concerning (1) the velocity allowing you to fulfill the navigation task, (2) the fun factor of the application, (3) the requirements defined by the system to you (4) the subjective feeling being on the right track, (5) the overall feeling being guided) confirm the results of the NASA RTLX and the SUS Score. The participants rated augmented reality significantly more negative (*Chi-squared* = 16.333, $p < 0.001$) regarding the overall navigation by the application. Again the differences between the other two navigation technologies did not show any significant differences in regard to overall navigation. The three different age groups participating in the field tests did not show any significant differences regarding the rating of the demands imposed by the technologies, the usability of the technologies or the overall rating.

1.6 Discussion

Our study results show that pedestrian navigation assistance with digital map and voice-only interfaces leads to similar results regarding navigation performance and user experience. In terms of completion time, voice has a slight edge over digital map. However, in terms of stops, participants performed better with a map-based interface. One reason is that people are used to navigate with digital maps and do not feel comfortable if the map is replaced by a voice-only interface. It is worth to mention that the authors spent in-depth research in voice-only guidance in the context of pedestrian navigation in the past. This is the reason why the reported results are generally better than previous findings (e.g. (Goodman, Brewster & Gray 2005). However, several authors came up with the issue that a number of questions have to be answered in the context of voice-only guidance. Some of these questions have been addressed over the last years. Thus we conclude that the implementation of both interface technologies is on similar levels since also the implementation of the digital map interface is based on best practice guidelines published in related work. Although the field study reveals similar navigation performance and user experience of both interface technologies, it is worth to mention that individual participants either prefer the one or the other technology and most of the participants suggest a combination of both interfaces.

In contrast to voice and digital map guidance, augmented reality shows significant drawbacks regarding basically all assessed measures. Participants needed longer to complete the routes, more and longer stops have been observed and the imposed task load was rated higher whereas the system usability was rated lower. A possible reason why study results are not in favor of augmented reality could be that participants had no previous experience with this technology. Also participants who had experience with smartphones in general and in the context of pedestrian navigation became uncomfortable while using the augmented reality interface as they were not familiar with the technology. Especially the handling of the smartphone while navigating with the augmented reality interface revealed severe problems, e.g. some test persons continuously looked through the camera view of the device and thus had a high risk to stumble because of missed obstacles. Maybe in subsequent studies the user interface can be improved in a way that it becomes clear for navigating persons that it is not necessary to permanently look through the camera.

Another important aspect which we assessed in our study is the effect of inaccurate GPS signals of the different user interface technologies. The analysis of reasons for stops *(Figure 1.4)* clearly shows that noisy GPS signals do nearly not have any effect on navigation performance with a digital map. The reason is that on the one hand the GPS noise is filtered by means of map matching algorithms and on the other hand the digital map provides a good overview which is rather robust to positioning inaccuracies. Voice-only guidance is not that robust, since inaccurate GPS signals most likely lead to missing, too early or too late voice instructions. However, with varying radii of sensitive areas at decision points, this shortcoming can be widely compensated. The most error-prone technology is augmented reality, since it requires accurate GPS positions as well as accurate values from the magnetometer and the tilt sensor in order to correctly overlay route information over the live camera view. Effects like GPS bouncing become noticeable for participants and thus had negative effects on users' navigation performance. As a consequence, inaccurate signals caused a high number of stops (see stops with the reason GPS positioning and orientation in *Figure 1.4)*.

Finally, it is worth to mention, that despite of the quantitative results, qualitative finding (SUS, RTLX and final questionnaire) showed that voice and digital map interfaces have a significant edge over augmented reality regarding almost all aspects, including also the assessment of the subjective walking speed.

1.7 Conclusion and Future Work

In this paper we compared navigation performance and user experience of the three user interface technologies augmented reality, voice, and digital map in the context of GPS-enhanced pedestrian navigation. In general, participants succeeded in using our proposed user interfaces and could complete all test routes. We conclude that digital map interfaces are commonly used and thus lead to good navigation performance and user experience. Voice-only navigation, although rated poorer by several authors, leads to similar results given that voice instructions are carefully composed. Navigation with an augmented reality interface leads to significantly worse results, either in terms of navigation performance as well as in terms of user experience. When concluding on these results we have to consider, that people are inexperienced in using augmented reality technology and thus the results could basically mirror this inexperience. Another interpretation is that further research needs to be directed at this technology in order to catch up with best practices of other interface technologies. A third conclusion is, that augmented reality technology still suffers from technological drawbacks of current smartphones such as positioning inaccuracies or inaccuracies of the magnetometer. Image-based augmented reality, when ready for widespread outdoor use, could solve these problems.

In future work we will undergo a second iteration in the user-centered design approach. We will consider open issues as well as further improvements of the user interfaces with the qualitative user feedback from the field test. E.g. we will re-evaluate the digital map design and plan to improve landmark illustrations for better cognition. Concerning augmented reality technology, the most promising direction to follow is a 3D approach, such as indicating the route with 3-dimensional footprints. In this case, a critical issue is to overcome the mentioned GPS problems as well as the intersection of the footprint with 3D building models.

References

Arikawa, M., Konomi, S. and Ohnishi, K. Navitime (2007) Supporting pedestrian navigation in the real world. IEEE Pervasive Computing 6, 3, 21–29

Agrawala, M. and Stolte, C. A (2000) Design and Implementation for Effective Computer-Generated Route Maps. AAI Symposium on Smart Graphics

Baus, J., Cheverst, K. and Kray, C. (2005) A survey of map-based mobile guides. Map-based Mobile Services – Theories, Methods, and Implementations, Springer, 193–209

Brooke, J. SUS (1996) A "quick and dirty" usability scale. Usability Evaluation in Industry, Taylor and Francis

Chittaro, L and Burigat, S. (2005) Augmenting Audio Messages with Visual Directions in Mobile Guides: an Evaluation of Three Approaches. Proc. of the 7th Int. Conference on Human Computer Interaction with Mobile Devices & Services, 107–114

Coors, V., Elting, C., Kray, C. and Laakso, K. (2005) Presenting route instructions on mobile devices: From textual directions to 3D visualization. Exploring Geovisualization, Elsevier, 529–550

Darken, R. P. and B. Peterson (2002) Spatial orientation, wayfinding and representation. Handbook of Virtual Environment Technology. Lawrence Erlbaum Associates

Dillemuth, J. (2005) Map Design Evaluation for Mobile Display. Cartography and Geographic Information Science 32, 285–301

Gartner, G. and Radoczky, V. (2007) About the Role of Cartographic Presentation for Wayfinding. Geographic Hypermedia Concepts and Systems, Lecture Notes in Geoinformation and Cartography, Springer, 381–398

Gartner, G. and Uhlirz, S. (2005) Cartographic Location Based Services. Map-Based Mobile Services – Theories, Methods, and Implementations, Springer, 159–169

Goodman, J., Brewster, S. and Gray, P. (2005) How can we best use landmarks to support older people in navigation? Behaviour & Information Technology 1, 3–20

Hart, S.G. and Staveland, L. E. (1998) Development of NASA-TLX (Task Load Index): Results of empirical and theoretical research. Human mental workload, Elsevier, 139–183

Hegarty, M., Richardson, A.E., Montello, D. R., Lovelace, K., Subbiah, I. (2002) Development of a self-report measure of environmental spatial ability. Intelligence 30, 425 – 447

Heuten, W., Henze, N., Boll, S. and Pielot, M.: Tactile Wayfinder (2008) A Non-Visual Support System for Wayfinding. Proc. of the 5th Nordic Conference on HCI: building bridges, ACM, 172–181

Hile, H. and Borriello, G. (2008) Positioning and Orientation in Indoor Environments Using Camera Phones. IEEE Computer Graphics and Applications 28, 4, ACM Press, 32–39

Ishikawa, T., Fujiwara, H., Imai, O. and Okabe, A. (2008) Wayfinding with a GPS-based mobile navigation system: A comparison with maps and direct experience. Journal of Environmental Psychology 28, 74–82

Ishikawa, T., and Yamazaki, T. (2009) Showing Where To Go By Maps or Pictures: An Empirical Case Study at Subway Exits. COSIT 2009, Springer, 330–341

Klippel, A., Dewey, C., Knauff, M., Richter, K.-F., Montello, D. and Freksa, C. (2004) Direction Concepts in Wayfinding Assistance Systems. Workshop on AI in Mobile Systems, 1–8

Klippel, A., Tappe, T., Kulik, L. and Lee, P. (2005) Wayfinding choremes – A language for modeling conceptual route knowledge. Journal of Visual Languages and Computing, 16, 4, 311–329

Klippel, A., Richter, K.-F. and Hansen, S. (2009) Cognitively ergonomic route directions. Handbook of Research on Geoinformatics, Idea Group Inc., 230–238

Kolbe, T.H. (2004) Augmented Videos and Panoramas for Pedestrian Navigation. Proc. of Symposium of Location Based Services and TeleCartography, Springer

Raubal, M. and Winter, S. (2002) Enriching Wayfinding Instructions with Local Landmarks. Proc. GIScience 2002, Springer, 243–259

Rehrl, K., Gartner, G., Leitinger, S., and Ortag, F. (2009) An analysis of direction and motion concepts in verbal descriptions of route choices. COSIT'09

Rehrl, K, Häusler, E. and Leitinger, S. (2010) Comparing the Effectiveness of GPS-Enhanced Voice Guidance for Pedestrians with Metric- and Landmark-Based Instruction Sets. GIScience 2010, Springer, 189–203

Reiter, E., and Dale, R. (1997) Building applied natural language generation systems. Natural Language Engineering 3, 57–87

Schmalstieg, D. and Reitmayr, G. (2007) Augmented Reality as a Medium for Cartography. Multimedia Cartography, 2nd edition, Springer, 267–282

Schmid, F., Kuntzsch, C., Winter, S., Kazerani, A., and Preisig, B. (2010) Situated Local and Global Orientation in Mobile You-Are-Here Maps. Proc. MobileHCI ′10, ACM Press, 83–92

Shi, H., Mandel, C., and Ross, R. (2007) Interpreting Route Instructions as Qualitative Spatial Actions. Spatial Cognition V, Springer, 327–45

Stark, A., Riebeck, M., and Kawalek, J. (2007) How to Design an Advanced Pedestrian Navigation System: Field Trial Results. Proc. of IEEE International Workshop on Intelligent Data Acquisition and Advanced Computing Systems: Technology and Applications

Walther-Franks, B. (2007) Augmented Reality on Handhelds for Pedestrian Navigation. Master's thesis, University of Bremen

Werner, S., Krieg-Brückner, B., and Herrmann, T. (2000) Modeling navigational knowledge by route graphs. Spatial Cognition II. Integrating Abstract Theories, Empirical Studies, Formal Methods, and Practical Applications, Springer, 295–316

Chapter 2

Mobile Augmented Reality for Tourists – MARFT

Patrick Luley[1], Roland Perko[1], Johannes Weinzerl[2], Lucas Paletta[1] and Alexander Almer[1]

[1] Joanneum Research Forschungsgesellschaft mbH
{patrick.luley, roland.perko, lucas.paletta, alexander.almer}@joanneum.at
[2] c.c.com – Andersen & Moser GmbH
jweinzerl@cccom.at

Abstract

The aim of the project MARFT is to demonstrate the next generation of augmented reality targeting current mass market mobile phones. MARFT sets out to launch an interactive service for tourists visiting mountainous rural regions. During local trips they will be able to explore the surrounding landscape by pointing the lens of the smart-phone camera towards the area of interest. As soon as the view-finder shows the area of interest, the tourist will be able to choose between two products: (i) an augmented photo superimposed with tourist information like hiking tours or lookout points or (ii) a rendered 3D virtual reality view showing the same view as the real photo also augmented with tourist objects. The outstanding step beyond current augmented reality applications is that MARFT is able to augment the reality with *cartographic accuracy.* In addition to the benefit of presenting reliable information, MARFT is able to consider the visibility of objects and further to work completely offline in order to avoid roaming costs especially for tourists visiting from abroad.

Keywords: Augmented Reality, Computer Vision, 3D Rendering, Tourism

2.1 Introduction

The aim of the project MARFT, which is funded by the Austrian Federal Ministry for Transport, Innovation and Technology (BMVIT) and by the European Union's EUREKA Eurostars Programme, is to demonstrate the next generation of augmented reality targeting current mass market mobile phones with enough computing power

G. Gartner, F. Ortag (Eds.), *Advances in Location-Based Services*, Lecture Notes in Geoinformation and Cartography, DOI 10.1007/ 978-3-642-24198-7_2,

Fig. 2.1. Augmented landscape with cartographic accuracy by the MARFT system

to render a 3D-world on their display in real-time. MARFT sets out to launch an interactive service for tourists visiting a dedicated test region in Upper Styria. During local trips through the rural mountainous region, they will be able to explore the surrounding landscape by pointing the lens of the smart-phone camera towards the area of interest. As soon as the view-finder shows the area of interest, the tourist will be able to choose between two products: (i) an augmented photo superimposed with tourist information like hiking tours or lookout points (see *Figure 2.1)* or (ii) a rendered 3D virtual reality view showing the same view as the real photo also augmented with tourist objects. This choice will be given for the case, that the view is obstructed by clouds or other obstacles. Further the information overlay is kept interactive. By pressing the symbols and lines of the overlay, a new window will be opened showing details on the selected tourist information like any hiking track, look out points or different mountain peaks.

The outstanding step beyond current augmented reality applications is that MARFT is able to augment the reality with *cartographic accuracy.* This is achieved by matching a locally rendered virtual landscape (virtual 3D view) according to the current position and orientation with the corresponding real camera picture by the use of computer vision algorithms. In addition to the benefit of presenting *reliable information,* MARFT is able to consider the *visibility of objects* and is able to work *completely offline* to avoid roaming costs especially for tourists visiting from abroad.

Additionally the MARFT application is integrated with different social communities on the internet like Facebook to share outdoor experiences and augmentation results produced by MARFT. The user can add his own comments to any point of interest and can (i) share them with the whole MARFT user community and (ii) post them to the social community platforms he is using to share the information also with friends that are not using the MARFT client.

A prerequisite for augmentation with cartographic accuracy is that the spatial referenced tourist information has to be of high quality and accuracy. To ensure this, MARFT further offers a data management tool, which can be used by regional tourist boards to add and manage their own tourist information they want to present to the users of the MARFT system.

2.2 Initial Situation

When Nokia researcher developed their first mobile augmented reality system in 2006 called MARA[1] they were using external sensors attached to their S60 phone, namely accelerometers in all three axes to determine orientation, a tilt compensated compass for heading, and GPS for positioning. This prototype application overlays the continuous viewfinder image stream captured by the camera with graphics and text in real time, annotating the user's surroundings. Since the appearance of the iPhone[2] and the introduction of the first Android phone[3] (G1) it is state of the art that all sensors needed for augmented reality applications are integrated in modern smartphones.

A main prerequisite for the MARFT system is firstly the knowledge of the high accurate three dimensional position of the mobile device – the position where the picture was taken by the user with the mobile phone. And secondly the spatial orientation of the mobile phone in order to know exactly in which direction the phone was pointed at the time of taking the picture. The spatial position combined with the orientation is used to define the line of sight within a three dimensional virtual digital landscape model. From the line of sight a virtual view can be calculated, which will show nearly the same picture as the real view of the mobile phone's build-in camera. The picture taken by the user and calculated virtual picture out of the digital landscape model will show nearly the same view.

One of the most computing-power demanding application has always been the 3D-visualisation of an interactive scene in real-time. A complex 3D-scene can consist of millions of vertices and polygons, which are all mathematical products of simple, but yet many operations. Most of these operations are floating point operations, which can be easily parallelized. Processors used in mobile phones have gotten powerful floating point units during the last year like the Samsung Nexus S or Galaxy SII. Due to the fact that many semiconductor manufacturers have shifted their production to 45nm wafer production, the result was massive computing power with little power consumption. That meant dedicated, embedded 3D GPUs

[1] http://research.nokia.com/research/projects/mara/index.html
[2] http://developer.apple.com/iphone/
[3] http://code.google.com/android/

(graphical processor units) with a new generation of processors for mobile phones. The industry for 3D-application on mobile devices is driven by the gaming industry, because their business model is clear and massive revenue streams can be generated. This leads to even more powerful graphic capabilities in mobile devices, which of course can be used for other application as well, such as MARFT.

Image analysis has recently made significant advances through the application of Artificial Intelligence (AI) enabled methodologies, such as machine learning. Nowadays, even outdoor vision systems, such as visual surveillance, have been successfully applied and efficiently cope with uncertain and noisy information to gain robustness in the automated analysis of videos. Currently, the R&D is turning attention to image analysis of mobile imagery, in particular, to make sense of billions of mobile image captures stored and sent around worldwide, and hence to benefit from the mass market of camera phones. Successful analysis of mobile imagery has been proved to be successful but under certain constraints, such as, standard illumination conditions, moderate changes of viewpoints with respect to reference images, discriminative textures and objects, and so forth.

However, current applications using mobile image recognition are still of prototypical nature and have not yet reached a large segment of the mass market. Software for mobile vision has so far been presented for sample services, such as, mobile tourist guides (Tollmar et al. 2004, Fritz et al. 2006, Amlacher et al. 2008), image based barcode reading (Adelmann & Langheinrich 2007), mobile games (Vodafone 904SH Mobile und Samsung G800 using face recognition), mobile marketing (Daem Interactive, Spain) and navigation (Cuellar et al. 2008). Nokia has launched a project towards so-called hyperlinked reality applications (MARA 2006, Che et al. 2007) that associate multimedia content with geographic locations, can serve as mobile guides for cultural heritage, education, sightseeing and geo-caching or raise public awareness by informing about emergencies and infrastructure deficiencies on-site.

MARFT has the objective to enable a maximally precise, image based estimation of orientation in rural areas. Currently, mobile image based services have actually been designed for urban environments but may serve in the function of a search key for rural areas as well. With regard to MARFT visual information in the mobile image contains cues about the position and the orientation of the mobile user like the specific profile of the horizon and/or known geo-referenced landmarks.

2.3 The MARFT Approach

The following chapter describes a technical sight of the MARFT system. Starting with the system architecture and the deduced functional modules, processes and workflows will be described in detail. The architecture was designed to match the requirements of the use cases of MARFT –especially the main use case, which is the augmentation of pictures with cartographic accuracy.

2.3.1 System Architecture

Basically the MARFT system is designed as loose client-server architecture. "Loose" means that an established connection between client and server is not a prerequisite to run the core functionalities on the client side. As MARFT should be used in rural and mountainous areas, where the availability of a wireless (broadband) internet connection over the cell network cannot be guaranteed, an update mechanism is needed to push the necessary data to the client prior to the usage of the system outdoors. This update functionality will be automatically started the moment a broadband internet connection is available on the mobile device.

Updates will be used for huge amount of data which is nearly static and valid for a long time period like maps, 3D model and multimedia meta-data on POI's and tracks. All client-server data communication will be achieved by a direct link over an internet connection without any proxy module. Beside updates for the huge amount of static date a second bidirectional communication channel will be established for highly dynamic information, which has low data amounts, like textual comments on objects, user positions and tracks. Maybe user pictures in a reduced resolution will also be collected to this kind of dynamic information.

On the server side a central database will be used to collect all relevant data for the MARFT system and also user content. The data update for the huge amount of static geo-data to the client is handled by the provider module, while dynamic data is bidirectional synchronised by the sync module. Different server modules will interchange data with relevant social network communities and geo-data platforms (external data providers) on the internet. This concept guarantees that MARFT users can share their information directly with users in other communities. The server module for user- and geo-data management can be seen firstly as an interface to the database, which can be used by administrators to insert and maintain information and secondly as a corresponding module for the client intended for user management. The users profile and account can be managed over this module. View alignment is a special server module with direct connection to the client framework and serves as a backup, if the client is not able to carry out the matching process between real and rendered views.

The client software is designed after a controller-view-modules architecture and will be implemented on a mobile device platform. A local database can be filled with regional static geo data by the update process from the server and guarantees the autonomous usage of the MARFT client. The update is handled by the same named module on the client side. Dynamic community and user data like user comments and current positions as well as new MARFT augmentations requests by users are synced on demand via the sync module to spread the new peace of information to the linked communities like social networks.

Any user interaction on the client side is covered by the graphical user interface (GUI) module. The GUI, which triggers all functionalities and visualises augmentation results or any demanded object meta-information, will be implemented in this client module. The GUI is the start and end point for each use case.

The logical implementation of use cases is done within the workflow controller module, which can be seen as the business logic of the whole client. The workflow controller requests any necessary information needed to process the users request according to the defined workflow from software modules available on the client side.

Functional modules on the client side are 3D rendering, multi-sensor data processing and view alignment. All sensors for orientation and position of the mobile phone as well as the integrated camera can be queried and processed within the multisensory module. According to the current spatial geometry of the photo-snapshot point the 3D rendering module can process a virtual landscape model based on maps and height-models from the local database. Further an information overlay can be rendered, taking into account the line of sight. The view alignment module is able to match the virtual view and with the real image of the landscape to achieve highest accuracy for superimposing the real picture with information overlays. As a backup for this complex image processing task a corresponding server module is implemented to provide assistance if necessary.

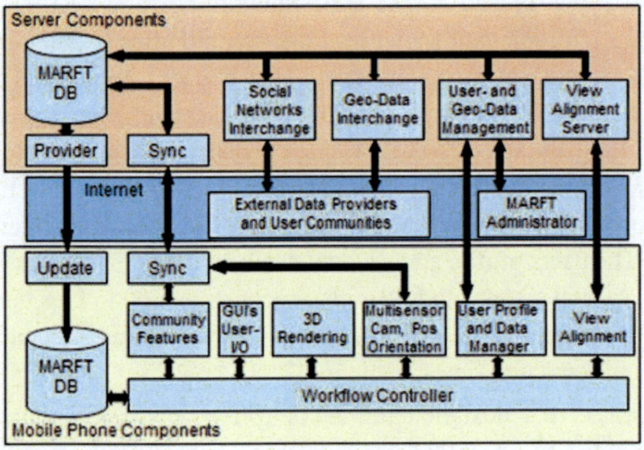

Fig. 2.2. System architecture of the MARFT system.

As the MARFT system is designed as a personalised system, the user profile and data manager module is intended to handle the user authorisation and filtering data according to the user's personal profile. A direct connection to the corresponding server module is necessary for the first login process and a profile update.

2.3.2 Modules of the system architecture

Server Modules

Table 2.1. Description of the MARFT server modules

MARFT database	This is the central database for the MARFT system on the server side. It will store any geo-referenced multimedia information as well as maps and a digital elevation model. Beside this any user data like profile, history and self-generated content can be added to the repository. Any data object can be linked to external databases and social networks.
Provider	This data provider handles the access to the database for the update process towards mobile phone clients. The update process transfers all relevant almost static data like maps, object information and 3D models, filtered for a specific region, to a MARFT end device. This update enables the use of basic MARFT services autonomous on the mobile device without the need of an internet connection (offline usage in rural areas). Due to the huge amount of data to be transferred during an update it will only be executed on demand and when a broadband internet connection is available.
Sync	The sync module is dedicated to synchronise highly dynamic data like user comments or user positions bidirectional between client and server. The amount of data transmitted is fairly low. The synchronisation can run on demand or automatically when a internet connection is available (even with low bandwidth).
Social networks interchange	This module interchanges user specific information with relevant social networks used by the user. User generated content or user requested results of the MARFT system can be shared with the whole community on the social network or just with friends.
Geo-data interchange	This module is used to connect MARFT to relevant geo-data internet platforms, which have a data repository of tracks (any tracks for outdoor activities like hiking, biking) or POI's (relevant to the touristic user like restaurants, markets, sights, infrastructure...) within the geographic region of MARFT client usage. All gathered geo-referenced information can be integrated in augmentation process of the MARFT system.
User- and geo-data management	This module keeps track of the user's profile, which stores beside personal information also his interests or any other favours deduced from his/her behaviour. The interest profile can be used to filter the relevant information for individual user needs. Privacy, security and access rights can also be managed with this module, which has a direct communication channel to the mobile phone client for any update to the user's profile.
View alignment server	This module serves as a helper to the view alignment on the client side, which is based on complex image processing technologies. View alignment is the matching process between the real landscape photo taken by the user and the artificial rendered view from the 3d model. It will be used on demand and only in special cases directly by the corresponding client module.

Mobile Phone Client Modules

Table 2.2. Description of MARFT software modules

MARFT database	This is the local data repository on the mobile device. It is designed to store all necessary data to guarantee a basic MARFT service to run even if the mobile device is not connected to the internet.
Update	This module is the client side interface for the update process. During the update process a huge amount of static data like maps, 3D model, POIs and tracks are transferred from the server to the client in a single step. All data is valid for a long time period and is bundled after a specific geographic area.
Sync	The sync module on the client side is, like its corresponding module on the server side, the interface for the bidirectional synchronisation of dynamic data like user comments on POIs or tracks. MARFT generated augmented photos can also be shared to friends on social networks over this interface.
Workflow controller	The workflow controller implements the core business logic of the MARFT client. After MARFT functionalities are triggered by the user, the workflow controller has to manage the whole processing chain till the requested result for the user is generated and displayed. It has direct connections to all functional modules on the client, requests intermediate results on demand and integrates them to the end result for the user.
Community features	This module handles any sharing of personal user information. User data like comments or individual augmented pictures can by synchronised to the server side and broadcasted to social networks.
GUI's and user I/O	According to the MARFT use cases different graphical user interfaces are needed to fulfil user interaction demands. Triggering MARFT services and viewing personal augmentation results and meta information on touristic POIs or tracks is only one perspective of useful user interfaces. From the current point of view at least the following user interfaces have to be implemented: (i) Camera Client, (ii) Photo Gallery, (iii)Image Annotation Viewer, (iv) Map Viewer, (v) Multimedia Object Details Viewer and (vi) Commenting Interface.
3D rendering	This functional module will be able to render artificial landscape views based on a 3D model and maps according to the actual position and orientation of the mobile phone corresponding to the picture taken by the user. Further an information layer matching the artificial landscape view is rendered, which shows relevant touristic information taking the line of sight into account. In the final MARFT service this overlay will be placed over the real user picture with highest accuracy.
Multisensor: camera, position, orientation	This module keeps track of data provided by any available integrated mobile phone sensor. The determination of positioning and orientation of the mobile phone with highest accuracy is curial to the whole MARFT system and will be achieved by the use of the internal GPS (in combination with cell and WIFI positioning), accelerometer and the digital compass. The interface to the internal camera will also be realised within this module.
User profile and data manager	This module will handle all issues in connection to the user's authentication, profile and access rights. The profile can be adopted on the client to any time and will have influence on the individual information filtering.
View alignment	This module is focused on complex image processing implementations for matching the real picture taken by the user with the artificial 3d rendered landscape view for the current photo position and geometry. The goal is to calculate a spatial transformation, which will be applied on the rendered information layer in order to get a high accurate overlay for the real picture.

2.3.3 Use Cases of the MARFT System

Within the MARFT system we have to distinguish five different users described in *Table 2.3*.

Table 2.3. Different user roles of the MARFT system

Name	Profile: Background and Knowledge
Mobile user	Authenticated person, who uses the MARFT application on the mobile device.
Web user	Authenticated person, who uses the MARFT web site to browse touristic content.
Guest user	Anonymous person, who uses the MARFT web site to browse the touristic content.
Content administrator	Person, who manages the touristic geo-referenced content in the MARFT system.
System administrator	Person, who administrates and maintains the MARFT system.
System	Analyses the user's behaviour and processes user requests.

There are five different use cases of the MARFT system which are shown in *Figure 2.3* and all of them are triggered by different types of users.

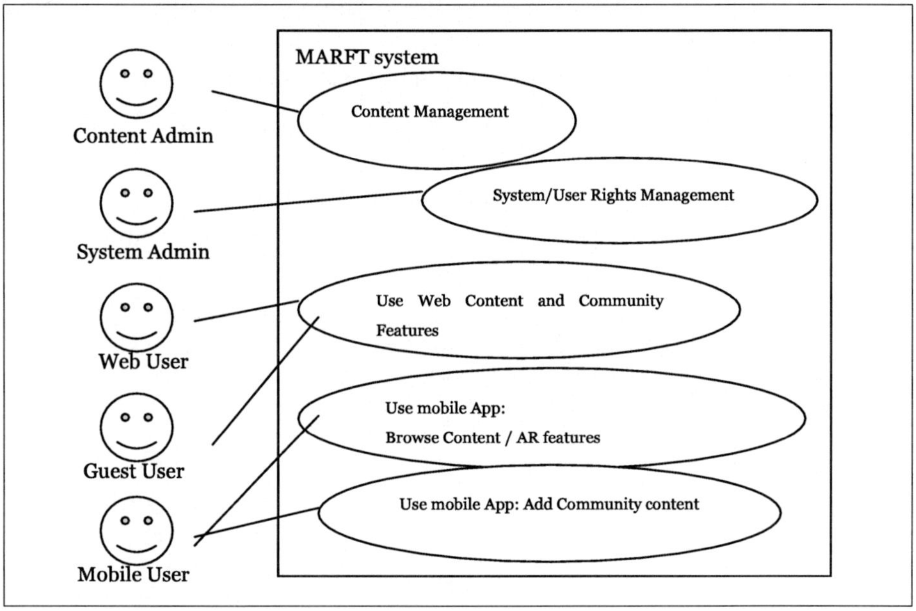

Fig. 2.3. Use Cases of the MARFT system

2.3.4 Main Use Case and Workflow

Description. The *Mobile User* is taking a picture with MARFT client on the mobile phone and wants to get it annotated by the *System*

Goal. The *Mobile User* wants to get an annotation with touristic information of the picture he has just taken of the landscape in a rural outdoor scenario.

Preconditions

1. Mobile MARFT client is ready to use on the mobile phone
2. Mobile User has successfully authenticated
3. GPS position and orientation of the mobile phone is available.

4. Geo-Information and 3D landscape model is locally available on the mobile phone.

Basic Course

1. Use case begins when the mobile user takes a picture of the landscape with the MARFT application on the mobile phone.
2. System detects position and orientation of the mobile phone
3. System calculates synthetic view and an information layer from the 3D model corresponding to the picture geometry
4. System calculates correction matrix between picture and rendered view by computer vision techniques to enable high accuracy augmenting
5. System augments real picture with the rendered information layer by applying the correction matrix
6. Mobile client shows the high resolution augmented picture with an interactive information layer
7. Mobile User browses touristic information
8. Mobile User comments and rates touristic information
9. Use case ends when the mobile user closes the augmented picture on the mobile phone

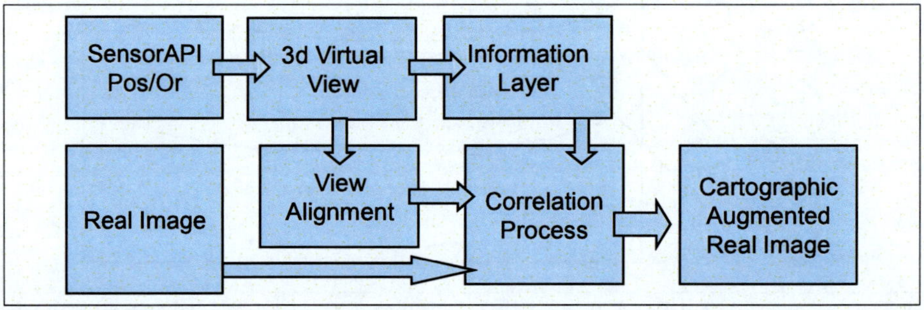

Fig. 2.4. Workflow for augmenting pictures

2.4 Key Technologies

2.4.1 3D Rendering

The 3D rendering module is able to render artificial landscape views based on a 3D model and aerial pictures according to the actual position and orientation of the mobile phone corresponding to the picture taken by the user. *Figure 2.5* illustrates an example of a real photo taken from an actual location and its corresponding rendered virtual view.

Fig. 2.5. Comparison of a real landscape photo with the corresponding virtual view

The MARFT Rendering Engine is made out of two parts: The Nütex 3D Engine and the Terrain Engine, which relies on the 3D Engine that is optimized for speed and size. It is implemented in C++ in order to make it portable. Another advantage of implementing the 3D Engine in C++ is that for example the code may be compiled in a way that it is optimized for chips with specific features like NEON.

The minimal requirement of the 3D Engine is OpenGL ES 2.0, since it includes support for VBO (Vertex Buffer Objects). VBO may be used to upload the vertices to the video device memory without passing the same data to the video device, which means that the overhead to render every frame is kept low and the frame rate is increased. The VBO may contain 3D coordinates, vertex colour, texture coordinates and the "normals", used for lighting computations. Another reason to choose OpenGL ES 2.0 as the minimal requirement is the need of a Framebuffer Object, which is used to render the scenery and save it to a texture for further processing.

The "Terrain Rendering Engine" uses the shader functionality of OpenGL ES 2.0 to render the same scenery in black and white. Shaders are also used to add more realism to the rendered scenery by simulating fog which makes the distant mountains look like they are fading away.

OpenGL ES has its own way of eliminating triangles which are not visible. The 3D Engine implements frustum culling based on bounding boxes, which are axis-aligned and they enclose a 3D object, in order to optimize the rendering process. The 3D engine camera function contains the position, the viewing direction and six planes. The camera can then decide whether a 3D object is visible by verifying that at least one of the bounding box corners is contained in the frustum, which is defined by the six planes of the camera. Moreover, the 3D Engine also includes code to generate “Mipmaps” in order to increase the rendering speed and reduce the aliasing artefacts.

As the range of vision is sometimes more than 25 kilometres and as the scenery should contain as much detail as possible, the Terrain Engine uses a quad-tree to split up the world in tiles. The tiles nearby are very detailed, which means that the geometry of the closest tiles is more detailed and moreover textures at a higher zoom level are used. The higher the distance from the point of view within the scene, the less detailed are the textures used for the tiles. By using this approach the memory usage gets reduced dramatically without sacrificing the detail of the scenery.

The Terrain Engine uses pictures to generate the geometry of the 3D World. The geo-referenced 3D model (Height-Map) of the terrain is represented by greyscale pictures, in which each pixel and its colour value represents the height at a particular spatial position. This Height-Map is used for each tile of the 3D world to generate the triangles, used to build up the 3D model. Storing Height-Maps as pictures results in less storage space, compared to storing the world as 3D coordinates. Another very practical side effect is that the Height-Map could be used to generate different Levels-Of-Detail of the same 3D World tiles. This means that in case one tile of the 3D World is too far away, the mesh with fewer details of the same tile would be rendered and that unloads the video device.

Another important part of the 3D Engine is the built-in Font Manager. However, there is only support for Bitmap-Fonts as they do not require much computational power. Each font is stored as a picture. The size and position in the bitmap of each glyph are stored in an additional file.

In order to prevent the 3D Engine from relying on “Dalvik virtual machine” to access content (e.g. Textures) in zip files, there is built-in support for reading and writing zip packages by the library “libzip”.The Math Library which is part of the 3D engine provides speed optimized routines for matrix and vector operations.

Another key feature of the 3D engine is the capability of being able to run the native code on a desktop computer. There is no need to install it on a mobile phone or to use the Android emulator, which is always very slow and limited when using 3D, in order to test the 3D code. Running the native code on a desktop computer shortens the development time. The format of pictures is currently limited to PNG

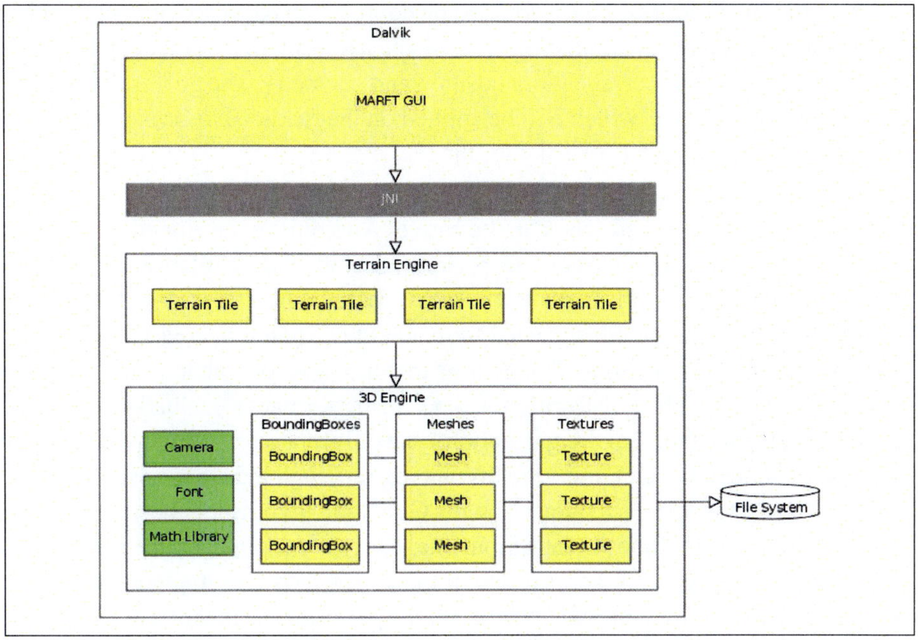

Fig. 2.6. 3D Rendering Engine Architecture on Android OS

which is implemented directly in the 3D engine. Support for JPEG could be added at a later time.

Figure 2.6 illustrates the architecture of the 3D rendering engine running in the "Dalvik Virtual Machine" on the Android operating system. The Java Native Interface is used to combine the Java GUI with the 3D engine written in native code.

The NüTex 3D-engine uses new vertex buffer/VBO to reduce frame rate and reduce main processor load. ETC1 texture compression is used because gains in rendering and loading speed are quite large (ratio 4:1 compared to commonly uses RGB565 formats).

2.4.2 View Alignment

One crucial topic of MARFT is to develop and implement image analysis algorithms to overcome the lack of augmentation accuracy in state-of-the-art applications in the scope of augmented reality. JOANNEUM RESEARCH has built up comprehensive experience in the field of image processing and computer vision during the last two decades. Industry leading applications in the scope of environmental analysis, robotics, near field geometry, navigation, object recognition and semantic attention analyses have been rolled out. Therefore an image processing software framework has been built up called IMPACT (Image Processing and Classification Toolkit).

View alignment for MARFT is focused on complex image processing implementations for matching the real picture taken by the user with the artificial 3d rendered landscape view for the current photo position and geometry. The goal is to calculate a spatial transformation, which will be applied on the rendered information layer in order to get a high accurate overlay for the real picture.

Therefore the potentials of the mobile platform towards the needs in the project MARFT had to be checked. Thus, in the case of Android we developed a few small software-prototypes to measure the performance of such a platform regarding memory usage and timing. These experiments include the first step of the horizon extraction algorithm, which is needed to match the virtual rendered view with the real picture taken by the user. The goal of this test was to evaluate the feasibility of such tasks on a mobile platform as well as the performance evaluation regarding user requirements in terms of e.g. reaction-time. Given a set of 30 images captured from our first demonstrator site – the Styrian region "Joglland" the tests involve the following processing steps: (i) Based on the GPS and INS sensor data, a synthesised view was generated offline for each input image, (ii) camera images and synthesised views were feed into a horizon extraction module, which was implemented in the NDK framework of the Android OS and (iii) the performance in terms of processing time and memory consumption was measured over multiple runs for different input images.

The horizon extraction module is built upon a global optimization where a cost function is minimized employing dynamic programming for speedup. The cost function itself holds information like edge orientations and magnitudes, image location and penalties for gaps. The algorithm enables the extraction of a smooth continuous horizon line, also in cases when edge information is missing or weak. *Figure 2.7* depicts an example image from the demonstrator site, while the corresponding edge image is shown in *Figure 2.8.* Finally the extracted horizon is pictured as an overlay in the original image and depicted in *Figure 2.7* again.

Fig. 2.7. Depict the extracted horizon superimposed on the original example image

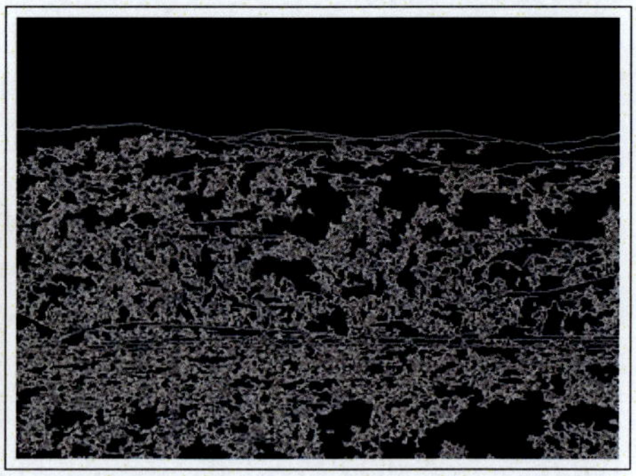

Fig. 2.8. Edge image extracted using the test program within the NDK

The performance for edge extraction was measured during ten runs on a Google Nexus One with Android 2.2 OS. The average processing time for extracting the horizon was 52.6 ms +/- 11.8 ms. The required memory for executing the tests was approx. 30 kByte. This is not very significant due to the fact that the Android OS (Dalvic Java VM) uses a garbage collector that administrates the required memory on its own, without being controlled by the program itself.

The experiment figured out that image processing algorithm can be developed on "state-of-the-art" smartphones nowadays, making offline image processing applications available for users of the MARFT service.

2.5 Conclusion

In this paper, we have presented the basic idea behind the project "Mobile Augmented Reality for Tourists" (MARFT) and how it will be technically realized. The technical challenges beyond the current state-of-the-art are identified as: (i) the augmentation of rural landscapes with cartographic accuracy, (ii) the goal of presenting reliable augmentations so that the user can trust that points of interest are really located at the shown position, (iii) avoiding information overload and confusion by showing only really visible objects by making line of sight intersections with 3D landscape models and (iv) making the service locally available on mobile devices without the need of an internet connection to avoid roaming costs for tourists and to overcome the lack of network coverage in rural and mountainous areas.

Current available augmented reality applications like Layar[4] or Wikitude[5] are interfaces with a more playful attempt. They are not able to present reliable information to their users, because the augmentation is only based geometry calculations deduced from GPS/Wifi/cell positioning and the orientation estimate of the digital compass. Those sensors deliver data with little accuracy, which lead to an error in augmentation up to several hundred meters and even more when the distance to the augmented object is high, like in the MARFT scenario of augmenting pictures of the surrounding landscapes with distances of several thousand meters. MARFT uses computer vision algorithms and 3D landscape reconstructions to overcome those weaknesses with the aim to present reliable augmentations.

Taking into account the sensor accuracy of current smart phone sensors for position and orientation measurements the deviation of the augmentation from the real position in the taken picture could be up to +/- 110 pixels (according to our expertise based on field trials) on a display with a resolution of 800x600 pixel. MARFT is aiming for a maximum deviation of +/- 10 pixels which is 11 times better than current AR attempts.

[4] http://www.layar.com/

[5] http://www.wikitude.org/de/

The presented results of current developments towards this challenging goal are, as shown in this paper, very promising and will lead to an integrated prototype for an on-site live demonstration in the Styrian region "Joglland" in spring 2012.

Acknowledgements

This work is part of the project MARFT and has been funded by the research program EUREKA/Eurostars[6], which is a co-funded program by the EU 7th framework program and the Austrian Ministry for Transport, Innovation and Technology (project number E! 4802 MARFT).

References

Adelmann, R., and Langheinrich, M. (2007) Rapid Prototyping Platform for Mobile Phone Based Services on Retail Products. Proc. International Symposium on Ubiquitous Computing Systems, UCS 2007, Akihabara, Tokyo, Japan, November 2007.

Amlacher, K., Fritz, G., Luley, P., Almer, A., and Paletta, L. (2008) Mobile Object Recognition Using Multi-Sensor Information Fusion in Urban Environments, Proc. IEEE International Conference on Image Processing, ICIP 2008, San Diego, CA, 12–15 October, 2008.

Chen, W-C., Xiong, Y., Gao, J., Gelfand, N., Grzeszczuk, R. (2007) Efficient Extraction of Robust Image Features on Mobile Devices. In Proc. ISMAR 2007.

Cuellar, G., Eckles, D., and Spasojevic, M. (2008) Photos for Information: A field Study of Cameraphone Computer Vision Interactions in Tourism, Proc. Conference on Human Factors in Computing Systems, CHI 2008, Florence, Italy.

Fritz, G., Seifert, C., and Paletta, L. (2006) A Mobile Vision System for Urban Object Detection with Informative Local Descriptors. Proc. IEEE International Conference on Computer Vision Systems, ICVS 2006, New York, NY.

Kumar, S. and Hebert, M. (2003) Man-made structure detection in natural images using a causal multiscale random field. Proc. IEEE Conference on Computer Vision and Pattern Recognition, Madison, WI, June 2003.

[6] http://www.eurostars-eureka.eu

Chapter 3

I'm feeling LoCo: A Location Based Context Aware Recommendation System

Norma Saiph Savage[1], Maciej Baranski[1], Norma Elva Chavez[2] and Tobias Höllerer[1]

[1] Computer Science Department, University of California, Santa Barbara, USA
{saiph, mbaranski, holl}@cs.ucsb.edu
[2] Computer Engineering Department, National Autonomous University of Mexico
normaelvacr@gmail.com

Abstract

Research in ubiquitous location recommendation systems has focused on automatically inferring a user's preferences while little attention has been devoted to the recommendation algorithms. Location recommendation systems with a focus on recommendation algorithms generally require the user to complete complicated and time consuming surveys and rarely consider the user's current context. The purpose of this investigation is to design a more complete ubiquitous location based recommendation algorithm that by inferring user's preferences and considering time geography and similarity measurements automatically, betters the user experience. Our system learns user preferences by mining a person's social network profile. The physical constraints are delimited by a user's location, and form of transportation, which is automatically detected through the use of a decision tree followed by a discrete Hidden Markov Model. We defined a decision-making model, which considers the learned preferences, physical constraints and how the individual is currently feeling. Our recommendation algorithm is based on a text classification problem. The detection of the form of transportation and the user interface was implemented on the Nokia N900 phone, the recommendation algorithm was implemented on a server which communicates with the phone. The novelty of our approach relies on the fusion of information inferred from a user's social network profile and his/her mobile phone's sensors for place discovery. Our system is named: I'm feeling LoCo.

G. Gartner, F. Ortag (Eds.), *Advances in Location-Based Services*, Lecture Notes in Geoinformation and Cartography, DOI 10.1007/ 978-3-642-24198-7_3,

Keywords: Personalization, Recommendation Systems, Pervasive computing, Human Computer Interaction, Context Aware Recommendation Engines, Automatic Travel Guides.

3.1 Introduction

Personalization (see e.g., Mulvenna S. M and Buchner A., 2000) is a key component of modern location based services (LBS). To achieve personalization the system builds a model representing the user's needs and preferences. In the literature this process is generally called a user model (UM) (see e.g., Kobsa. A., 2001). Most LBS obtain this data by extended surveys. Yet responding to long questioners can be cumbersome and imposes a cognitive burden for the user.

Recommendation systems (see e.g., Resnick H. V. P., 1997) are a particular type of personalized system which filters information and presents only what is relevant to an individual. There have been many attempts to build systems, which recommend places to visit and facilitate decision-making. For example, Rinner and Raubal (2004) designed a service named Hotel Finder which by considering a user's location, spatiotemporal constraints and preferences, recommended suitable hotels. Espeter and Raubal (2009) extended Hotel Finder and created a system that aided cohort decision-making. Albeit decision making in unfamiliar environments was improved, as these system required users to manually input the majority of their preferences, the user experience in both cases may have deteriorated

According to the studies done by Regula and Bachman (1981), when responding to long questionnaires individuals are more likely to give identical answers to most or all of the items. Therefore in Hotel Finder and in the work of Espeter and Raubal (2009), because the user had to manually provide a series of personal preferences, it is likely many of the user' responses did not truly reflect their interests. From the study of Regula and Bachman (1981) it is clear that a system, which could automatically and accurately infer an individual's preferences, would dramatically boost the user experience because the user would not require to spend time completing surveys.

The work done by Sugimoto Takeuchi Y (2006) took a first approach in place recommendation without needing the user to pass through any survey phase. By utilizing information from a user's location history, the system in their study automatically suggested stores to visit. The drawbacks were that for logging the history of visited places, the system assumed that all locations were stores. The system stored no information related to the nature of the place the user was visiting: grocery stores, Mexican restaurants and pawn shops were treated as being the same category. The system simply calculated the probabilities that existed for moving from one position to another. It was therefore impossible to query the system and ask for suggestions of where to eat or where to buy an inexpensive pair of jeans, which are some of the typical questions a user might have for a location-based recommendation system.

A number of previous systems automatically discovered the significant places related to a user and also stored information related to these sites. The work done by Marmasse, N. and Schmandt, C. (2006) is an example of this type of system, which discovered the places a user frequently visited and additionally stored information the user manually annotated, such as notes and to-do lists for a particular site. The work of Marmasse, N. and Schmandt, C. (2006) suffered a problem similar to that of Hotel Finder. It forced the user to spend time in selecting information from a small output device and entering data through uncomfortable input interfaces such as a thumb keyboard or a stylus. Another limitation of Marmasse, N. and Schmandt, C. (2006) was that its employed recommendation algorithm was very basic. The recommendations were based mostly on "reminders". For example if the system detected that the user was near a place for which they had set a to-do list, the system suggested visiting this site and completing the to-do list. It was not able to suggest new places for the user to visit and it provided little aid in the decision-making.

A system, which provided more assistance in the decision-making, while leveraging user input was TripTip by Kim et.al (2009). Given the places the user had visited and their characteristics, TripTip recommended sites. All the data was obtained by mining a person's 43places.com profile and crawling the 43place.com website. Despite TripTip's improvements over recommendation systems that automatically inferred user preferences, TripTip still suffered several limitations: it recommended places only within walking distance of where the user was last seen. The system disregarded spatial-temporal constraints. Additionally, because TripTip was not a mobile system it could not automatically detect the user's current location. Therefore to receive recommendations, the user had to actively update where they were. This evidently damaged the user experience. Furthermore, TripTip only suggested places that had a similarity with the sites in a user's 43places.com profile. Therefore an individual, whose 43places.com profile held only information about visits to educational institutes, would most likely have a difficult time receiving restaurant recommendations.

A system, which sought to suggest significant places, despite having little user data is foursquare's recommendation engine. foursquare is a location-based online social networking website, which permits users to "check-in" to places by either visiting a mobile version of their website, text messaging or by using a smart-phone specific application. In late May 2011, foursquare began offering place recommendations. foursquare's recommendation engine considers the user's location, check-in history (the places the user had visited) and "popular" sites near the user (foursquare developed a metric for inferring place popularity based on the number of user's that have visited the place, as well as the number of visits that all users have made). Due to the popularity metric, foursquare's recommendation engine can suggest relevant places to visit, without needing extensive user information. Albeit foursquare's recommendation engine can potentially suggest relevant places to visit without having the user provide extensive amounts of information, the system's recommendations are less user tailored and more generic. Furthermore, in cases where foursquare does have sufficient data, the recommendation engine fails to consider context for the recommendation: the engine does not acknowledge that the user's current transportation mode could affect the type of places a user would want to visit. Additionally because the engine only considers the category associated with the places the user visits, foursquare's recommendation engine disregards the place's contextual information, such as: Does the restaurant cater organic healthy food? What type of people do visit the restaurant: business men, students, surfers?

The research on place recommendation systems has paid little attention to the integration of contextual information for the recommendation algorithm. Systems, which do consider context, require the user to complete extensive surveys and constantly update their contextual information.

The aim of this investigation is to design a ubiquitous location based recommendation system, which by considering time geography and similarity measurements, presents a more complete recommendation algorithm. Our algorithm takes an approach similar to that of Rinner and Raubal (2004), but unlike Rinner and Raubal (2004) is not restricted to an extensive questionnaire phase or does it require the user to constantly update their contextual information.

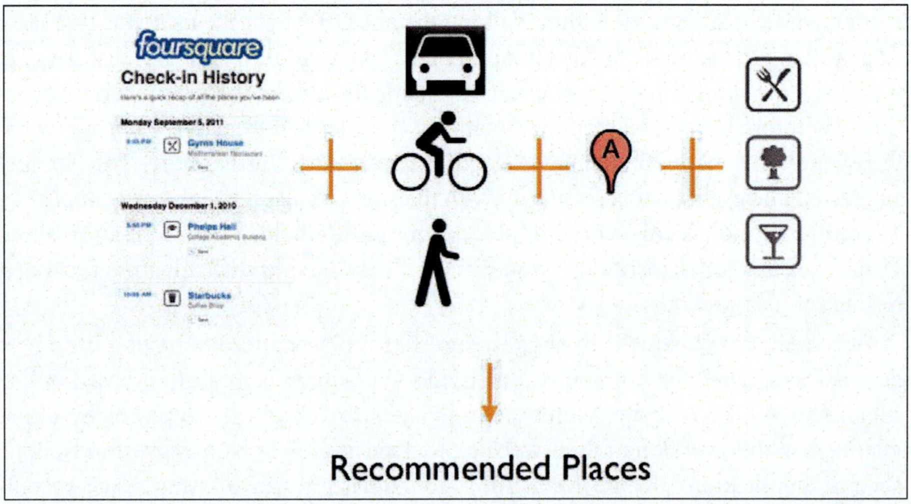

Fig. 3.1. Features considered for the recommendation algorithm: the user preferences (based on the user's foursquare check-in history), the user's current transportation mode, the user's current location and the user's mood(the type of places the user is currently interested in visiting).

Instead of including a survey phase, our system mines a person's social network profile and maps this information into user preferences. For inferring the user's preferences, our system, unlike TripTip and foursquare's recommendation engine, considers the contextual information related to the places the user has visited: tags and categories associated with a place are utilized for learning the user preferences. Our system additionally can also offer relevant place recommendations even under the circumstance that user data is lacking. Our system detects whether sufficient user data has been provided. If sufficient user content is not present, our system mines the information of the wikitravel page (http://wikitravel.org) of the city the user is in, and automatically finds the city's landmarks and adopts this data for the recommendation.

Furthermore, our system includes a mobile application, which automatically infers a user's current mode of transportation and utilizes this information to determine how far a person would be willing to travel to visit a location. In our approach the user is only required to input their mood. The user's mood is used to delimit even more the type of places, which will be recommended by the system. *Figure 3.1* shows the features utilized for recommendation. Our system is named: I'm feeling LoCo. Where LoCo is short for Location and Context.

In the following sections we present in greater detail each component of our recommendation system. First, we explain how personal spatiotemporal constraints and preferences are automatically inferred and used as features for our place recommendation algorithm. Our place recommendation algorithm is presented afterwards. Insights gained from our study are given subsequently. In the final two sections we describe additional evaluations obtained via cognitive walkthrough methodology, and present our conclusions.

3.2 Automatic Integration of a User's Spatiotemporal Constraints

Time geography considers that there are natural laws and social norms, which determine a person's ability to be present at a specific location and time. Communication and transportation services aid individuals in trading time for space and play an important role in allowing people to be physically present at a certain location and time. The work of Raubal et.al (2004) showed the importance of integrating time geography to personalized LBS, in particular personal spatiotemporal constraints.

Our study follows this guideline and incorporates in the personalization process capability constraints. Capability constraints are a particular type of spatiotemporal constraints, which confine human activities in time and space to available resources. Personal capability constraints can be delimited by the individual's mode of transportation, because the form of transportation bounds the places the person can visit. Our recommendation algorithm utilizes this constraint for delimiting the list of suggested places: only places "near" the user are analyzed and nearness is defined by the mode of transportation. For example, if an individual is riding a bicycle and requesting a restaurant recommendation, the system will not suggest places, which are an hour biking distance away. Whereas, if the person is driving a car, a restaurant that is an hour away by bike could still be recommended. Our study also considers that requiring the user to constantly update their current form of transportation is uncomfortable. Therefore a person's mode of transportation is automatically detected. The detection is done on the user's smartphone. In this case it was implemented on the Nokia N900 phone.

The method for automatically detecting a user's form of transportation is similar to the process proposed by Reddy et al. (2010). Their mobile system discriminated between a person who was stationary, walking, biking, or driving. The classification was realized by a decision tree (DT) followed by a first-order discrete Hidden Markov Model (DHMM). The combination of the decision tree with a discrete hidden Markov model improves classification, because the decision tree is tailored to differentiate between the boundaries of transportation modes, and the discrete hidden Markov model helps in reducing noise by utilizing temporal knowledge of

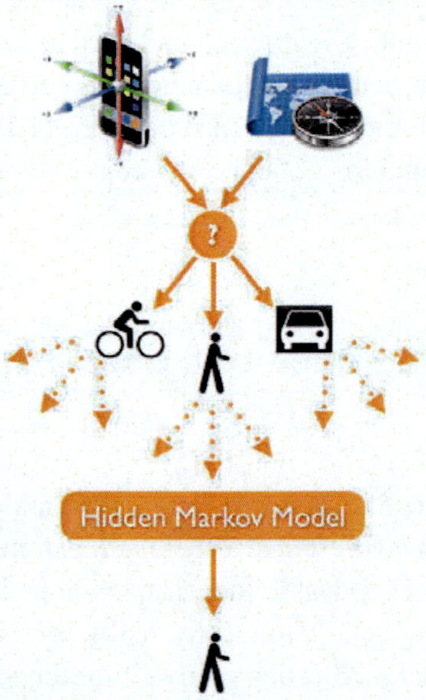

Fig. 3.2. Overview of the algorithm utilized for the detection of the transportation mode: The accelerometer variance and the GPS speed data from the n900 are the decision tree's inputs. The decision tree classifies the input to an activity: biking, walking or driving. A hidden Markov model is fed a series of activities encountered by the decision tree. Based on the presented pattern, the hidden Markov model determines the final activity classification.

the previous transportation mode that were detected. There are times for example, when the user is driving and due to traffic or stop lights their speed is decreased. This may cause the decision tree to classify the user's activity as biking, but because a transition from driving a car to biking is unlikely, the discrete hidden Markov model corrects the classification. As feature vectors, we used the variance of the accelerometer signal and the GPS speed data.

The approach of Reddy et al (2010) was selected because their system was capable of running without strict orientation or position requirements, which is fundamental when doing classification on smartphones, since the form in which individuals carry their phones varies widely. Some, for example, keep their mobile device in their backpacks, while others place them on their belts. This method was also selected, because the authors demonstrated that historical user pattern data is not required. It can therefore be immediately utilized by a person, without needing a prior training phrase. *Figure 3.2* presents an outline of the algorithm.

3.3 Automatic Recollection of User Preferences

For limiting the time spent on questionnaires, user preferences are automatically obtained by mining the user's social network profile. The utilized social network was foursquare. From a user's GPS coordinates, foursquare returns a list of possible places the person could be in. Each item on the list has an associated name, category

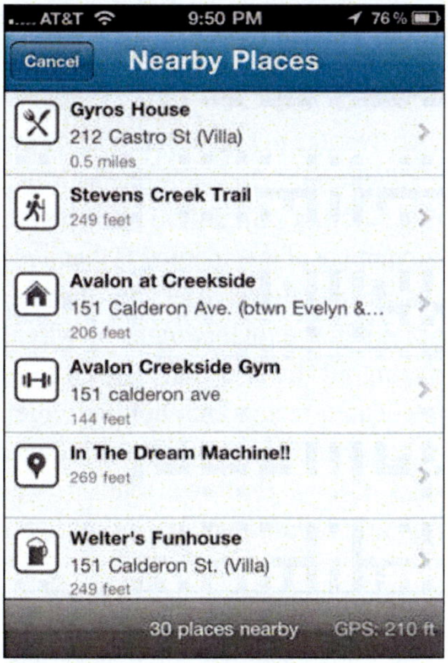

Fig. 3.3. foursquare interface to check-in to a place.

and relevant tags which define the place. An item on the list might be for example "Ruben's Tacos," with the category of restaurant and tags such as: Mexican food, burritos, mariachi, margaritas etc. The user, with a simple click or tap, can select the place they are currently in, and foursquare will log to the user's profile, the place along with all the associated tags. *Figure 3.3* shows the foursquare user interface with a list of places the person can check in to.

The foursquare API permits the retrieval of all of the user check-ins along with the information conjoined with the places the user has visited, such as tags, name of place and category. We used the retrieved information to build a user model that holds contextual information related to the type of places the user visits. The user model is in essence a document, which holds a series of words. Each time a user visits a place, its name, category and tags are added to the end of the document. In our implementation, a server manages the creation of the user model and has a daemon, which periodically checks the user's foursquare profile for new check-ins, and performs an update if necessary.

3.4 Place Recommendation Algorithm

Recommendation algorithms are generally divided in two types: algorithm that utilize collaborative filtering and algorithms that utilize content based filtering (see e.g., Baudisch, 1999). Collaborative filtering establishes that personal recommendations can be computed by calculating the similarity between one user's preferences and the preferences of other individuals. In collaborative filtering the preferences of a large user group is registered. Given a user A that is seeking recommendations, similarity metrics are utilized to find a subgroup of people that present preferences similar to that of user A. An average of the preferences of that subgroup is computed. The returned preference function is what is utilized for user A's recommendations. By contrast, content-based filtering utilizes the information about an item itself for recommendations. The advantage of this method is that it is not limited to suggesting options that have previously been rated by users. Furthermore content based filtering can provide the user with a better explanation as to why option X was suggested. For example, the system can tell the user that 'Ruben's Tacos' was recommended, because the user had frequented restaurants before, which serve Mexican dishes. Content based filtering recommendation algorithms hold a set of items denoting the user's preferences. The task of the algorithm is to classify an unseen item (an option the user has not expressed any opinion about), as something relevant or irrelevant for the user.

Due to the nature of the data we were handling, it was decided to utilize a content based filtering approach for our recommendation algorithm. In this study, the items denoting the user preferences are the restaurants visited by the user along with their associated information: tags and assigned category. The unseen items to be classified are places "near" the user that they have never before visited. Because the space of places to analyze is immensely large, our system utilizes the user's capability constraints, preferences and mood, to delimit the search. Capability constraints influence the outcome of the algorithm as follows: given the current location of the user, the foursquare API is utilized to return all of the places within a certain radius to where the user is. The size of the radius depends on the user's mode of transportation: the faster the user moves, the larger the radius. A larger radius generally implies that more places will be considered for the recommendation. The radius size was empirically calculated. From this list of places, which are around the user, a second filtering step is performed. The filtering is now based on how the user is "feeling": foursquare labels every place with a category. As of December 2010 there were seven different categories: Arts & Entertainment, College & Education, Food, Work, Nightlife, Great Outdoors, Travel and Shops. Each of these categories was mapped to a particular feeling:

- Arts & Entertainment = "feeling artsy"
- College & Education = "feeling nerdy"
- Food = "feeling hungry"
- Home/Work/Other = "feeling workaholic"
- Nightlife = "feeling like a party animal"
- Great Outdoors = "feeling outdoorsy"
- Shops = "feeling shopaholic"

Our system provides the user with an interface through which they can select with a tap one of the moods mentioned above and portray to the interface their feelings. *Figure 3.4* shows the 'I'm feeling LoCo' interface. The selected mood delimits even more the places considered for recommendation: only places labeled with the category to which the chosen feeling is mapped to are selected. For example, if the user stated they were "feeling nerdy" only places labeled with the category of College & Education are acknowledged.

Fig. 3.4. I'm feeling LoCo interface. The interface displays the detected transportation mode of the user and presents eight buttons from which the user can select their current mood.

Aside from the "LoCo" mood, all the other user moods consider the contextual information associated with the places the user visited for the recommendation procedure: from each place in the newly filtered list of places, its associated tags are obtained. A set of words containing the intersection between the tags of the user and the tags of the particular place is created. For each term in this set of words its log frequency weight is obtained. The log frequency weight of a term t in a set d can be defined as a function F(t):

$$F(t) = \begin{cases} 1 + \log tf_{t,d} & tf_{t,d} > 0 \\ 0 & otherwise \end{cases}$$

where tf_{t,d} represents the number of times t occurs in d. In this case, d is the document, which holds all of the words associated with the places the user has visited, and t refers to one particular term or word present in the document. Once the weights for all of the tags of a particular place are calculated, a summation over all of the weights is done. This summation represents the log frequency weighting score of a particular place. The K places with the highest log frequency weighting are selected and are what is recommended to the user. K is a design parameter, which can be chosen arbitrarily. In our study, for visualization purposes, we set K to a value of 4. This list of K places represents the places that best match the user's personal preferences and spatiotemporal constraints. The list contains for each place, its name, the distance from the user's current location and the GPS coor-

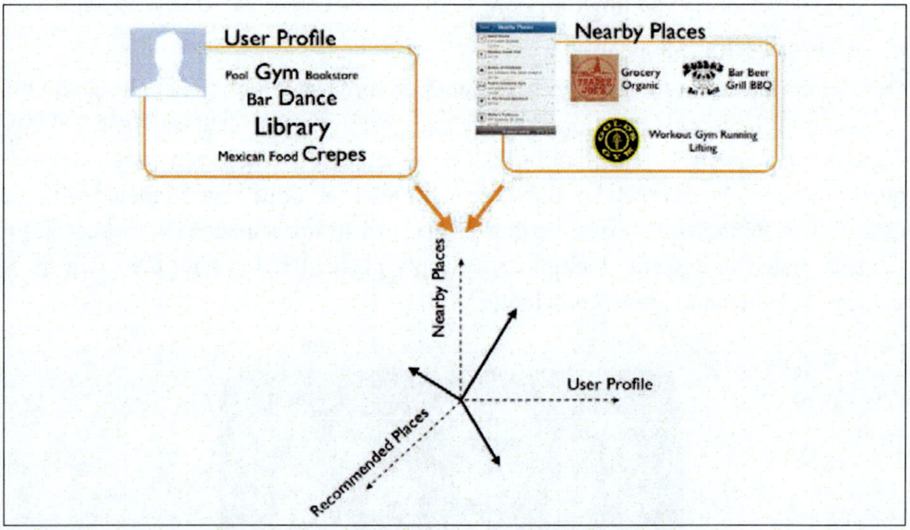

Fig. 3.5. Outline of I'm feeling LoCo's recommendation algorithm. Depending on the user's profile and the places near the user, the algorithm recommends sites to visit. Only nearby places are considered for the recommendation. The definition of nearness changes according to the user's transportation mode. Places nearby are then further filtered based on the category they present and the similarity they have with the other places the user has previously visited. The K places with the highest similarity score are what is suggested to the user.

dinates of the place. The phone displays the recommended places on Google Maps, allowing the user to select the place he or she wishes to visit. For the particular case of when the LoCo mood is selected, the system retrieves all venues that are within a convenient distance to the user, nearness to the user being the only considered factor. The top K suggested places are picked randomly by the system.

Because our recommendation algorithm is content based, it depends on the user's foursquare check-ins to generate place suggestions. Relying on how active a user is on a social network can be problematic, especially in cases where the user rarely utilizes the social network site. For this reason, we developed a metric that will recommended meaningful venues to visit, regardless of the user provided content. The metric functions as follows: given the city the user is in, we mine the City's wikitravel page (wikitravel.org) for the city's iconic places or landmarks. Each landmark is then searched on foursquare, where its address and associated category is retrieved. If the landmark is conveniently near the user and has the category the user requested, the landmark is suggested to the user. This metric permits the system to recommend significant venues without requiring excessive user generated content. *Figure 3.5* presents an outline of the final recommendation algorithm and *Figure 3.6* of the entire system.

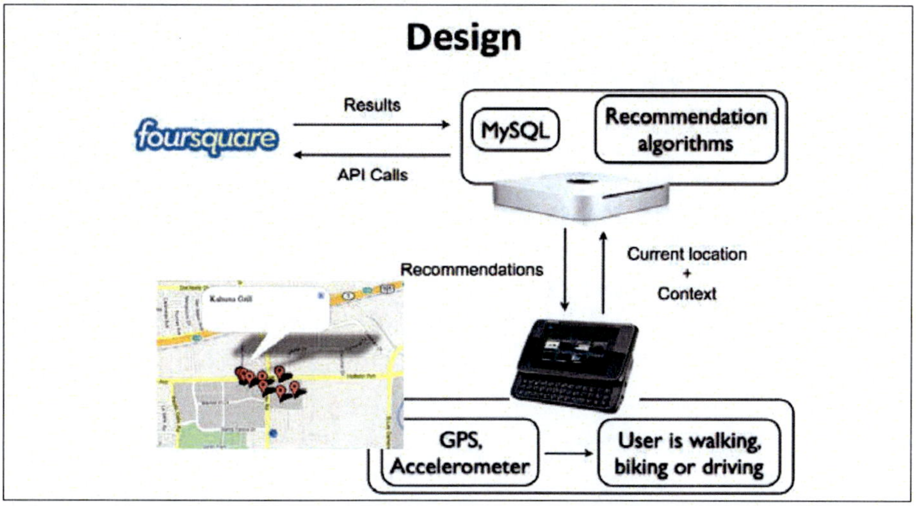

Fig. 3.6. Overview of I'm feeling LoCo system operation. Through foursquare API calls, a server is constantly updating an individual's user model, and storing in a database this information. On the other hand, a mobile application is continuously detecting the user's transportation mode as well as offering an interface through which the user can request place recommendations. When the user queries the system for a recommendation, the mobile phone submits to the server, the user's location and contextual information: their transportation mode. The server inputs this data, along with the generated user model to the recommendation algorithm, that utilizes these features to decide what places are the most suitable to be recommended. The server returns to the mobile phone a list of the best K places for the user to visit. The mobile phone then displays these places on a virtual map.

3.5 Iterative Design

During the course of our work, which we tested throughout by frequent use by the authors and several volunteers in informal formative design evaluations, a number of limitations were encountered. In the following, we mention a few of the most interesting ones: Foursquare returns only places within 400 meters radius from the user's current location. But our place retrieval algorithm requires a radius, which varies according to person's mode of transportation. An algorithm for automatically increasing the considered radius had therefore to be implemented. The functionality of this new feature is presented in *Figure 3.7*.

Another issue was that the places returned by the foursquare API only contained the name and categories associated with the place, its corresponding tags were not given. Therefore additional API calls were required per place. These extra API calls led to exceeding the permitted number of foursquare API calls a user is allowed to make within an hour, which is 200. In order to overcome this limitation for the purpose of facilitating a meaningful evaluation (the cognitive walkthroughs described in *Section 3.6),* we resorted to the following temporary work-around, which is not adhering to foursquare's usage policies and which we don't advocate for any real use, but which was workable for the duration of our experiments: The API calls for attaining tags associated with a given place do not have to be authenticated by the user who is currently using the "I'm Feeling LoCo" application. Hence a number of foursquare "helper" accounts were created, for the purpose of making additional API calls. An automatic account switcher was implemented. This account switcher tracked per "helper account" the number of foursquare API calls which had been made in the last hour. When the number of calls was approximating the limit, it gracefully switched to another helper account: one that had not been utilized for the last hour. On top of account switching, a server-side caching strategy was also implemented to store the data associated with places that had previously been retrieved from foursquare. Before making an API call to retrieve

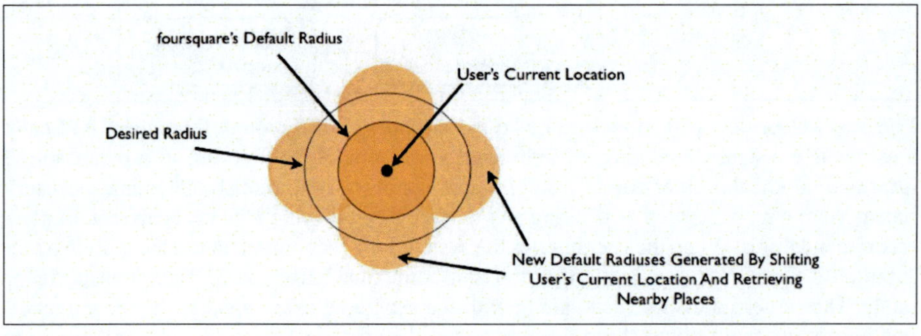

Fig. 3.7. This diagram describes how the radius of places near the user is incremented beyond foursquare's limit.

the tags of a place, it was determined whether or not that specific place had already been added to the database.

We believe that an important lesson to be learned from our studies is that although integrating a social networking site into a system can allow for a quick profiling of an individual, there will be always certain shortcoming from the offered API, mainly because it is not tailored to the specifics of the system that is integrating it.

3.6 Usability Inspection of I'm feeling LoCo

In this section, we inspect the usability of I'm Feeling LoCo using cognitive walkthrough methodology. The cognitive walkthrough is a practical evaluation method based on the work of Polson, Lewis et al. (1992) in exploratory learning. In a cognitive walkthrough the user explores the interface and with the system tries to accomplish a series of assigned tasks. The user first searches the interface, analyzing what the system enables them to do. Next the user selects the specific actions that appear to aid them in achieving their final assigned goal. The cognitive walkthrough helps identify the ease of learning, use and usability of an application. The 'I'm Feeling LoCo' usability inspection was done in different US cities (Portland, OR; Beaverton, OR; Santa Barbara, CA; and Goleta, CA) by eight different users under diverse transportation modes. Each user utilized the system on at least two separate occasions.

3.6.1 Users

The people selected for the cognitive walkthrough were foursquare users that had at least 20 check-ins. All of the participants owned a Smartphone and had utilized the navigation assistant equipped in their phone. Only two of the users in the study had utilized a personalized travel guide. Five of the users obtained their place recommendations from casual conversations with acquaintances and reading online reviews in sites such as Yelp.

3.6.2 Tasks

The tasks we requested each user to perform were:

- Find a place to eat while walking in downtown Santa Barbara or Portland.
- Find a place for celebrating with friends while being a passenger and navigator in a car near Santa Barbara and Portland.
- Find a place for studying while biking in Goleta, CA.

Due to resource limitations, only three users did the task involving biking.

3.6.3 Results

The findings of the cognitive walkthrough were divided in two: the results obtained for the main menu and the results obtained for the mobile map in which the users visualized the recommended places.

3.6.3.1 Main Menu Usability

The main menu presents the user's detected transportation mode as well as a series of buttons denoting possible moods for the user to select. The observations and feedback received for the main menu was very positive. All of the users in the study were able to correctly select the mood associated with the type of place they were requested to find.

Additionally all of the users were inquired about what option they would select if they wanted to:

- Find a place for studying.
- Find a place for doing cultural activities.
- Find a park for walking.
- Find a place for shopping.
- Find a few of the corporations that had offices in the area.

All of the users selected the correct mood for each of the above tasks. It was straightforward for them to understand the mapping between their selected mood and the type of places returned by the system. All users appreciated the whimsical names selected for the moods. Furthermore, the users were surprised that I'm feeling Loco could flawlessly detect their current transportation mode. All users made positive comments about this feature. The users also made positive comments about the 'I'm feeling LoCo' button. They liked that this option allowed them to discover places that were not within their normal pattern. The users viewed this option as a fun component of the system that enabled them to explore their surroundings more. Three users suggested an option for querying the system for directions to a specific place. They mentioned that many times while shopping they had specific stores they wanted to visit. Therefore an alternative for directly searching for a particular place would be beneficial.

3.6.3.2 Mobile Map Usability

The participants were overall satisfied with the places returned by the recommendation system. For the first users participating in the study, the place recommendation requests were done in small US towns, such as Beaverton, OR; Hillsboro, OR; and Goleta, CA. Because these small towns offered a very limited selection of places, the system's suggestions were not very relevant for the user and did not fully portray the person's interest. It was due to this situation, that we opted to perform the cognitive walkthrough in downtown Portland and downtown Santa Barbara, where we could guarantee that a larger subset of places would be present, and better results would be obtained. When the study was done in downtown Portland and Santa Barbara, all the users expressed that the suggestions were places they would be interested in visiting. All the returned places were relevant to the user and were located at a distance reachable to the user.

An interesting pattern we observed was that on the second day of utilizing the system, the foursquare usage of all participants had incremented. We believe that after noticing that the application considered their check-ins for the recommendation, the users felt motivated to check-in to more places they visited and obtain therefore a far more tailored recommendation. On the second trial of the system, the users were pleased with how the recommendations better matched their personal preferences. We believe this shows that our system promotes the usage of location based social sites.

While visiting the downtown area many users were surprised that a few of the most popular and typical restaurants were not recommended by the system. This made us consider that in a location recommendation system, there are places that all users should be exposed to. Explicit serendipity should be enabled at all times and not only when there is not sufficient user content (Our current system explicitly recommends city landmarks, only when there is a lack of user data).

The participants enjoyed the idea that the recommendations changed accordingly to their transportation mode, but two of the users expressed difficulty in reaching the destination (These two users were the only ones that were not locals in the city where they tested our application). The difficulty arose because I'm feeling LoCo simply displays markers with the top recommendations, but offers no instructions as to how the destination can be reached. We conclude that this feature be integrated to the interface.

Additional suggestions that we derived from user feedback included modification of the type of presented map, providing detail based on the user's transportation mode; for example: show bike paths when biking. We also considered that eyes-free interaction with the system could be beneficial, especially while biking or driving. We plan on integrating an eyes-free approach similar to that of Savage et al. (2010). In summary, the cognitive walkthroughs demonstrated a good level of usability, yielded positive reactions from the participants, and generated ideas for further improvement.

3.7 Conclusions

In this study we presented a novel personalization system, which considers automatically inferred user preferences and spatiotemporal constraints for location recommendation.

This system can serve as an early research example, providing an outlook on future developments in personalized LBSs, in which the majority of the data utilized for generating the recommendations is automatically collected from different information sources, freeing the user from completing exhaustive surveys or manually updating their current state. Our system models personal spatiotemporal constraints by automatically discovering the user's mode of transportation. To infer personal preferences we proposed using a bag of words approach. We also presented the results from cognitive-walkthrough-style evaluations of our location recommendation system, comprising results from eight individuals, searching for places that best satisfied their personal priorities in the US cities of Goleta, CA; Beaverton, OR; Santa Barbara, CA; and Portland, OR. The cognitive walkthrough sessions demonstrated that our proposed system can be utilized to deliver useful location recommendations.

We believe that, the recommendation procedure could be improved by integrating other sources of user information, such as a person's Google calendar. The system would now consider the fact that the user has appointment X at time Y in location Z, creating a new spatiotemporal constraint. The inclusion of semantics may also provide additional useful information about the user. It would be interesting to analyze if meanings to certain visits could be inferred and effect the user model. For example, is there a meaning related to going to a nightclub every weekend or visiting a church every Sunday?

As this study concentrated on improving the user experience when utilizing a location recommendation system, little attention was paid to the energy efficiency. In our future work, we envision implementing a duty cycle on the mobile device, which could significantly extend the battery life. Instead of running the GPS and the accelerometer at all times, usage could be based on a duty cycle from the user's behaviors. We believe that if the user is actively using foursquare, it would be possible to construct a model of user behavior and predict when the user is at home, at work or sleeping. The system could predict user activity and therefore turn off sensors utilized for determining user activity.

Other areas of future research, which could provide a better user interaction with our system, are: changing the user interface accordingly to the person's form of transportation. For example, if the user is driving, an eyes-free interface could be presented and allow the user to keep their eyes on the road rather than on the recommendation system. Furthermore, if the user is biking, for example, the system could display routes of bike paths.

Our current system, due to foursquare's lack of user ratings to venues they visit, assumes that constant visits to a site means the user likes the place. But this is not always true, a user could frequent a place yet not enjoy it. In the future we hope to integrate to our study other social networks that do offer rating information (such as yelp). This would allow our system to return much better recommendations.

Despite its preliminary character, the research reported here indicates that it is possible to construct an adequate location recommendation system without requiring the completion of extended and complicated surveys. Furthermore, this study shows it is viable to integrate contextual information of the environment surrounding the user and user activities into the location recommendation engine.

Acknowledgments

This work was partially supported by a UC MEXUS-CONACYT fellowship, as well as NSF grants IIS-0747520 and IIS-1058132.

References

Bao L. and Intille S (2004) Activity recognition from user-annotated acceleration data. ACM Trans. Sen. Netw, pages 1–17.Bertin J (1967) Semiology of graphics: diagrams, networks, maps. University of Wisconsin Press, 1983 (first published in French in 1967, translated to English by Berg WJ in 1983)

P. Baudisch (1999) Joining Collaborative and Content-based Filtering. In: Proceedings of the ACM CHI Workshop on Interacting with Recommender Systems. ACM Press.

Espeter Martin and Raubal Martin (2009) Location-based decision support for user groups Journal of Location Based Services, Vol. 3, Iss. 3

Herzoh A. Regula and Jerald G. Bachman (1981), Effects on Questionnaire Length on Re-sponse Quality, The Public Opinion Quarterly, Vol. 45, No. 4 (Winter, 1981), pp. 549–559

Kim Jinyoung, Kim, Hyungjin, and Ryu Jung-hee (2009). TripTip: a trip planning service with tag-based recommendation. In Proceedings of the 27th international conference extended abstracts on Human factors in computing systems (CHI EA '09). ACM, New York, NY, USA, 3467–3472

Kobsa. A. (2001) Generic user modeling systems. User Modeling and User-Adapted Interaction, 11(1–2):49–63.

Marmasse, N. and Schmandt, C (2000). Location-Aware Information Delivery with ComMotion. In: Proceedings of the 2nd international Symposium on Handheld and Ubiquitous Computing (Bristol, UK, September 25–27, 2000).

Mulvenna S. M and Buchner A. (2000), Personalization on the net using web mining. Communications of the ACM, 43:122–125.

Polson, P.G., Lewis, C., Rieman, J., and Wharton, C(1992). Cognitive walkthroughs: A method for theory- based evaluation of user interfaces. International Journal of Man-Machine Studies 36, 741–773.

Reddy S., Mun M, Burke J, Estrin D., Hansen M and Srivastava M. (2010), Using mobile phones to determine transportation modes. ACM Trans. Sen. Netw, 69(2):13:1–13:27

Resnick H. V. P. (1997), Cross-representation mediation of user models. Communications of the ACM, 40(3):56–58.

Savage N.S., Ali S.R., Chavez N.E. and Savage R., (2010), Mmmmm: A multi-modal mobile music mixer. In Proceedings of the 2010 conference on New Interfaces for Musical Expression, NIME '10, pages 395–398, 2010

Seiler S, 2010, The impact of search costs on consumer behaviour : a dynamic approach. London School of Economics. Collaborative Management Talks.

Sugimoto Takeuchi Y, 2006, M. CityVoyager: An Outdoor. Recommendation System Based on User Location History. In Proc. UIC'2006, Springer Berlin

Rinner, C. and Raubal, M., 2004. Personalized multi-criteria decision strategies in location-based decision support. Journal of Geographic Information Sciences, 10, 149–156.

Chapter 4

A Mobile LBS for Geo-Content Generation Facilitating Users to Share, Rate and Access Information in a Novel Manner

Helmut Schrom-Feiertag[1], Patrick Luley[2] and Lucas Paletta[2]

[1] Dynamic Transportation Systems, Austrian Institute of Technology, Vienna
helmut.schrom-feiertag@ait.ac.at
[2] Remote Sensing and Geoinformation, Joanneum Research, Graz, Austria
{patrick.luley, lucas.paletta}@joanneum.at

Abstract

While mobile applications typically offer access to standardized 'basic' geo-content, there is evidence in the human sciences that people ac-tually prefer subjective information sources for decision making, e.g. personal stories about experiences by family and friends. The success story of content communities in web applications confirms the wide acceptance of innovative information systems that offer the potential to consume, to produce and to rate personalized data. Therefore in this paper an approach for a mobile, interactive and integrated system is presented that do not only deliver basic geo-referenced information, but also allow the users to create location aware information for themselves and other users: information like hints and personal experience will be geo-referenced, time stamped and annotated with text and keyword information. It is stored and exchanged between community members. This individual information creates information content which is continuously growing and updated. Image based and text based information retrieval in combination with location information are used to provide easy access to relevant information. In this paper we outline the system architecture and components that enable these new approaches also providing augmented reality navigation. The approach was tested in a field test study and results and open issues are given. The system worked well and could be applied to future experiments in order to gain more insight in the mobile users' behavior in real contexts.

Keywords: location-based, information, mobile, service, user generated content, tagging, rating, recommendation, object recognition, spatial-analysis, augmented reality navigation

G. Gartner, F. Ortag (Eds.), *Advances in Location-Based Services*, Lecture Notes in Geoinformation and Cartography, DOI 10.1007/ 978-3-642-24198-7_4, © Springer-Verlag Berlin Heidelberg 2012

4.1 Introduction

Internet enabled mobile phones have revolutionized the communication and with it the lifestyle of people. An increasing number of mobile phones allow people to access the internet wherever they are and whenever they want. In combination with GPS they can obtain on one hand information on nearby events like cinema, concerts or parties and on the other hand information on places like city maps, sights, restaurants, museums or hospitals by the use of location based service (LBS).

According to studies there is a huge interest in mobile and location-aware information regarding leisure and local services. While mobile applications typically offer access to standardized 'basic' geo-content, there is evidence in the human sciences that people actually prefer subjective information sources for decision making, e.g. personal stories about experiences by family and friends (Case, 2007). The success story of content communities in web applications confirms the wide acceptance of innovative information systems that offer the potential to consume and to produce personalized data.

Therefore in this paper we present the results of the project Captain Kirk, funded by the Austrian strategic initiative IV2Splus, which demonstrates a new approach for a mobile, interactive and integrated system that do not only deliver basic geo-referenced information, but also allow the users to create geo-referenced data for themselves and other users to provide hints and personal experiences: content with text, images and keywords can be created and is associated with current location and time. The information is stored and exchanged between community members. This individual information creates an information repository which is continuously growing and updated. Image and text based information retrieval plus item-based collaborative filtering in combination with location information is used to provide easy access to relevant information on mobile devices. In this paper we outline the system architecture and components that enable these new approaches. The proposed system was tested in a field study to identify the acceptance, the perceived benefit and ideas for future improvements.

The rest of this paper is organized as follows. *Section 4.2* sketches the state of the art. In *Section 4.3,* the modular system architecture of the proposed approach is outlined and the components for easy information access, creation and navigation support are explained in detail. *Section 4.4* describes the test set-up and the results for the field study. Finally, *Section 4.5* concludes with an outlook of future improvements.

4.2 Related Work

Finding the nearest service, accessing traffic news, getting help with navigation in unfamiliar environments or obtaining a local street map. These are just a few of the many location-based services. The market for information services is considered to be one of the most promising in terms of global revenue for developers and providers alike.

New powerful smartphones with high processing power and large storage capacity boost the transfer of services to be found primarily on the web to the mobile environment by the means of mobile applications. The major challenge is to support users with contextually adapted information through mobile devices. How to incorporate user's location in more intelligent ways has been ongoing research for several years (Beeharee & Steed, 2007; Meng, 2005).

4.2.1 Social Systems

The majority of the mobile social systems provide services similar to friend finder applications, but in different contexts. Some are for well known friends only, other give the opportunity to contact passers-by registered to the service.

Other services try to share knowledge in a location based way, some people sharing knowledge directly, some are also give access to people close to you which can help you to find the right way, shop or whatever.

The main approach of social services is to enable social interaction. Functions of mobile social networks are friend finders, see who is close to you and let you get in contact (i.e. Google Latitude, Facebook Places and so forth), or if someone need help or information some services let you get in contact with nearby users. Some services try to share knowledge in a location based way help finding answers to all questions related to a city, or share inside information on what's hot and what's not in a town within a community (i.e. Askalo.com). Also playful approaches are represented, i.e. FourSquare let people meet up with their friends and let them earn points and unlock badges for discovering new places, doing new things and meeting new people.

As the web opens up to more social, collaborative media creation, the content is becoming less restricted by corporate guidelines. Users are no longer just browse the web, but are the web and publish their own content.

4.2.2 User Generated Content

User generated content (UGC) refers to material on websites that is produced by the users of the website and has become tremendously popular. All digital media technologies are included, such as digital video, blogging, podcasting, forums, review-sites, social networking, mobile phone photography and wikis. Some of the most frequently visited sites on the Internet are primarily user generated (i.e. Wikipedia, YouTube). One of the challenges of user generated content is that it may be inaccurate and may not be appropriate for the purpose to obtain facts. To overcome these hurdles there are a variety of approaches. A very promising approach is the rating by user.

Another new way to organize content for navigation, filtering and search is "tagging" defined through new characteristics of social systems which allow users to mark content with descriptive terms, also called as keywords or tags. It describes a process by which users can add metadata in the form of keywords to shared content and has been an important research area in the field of human computer interaction (Golder & Huberman, 2005). Collaborative tagging on the web was gaining ground as a new paradigm for web information retrieval, discovering and filtering. Some websites include tag clouds as a way to visualize tags (Lamere, 2008).

4.2.3 Recommendation

In everyday life people tend to rely on recommendations from other people. In online services recommender systems support people to find the most interesting and valuable information for them. Recommendation systems make use of collaborative filtering techniques. The fundamental assumption of collaborative filtering is that users rated items similarly or having same behaviours will rate or behave on other items similarly. Collaborative filtering can be grouped into memory-based, model-based and hybrid techniques (Su & Khoshgoftaar, 2009). Memory based algorithms again can be split into user-based and item-based algorithms. User-based collaborative filtering calculates the similarity of users and creates predictions of preferences on new items based on similar users. In item-based techniques first users have rated the same items are searched and then to apply a similarity computation to determine the similarity between the co-rated items of the users (Sarwar, Karypis, Konstan, & Reidl, 2001). There are many different methods to compute similarity between users or items.

Model-based collaborative filtering methods are designed to learn and recognize complex patterns out of training data and compute predictions based on the learned models.

Hybrid collaborative filtering systems combines collaborative filtering with other recommendation techniques to improve prediction and recommendation performance, typically with content-based systems that are analyzing the content of textual information such as documents, messages or item descriptions to find regularities in the content (Pazzani, 1999). The main challenges collaborative filtering has to overcome by real-world applications are data sparsity, scalability, synonymy, privacy and so forth.

All these approaches have their weak and strong points and currently no cure-all solution is available and recommender systems are an active area of research.

4.2.4 Navigation

The increasing availability of powerful smartphones, which include multiple sensors for localisation like GPS, WIFI, network cell detection and accelerometers, enables a new quality of services in the field of pedestrian navigation. Currently the most popular application in this area is an add-on for the application Google Maps, which is preinstalled on all Android devices.

The system uses a three dimensional map view (perspective view) superimposed with the route, which rotates automatically according to the built-in electronic compass of the smartphone. In addition to this the inclination of the smartphone is used to change the perspective of the view (see *Figure 4.1)*. Beside street maps also aerial images could be used as a map overlay in order to increase the comparability of the map and the real world, which facilitates the navigation task for the user (Google Inc., 2010). Until now a real time augmentation of the camera view with navigation instructions is not offered.

The only available augmented reality navigation system for smartphones so far is the application Wikitude Drive (Mobilizy GmbH, 2010), which is optimised for car navigation. The system complements classical real time navigation applications for cars with an augmented camera view finder, which is shown in *Figure 4.2.*

This application lacks the support for pedestrian navigation. Despite the fact that the same technology could be adopted for pedestrian navigation purposes you

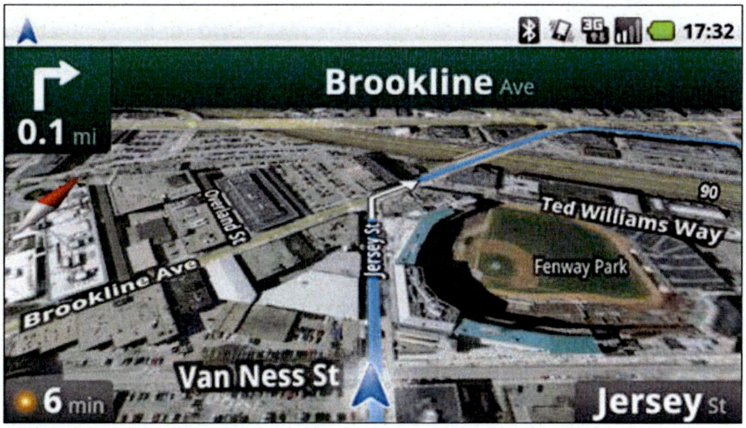

Fig. 4.1. Pedestrian navigation mode of Google Maps

Fig. 4.2. Wiktude Drive use case example

have to take into account that the needs and use-cases of pedestrians are completely different and an integration of public transportation facilities is indispensable.

The identified limitations of state-of-the-art navigation applications have been addressed in detail within the project Captain Kirk and led finally to an implementation able to match the requirements of pedestrians. This service will for example become highly useful for the application with specific social target groups that are not fully technology affine and from this favour highly intuitive user interfaces (Paletta et al., 2010).

4.2.5 Object Recognition

For human beings, the use of visual information for information access and decision making is most natural. With the upcoming of miniaturised cameras in mobile devices, it has been for the first time being possible to combine visual search with computational information processing in a most convincing and intuitive way. One of the first pioneers for mobile image based information retrieval has been (Tollmar et al, 2004), addressing the visual search in the real world for accessing information in the world wide web. In Europe, the EC funded project MOBVIS[1] has pioneered the combination of visual information with contextual awareness of pedestrians (Fritz et al., 2005 and 2006) in order to cut down otherwise untractable search spaces. At the same time, the service "Nokia Point & Find" has been developed that is limited to a specific set of Nokia smartphone devices. One can use this application to find information on movies by pointing the camera at movie posters and then view reviews, or find tickets at nearby theatres. Current uses include city landmark tagging, barcode scanning for comparison shopping and 2D barcode scanning and finding information related to products and services based on content provided by third party publishers. More recently, more services have been marketed, such as

[1] www.mobvis.org

Fig. 4.3. Google Googles mobile image recognition system. The service is available with all Android service platforms. However, it is sensitive with respect to image databases predefined in Google. This makes the service in general not applicable to all purposes, such as for the mobile service presented in this work.

from SnapTell, kooaba, JAGTAG and Google Goggles. Google Goggles *(Figure 4.3)* was developed for use on Google's Android operating systems for mobile devices. Google announced its plans to enable the software to run on other platforms, notably iPhone and BlackBerry devices. It is used for searches based on pictures taken by handheld devices. For example, taking a picture of a famous landmark would search for information about it, or taking a picture of a product's barcode will search for information on the product. However, the image databases as the basis for the image recognition service have to be defined by Google and in this sense the service is not applicable to arbitrary purposes.

In Captain Kirk, we present an image recognition service that is particularly suited to urban applications in the context of public transportation.

4.3 Proposed Approach

As of today, most of the mobile information services are unidirectional, providing information only. Extending the research to our domain reveals new ideas and designs that are adapted for the mobile applications and user interface concepts. The goal was to construct a mobile, interactive and integrated system that do not only deliver location aware information, but also allow the users to create location based information for themselves and other users to provide hints and personal experiences: content with text, images and keywords can be created and is associated with current location and time. The information is stored and exchanged between community members. The challenges therefore are to handle the information overload and provide ways to get accurate information.

In this section we describe the modular and general architecture that was used for the development of the interactive, multimedia, location-based applications, providing an extra level of service to the users. As with any application it's important to consider the target audience. In this case we made interviews with eight target audience members, persons between 18 and 50 years old and public transport users, to identify the requirements: easy information access, creation and navigation support.

Due to technical requirements the Android platform was chosen for the implementation. The Android platform provides an open basis, multi-sensor access (acceleration sensors, position sensors, electronic compass, GPS) and supports background processes that are necessary for the implemented services.

4.3.1 System Overview and Components

Android applications are basically composed of one or more application components (activities, services, content providers and broadcast receivers). As shown in *Figure 4.4* the client application has a modular system concept and consists of five components. Each component is independent of the others but work together to form a cohesive user experience.

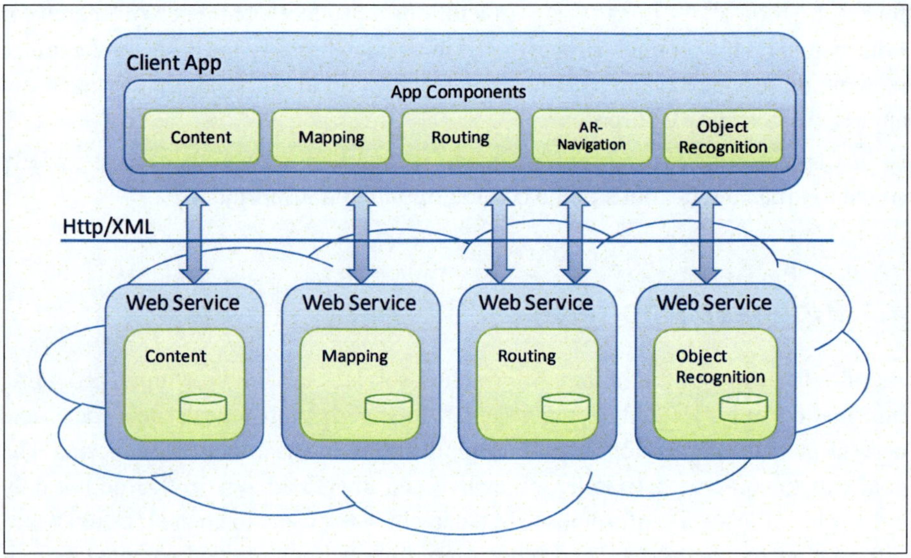

Fig. 4.4. Service oriented system architecture

First the **Content** component is responsible for the retrieval, creation, rating and recommendation of content. Second, the **Mapping** provides an interactive map and is used to show the current position, positions of geo-referenced content and also route information. Third, the **Routing** enables to query public transport routes using real-time information. Fourth, the **AR-Navigation** facilitates augmented reality technology to assist pedestrian navigation. Fifth, the **Object Recognition** enables image based information retrieval by taking photos of brand logos. Each of the components has its counterpart web service und communicates over the http-protocol and exchanging data in XML format.

4.3.2 Client Application

Location based services typically has to solve user requests like: Where am I? What is around me? How can I get there? We introduced a model for information retrieval to answer such questions with different options to improve the answer quality.

The normal information access is based on a map where users can identify their current location and query the space for information, events or objects in their surroundings, like the question answer model from (Heidmann, 1999).

The main interface to content in this system is the client application on the smart-phone. The application must be easy to use and provide easy access to relevant information. As shown in *Figure 4.5* the client provides access to information on events, attractions and community content. In order to alleviate the access it can be chosen between different filters. A location based search deliver information about

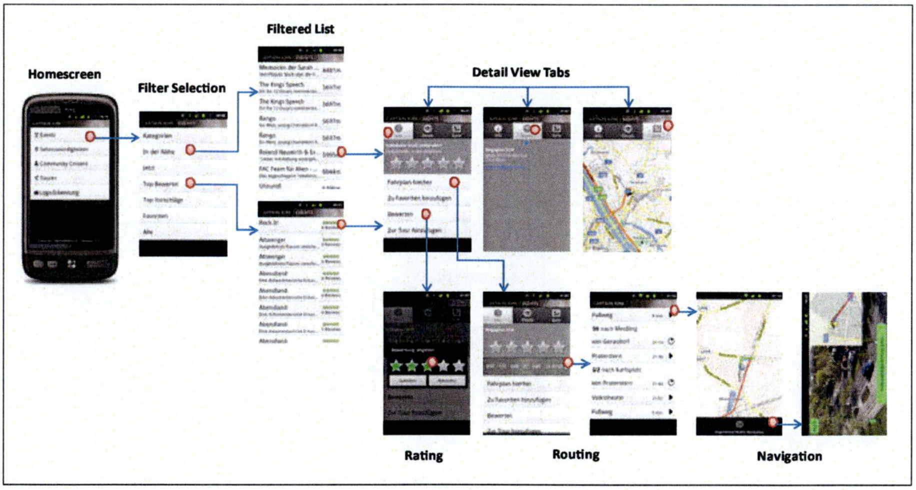

Fig. 4.5. Screen flow for accessing events, sights and community content

the surrounding and the results can be filtered on distance, rating and a recommendation score. Events get also filtered on their start time. Most interesting information can be added to a list for personal favourites or retrieved again over an access history. After selecting an entry from the filtered list the details are shown on a tabbed view, an "Info" tab for general information and a menu to retrieve a route, to rate the entry, add the entry to the favourites or add it to a personal tour. On the second "Details" tab the details like additional text, the address or an URL is shown. On the "Map" tab the location is shown in a map providing general manipulations like pan and zoom. If the user retrieves a route, a list shows the route segments and the transport lines. If one entry of the route is selected the location is shown on a map and walkways are shown with green lines and public transport with red lines. For walkways the augmented reality navigation can be activated.

The described handling with the client application is the same for events, sights and community content. First the type of content has to be selected, then a filter to be chosen and then one entry from the list. Finally the details of the entry are shown. Each entry has a location and a route with real-time information from the current position of the user to the target location can be calculated.

How new community content is created can be seen in *Figure 4.6.* A text for title, description and keywords has to be entered and an image can be taken using the camera application or chosen from the image gallery. To support the keyword selection over the "Keyword" button a pop up window can be opened showing the most recent keywords used in the close surrounding. Keywords can be selected by checkboxes and are applied to the keywords text field. This helps to find corresponding keywords and helps to improve the tagging system's effectiveness. Keywords can also be added by typing words and separate them by commas. Tags are used for text search and will improve the information discovery.

Each content entry can be assigned with a personal five start rating from 1 (bad) up to five (very good) were an average value from all ratings is calculated. Based on these ratings personal recommendations are calculated using an item-to-item

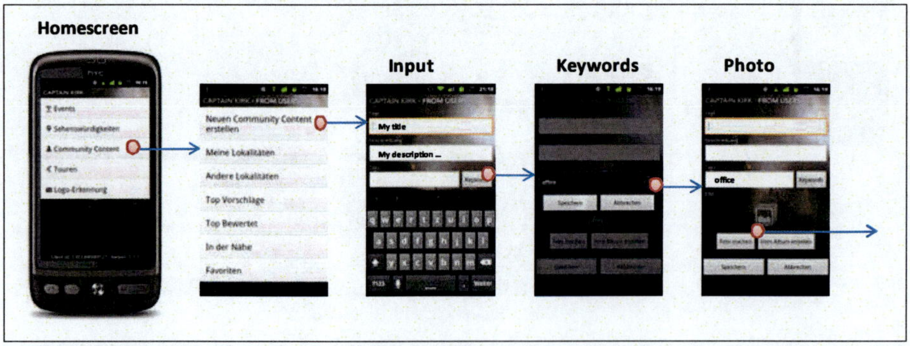

Fig. 4.6. Screen for community content creation

collaborative filtering recommender system (Horozov, Narasimhan, & Vasudevan, 2006). The used approach is an enhanced easy to implement collaborative filtering model based on average rating differential for high scalability and low latency and can compete against more expensive memory-based schemes (Lemire & Maclachlan, 2005).

4.3.3 AR-Navigation

This chapter describes the augmented reality navigation component for pedestrians developed within the project Captain Kirk by Joanneum Research. The aim was to develop and implement an assistive service, which facilitates augmented reality technology for the purpose of pedestrian navigation. Especially in situations of uncertainty with regard to navigation, like the junction shown in *Figure 4.7,* augmented reality could give a quick and intuitive answer to the question: Where should I go now?

In comparison to classical navigation solutions, Captain Kirk sets out to embed navigation instructions directly in the camera view finder of the smartphone specially rehashed for the needs of pedestrians. *Figure 4.7* shows a junction with three different possibilities of turn. In comparison *Figure 4.8* shows the view of the smartphone corresponding to this navigation situation with superimposed navigation instructions. The augmented reality view consists of the following components:

1. Camera view: Showing the camera view finder in real time in the background of the application user interface.
2. Distance Value: The current distance from the current position of the user to the next way point in metres is shown in the upper left corner.
3. Navigation instruction: The position of the next way point is shown with a blue marker. On the bottom there is a text box showing the current navigation instruction. When the next waypoint is not in the camera view, the system tells you by

Fig. 4.7. Example of a pedestrian navigation situation

displaying arrows in which direction you have to turn. If the next waypoint is a station also information on the public transport vehicle is given.

4. Rotating map: Additionally a traditional 2D map with the route is shown. The map is always centred on the current position of the user and rotates automatically according to the spatial orientation of the user. The colour of the route can change, whether it is a walkway or public transport.

Once the user has entered the public transportation system it doesn't make sense to show the augmented reality view since a deviation from the route is not likely. In that case the system switches automatically to a neat map view showing the route and the station where to change or exit the bus or train.

Figure 4.9 shows the overview information a user would get after entering the subway. Beside the current navigation instruction (Enter the U2 in the direction "Praterstern") it also shows the upcoming instruction just to make sure that the user

Fig. 4.8. Example of the augmented reality navigation assistance

Fig. 4.9. Map view while using public transport

can prepare for his next action. The map offers in this mode all common interaction possibilities like pan and zoom. The map mode is also used when there is no GPS reception, which can occur when the user is somewhere indoors like in subway stations where no free sight of GPS satellites is possible. One point to mention is that augmented reality navigation can only be used if the system has a very accurate position of the user. Augmented navigation instructions are not reliable anymore once the spatial geometry of the camera view is calculated on the basis of a wrong position. The consequence would be that wrong navigation instructions lead to confused and dissatisfied users, which has to be strictly avoided.

Sometimes, while using the navigation component, it could occur that the system is not able determine the current situation of the user to due problems with the localisation or for example when the routing is started while you are currently using public transport. In that case the system just asks the user about is current situation or position along the planed route.

By using WiFi access points for localisation purposes within subway stations, the system can assist the user by prompting instructions on how to switch from one vehicle to the other within the public transportation network. But, if a change is not detected, due to the lack of WiFi signals, then the system again just asks the user about his current situation.

Figure 4.10 illustrates how the system asks the user about his current situation. The question is: What are you currently doing? In this example the user has three different choices: (i) The user is currently using the subway number U1 towards Karlsplatz, (ii) the user is currently changing at the station Karlsplatz to the number U4 and (iii) the user is already using subway number U4 towards Pilgramgasse. Once the user defines his current situation, the system is able to initialise the navigation assistance along the planed route again.

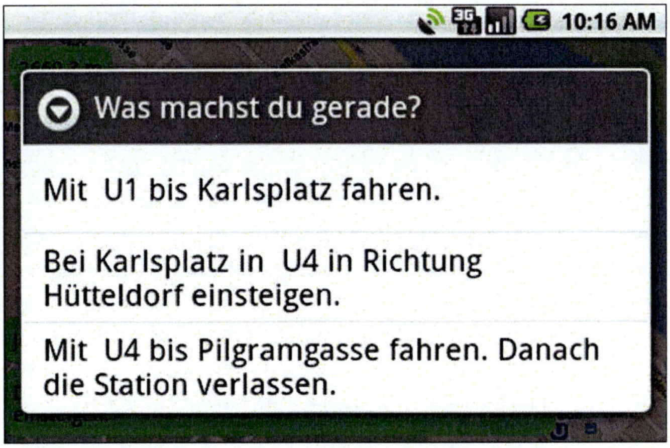

Fig. 4.10. Asking the user after his current situation

4.3.4 Object Recognition

The Object Recognition Component of the mobile service has been implemented in Captain Kirk on the basis of previous research results from Joanneum Research (Fritz et al., 2006, Amlacher et al., 2009). (Fritz et al., 2006) demonstrated that the use of informative local image descriptors for probabilistic object recognition is highly advantageous both in requirement on response time but also with respect to object recognition accuracy. (Amlacher et al., 2009) presented an efficient methodology to exploit the knowledge on geo-information to progress even more with respect to time complexity and computational accuracy in the results. Concretely, probabilistic knowledge about the local environment and its classification potential provides priors to impact the generation of distributions on class hypotheses.

For the application in Captain Kirk, company brands had to be localized in mobile phone based image material. Based on the recognition of brands in the image, annotation information is associated exactly to the image location where the brand had been localized by the automated image analysis methodology.

In order to detect brands in single image, we applied a straight forward methodology. Firstly, informative local image descriptors (i-SIFTs, Fritz et al., 2006) were extracted from the image and associated with a label of the most probable class. Descriptors of same labels are then grouped and a localization is provided by accepting firstly only those labeled descriptors that lie within Mahalanobis distance of the covariance of similarly labeled descriptor centers. Outliers are iteratively rejected and finally the mean center is computed to provide the final estimate of the complete brand centre.

Figure 4.11a depicts the complete collection of all 100 brands that have been individually attributed in the Captain Kirk project. *Figure 4.11b* shows the localization result (green marker) of the automated image analysis component in terms of the Mahalanobis distance with respect to the distribution characterized by the covariance of classified local image descriptors (Fritz et al., 2006). The distribution of *Figure 4.11c* visualizes how well the image based classifier can discriminate between the 100 object hypotheses, given the image information in *Figure 4.11b,* clearly favoring the correct object hypothesis #11.

In a thorough test phase on the 100-classes discriminator using the mobile image material with 1 Megapixel image resolution, a recognition rate of ≈94% was achieved. However, the processing time on this image resolution had to be reduced since the server on average took 2.5–4.5 minutes to process all extracted local image descriptors (ca. 70 000 per image). Therefore, a descriptor selection algorithm was applied, restricting processing only on most promising descriptors. Promising descriptors had been dedicated to be those that would not deviate too much in scale, thus too large and too small had been refused by the descriptor selection, arriving finally at an average processing time with 15–25 sec. per full resolution image, using ≈23 000 descriptors with the same recognition rate of ≈94%.

The brand based mobile vision service provides the user a very intuitive and fast way to access information about points of interest – here, brand and from this shop information. The service has been attributed by the users to represent a useful and natural interface for gathering in the urban environment.

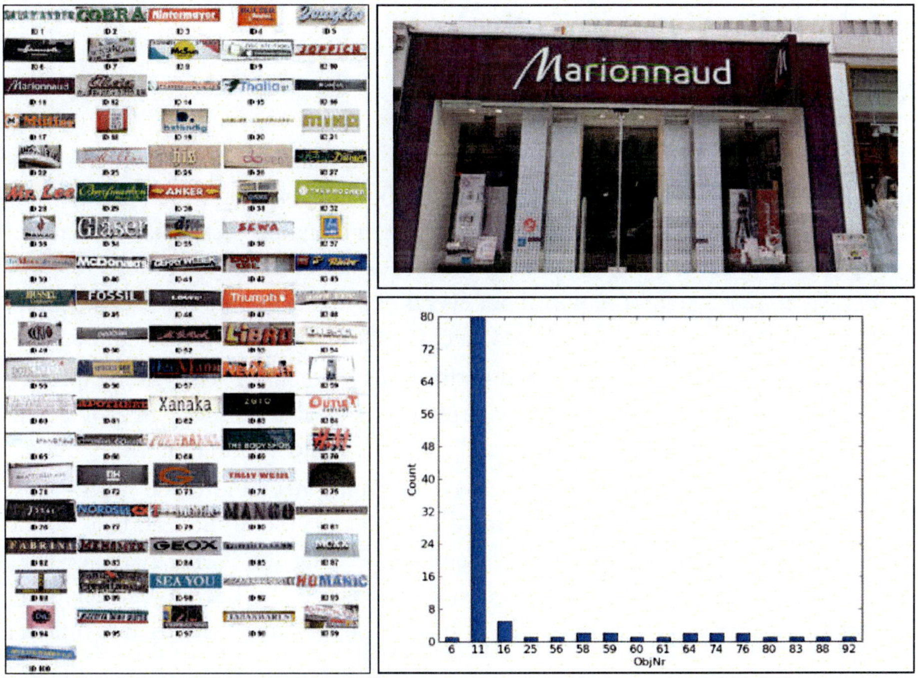

Fig. 4.11. Mobile service component for the recognition of company brands (logos). (a) Overview on the database of 100 brands, (b) sample correct localization of logo, (c) distribution of object hypotheses by number of local image descriptors voting for a specific object hypothesis and correct majority vote.

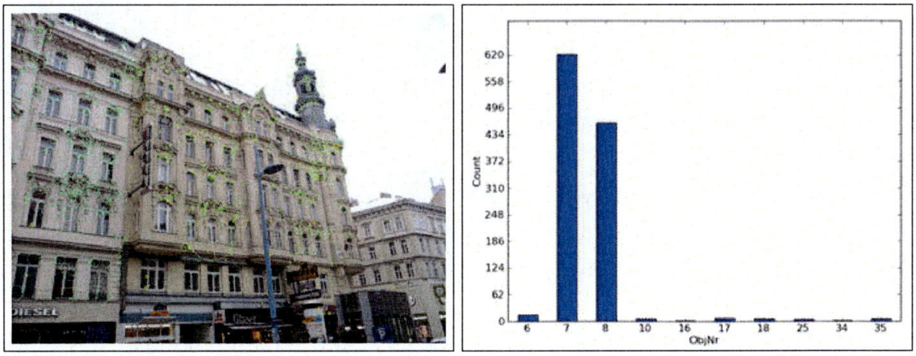

Fig. 4.12. Mobile service component for image based augmented reality for the annotation via recognition of facades.

Furthermore, the mobile vision service provided a component to access information about whole buildings, so as to get information about cultural information. As depicted in *Figure 4.12,* the service works similarly as for company brands.

4.4 User Studies and Evaluation

4.4.1 User Experience

The evaluation of the system was split into two phases. In the first phase the basic usability and the acceptance was tested in a laboratory setting which seems to be sufficient when studying user interface and navigation issues. Test users from target groups had to fulfil different tasks to get feedback about the different features. The tasks were to retrieve information about events and sights, to rate them, to assemble locations of events and sights to a route and finally to test the logo recognition.

The majority of all the users could perform the given tasks. Navigating through the application worked without major problems. But also areas for improvement, especially in categorization of events and real-time route information were found.

The logo recognition was perceived as something new and made fun to most of the users. Here too, there were some difficulties that could be lead back to the early state of implementation of the prototype. During the user experience tests the user feedback about problems and suggestions were collected and helped to identify necessary improvements. The significant suggestions were used to refactor the design.

4.4.2 Field Study

The research presented in this paper aimed to develop and test different aspects of location based services in the context of user generated content and pedestrian navigation. The hypothesis of the study was that there is evidence that people prefer subjective information sources for decision making and information systems such as content communities meets the requirements to consume and produce such personalized information.

From our research objectives the following questions that lead to the following set-up of our field study are derived. Will people share information and their experiences using a mobile location based service? Are the users supported properly by the developed system?

In the second phase field tests were conducted in the Mariahilfer Strasse in Vienna (6th district, the section between Neubaugasse and Zieglergasse), a very popular shopping area. In this area exemplary content about events, sights, reference logos for object recognition and community content was prepared. The test sessions took place in the last two weeks in April 2011.

To evaluate the application design test users were solicited to test the application on their personal smartphone and to provide their feedback in a final online questionnaire to gather information from the users about their skills, experience, preferences and opinion. The questionnaire was prepared with 48 questions about background information such as age, gender, preferred means of transport, frequency of public transport use, exiting experiences with smartphones and mobile applications, experiences made in the field test, feedback about the features and the perceived benefits.

The users were asked to complete the following tasks: 1. retrieve information about events, sights and community content, 2. make use of the routing service, 3. create at least 10 new content entries, 4. rate at least 20 entries, 5. test the object recognition with shop logos and finally 6. evaluate the augmented reality navigation. In *Figure 4.13* a heatmap over all keywords shows the spatial distribution of UGC during the pilot test.

From a pool of registered participants, 11 respondents installed the application on their smartphones and completed both the field test and the online questionnaire.

Keeping in mind the difficulties entering text on smartphones with touch screens, despite word completion, it was interesting to analyse the number of characters used for community content. In total 137 new entries in the community content were made and with 19 entries from pre-tests a total of 156 entries are in the database. In contrast a community content entry can have multiple keywords assigned. In total 221 keywords were entered.

Fig. 4.13. Heatmap over all keywords showing the spatial distribution of UGC

For the titles the shortest text had 2, the longest 30 and the average had 11.68 characters (median 11). The shortest descriptions had 3, the longest 102 and the average had 23.07 characters (median 20). In contrast a community content entry can have multiple keywords assigned. In total 221 keywords were entered, the shortest one had 2, the longest 24 and the average had 6.7 characters (median 6). To show the use of keywords 98 of the entries had 1 keyword, 23 had 2, 12 entries had 3, 9 had 4 keywords and only one had 5 keywords. *Figure 4.14* shows the tag cloud of the 149 unique keywords, the top keywords "bar" used 9 times, "restaurant" 8 times, "cafe" 7 times and so forth.

A critical characteristic of tagging systems that promote social navigation is their vocabulary, the set of tags used by members of the community. Instead of imposing controlled vocabularies or categories, tagging systems' vocabularies emerge organically from the tags chosen by individual members.

Personalization is the key to many successful services. The applied solution is to ask user to rate community content and make recommendations out of the ratings. Therefore it is a crucial point that users make a lot of ratings, not only for very good items, giving five stars, also to give lower ratings to find dislikes. *Figure 4.15* shows the distribution of the applied ratings with one (bad) to five stars (very good). The most ratings were made using five stars (very good, 83 times).

Finally in the test with end users, the object recognition component showed satisfying performance, both with respect to the time required for the server upload of images and return of recognition results (ca. 10–20 sec.) and the achieved recognition performance (≈ 93% of the uploaded images were recognized correctly).

4.5 Conclusion

The aim of the project was to develop a mobile LBS that enables users of a community to create, share, rate and access geo-content in a novel and improved manner. For the improvement of the information access rating and tagging was used to organize content and to overcome hurdles in user generated content. Object recognition was a further novel way to access information and showed satisfying performance. The experience gained for AR navigation showed that people using AR navigation in foreign environments have found their way faster than using a conventional 2D map visualisation. In general the sorting on entries based on their ratings, the recommendations and the AR navigation were perceived as very useful.

Some issues came up during the field tests not covered by the client application prototype. People creating community content will retain control over their entries; they will have the possibility to maintain their entries like changing or deleting. Also the possibility to control the access was mentioned, user want to control

Fig. 4.14. Tag cloud from the keywords

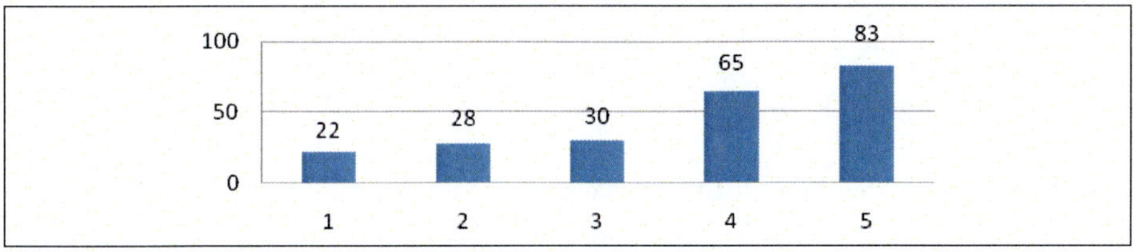

Fig. 4.15. Distribution of the rating values 1 to 5, total 228 ratings.

similar to Facebook or Google+ who can view their content, friends only, friends of friends, groups or set public to everybody. Some entries describing events, so there should be the possibility to set an expiration date that determines when the entry will be removed automatically from the repository. A critical characteristic of tagging systems is their vocabulary which emerges organically from the tags chosen by individual members. Here a solution that reduces the growth of keywords having a similar meaning could improve the performance for information retrieval.

A group of 12 test users will have a strong bias and cannot not be seen as representative but especially in usability research groups with 5 to 10 users will be large enough to discover the most significant difficulties and to give a glue of possible use cases and types of information people tend to share in communities.

One of the challenges will be to get people into a community and to keep them active for creating and sharing information to provide content interesting and up-to-date.

From the questionnaires it can be concluded that most of the participants rated the application as "useful" and are willing to use the application in future if some improvements will be made.

In conclusion the system worked well and could be applied to future experiments in order to gain more insight in the mobile users' behaviour in real contexts.

Acknowledgements

This work was part of the project Captain Kirk and has been funded by the research program line i2v within the framework of the Austrian strategic initiative IV2Splus "Intelligent Transport Systems and Services plus" under the project number 820714.

Project partners were the AIT Austrian Institute of Technology (LBS for community content and spatial analysis), Joanneum Research (object recognition and augmented reality navigation), Fluidtime (Android client application) and Wiener Linien public transport provider in Vienna.

References

Amlacher, K., Fritz, G., and Paletta, L. (2009). Geo-Contextual Priors for Attentive Urban Object Recognition, Proc. IEEE International Conference on Robotics and Automation, ICRA 2009, Kobe, Japan, May 12–17, 2009.

Beeharee, A., & Steed, A. (2007). Exploiting real world knowledge in ubiquitous applications. Personal and Ubiquitous Computing, 11, 429–437. doi:http://dx.doi.org/10.1007/s00779-006-0091-6

Case, D. O. (2007). Looking for Information, Second Edition: A Survey of Research on Information Seeking, Needs, and Behavior (2nd ed.). Emerald Group Publishing.

Fritz, G., Seifert, C., and Paletta, L. (2005). Urban Object Recognition from Informative Local Features, Proc. IEEE International Conference on Robotics and Automation, ICRA 2005, Barcelona, Spain, April 18–22, pp. 132–138.

Fritz, G., Seifert, C., and Paletta, L. (2006). A Mobile Vision System for Urban Object Detection with Informative Local Descriptors, Proc. IEEE 4th International Conference on Computer Vision Systems, ICVS 2006, CDROM. IEEE Computer Society 2006, ISBN 0-7695-2506-7, January 3–5, 2006, New York, NY.

Golder, S., & Huberman, B. A. (2005). The Structure of Collaborative Tagging Systems. cs/0508082.

Google Inc. (2010). Google Mobile Blog. http://googlemobile.blogspot.com/2010/09/walk-this-way.html. Accessed 14.01.2011

Heidmann, F. (1999). Aufgaben- und nutzerorientierte Unterstützung kartographischer Kommunikationsprozesse durch Arbeitsgraphik: Konzeption, Modellbildung und experimentelle Untersuchungen. GCA.

Horozov, T., Narasimhan, N., & Vasudevan, V. (2006). Using Location for Personalized POI Recommendations in Mobile Environments. Proceedings of the International Symposium on Applications on Internet (p. 124–129). Washington, DC, USA: IEEE Computer Society. doi:10.1109/SAINT.2006.55

Lamere, P. (2008). Social Tagging and Music Information Retrieval. Journal of New Music Research, 37(2), 101.

Lemire, D., Maclachlan, A. (2005). Slope One Predictors for Online Rating-Based Collaborative Filtering, In SIAM Data Mining (SDM'05), Newport Beach, California, April 21–23

Meng, L. (2005). Egocentric Design of Map-Based Mobile Services, 281–287. doi:10.1002/9780470979587.ch37

Mobilizy GmbH. (2010). Wikitude Drive. http://www.wikitude.org/en/drive. Accessed 10.11.2010

Mobilizy GmbH. (2010). Wikitude World Browser. http://www.wikitude.org/en/category/02_wikitude/world-browser. Accessed 10.11.2010

Paletta, L., Sefelin, R., Ortner, J., Manninger, J., Wallner, R., Hammani-Birnstingl, M., Radoczky, V., Luley, P., Scheitz, W., Rath, O., Tscheligi, M., Moser, B., Amlacher, K., and Almer, A. (2010), MARIA – Mobile Assistance for Barrier-Free Mobility in Public Transportation, Proc. 15th International Conference on Urban Planning and Regional Development in the Information Society, CORP 2010, Vienna, June.

Pazzani, M. J. (1999). A Framework for Collaborative, Content-Based and Demographic Filtering. Artificial Intelligence Review, 13, 393–408. doi:http://dx.doi.org/10.1023/A:1006544522159

Sarwar, B., Karypis, G., Konstan, J., & Reidl, J. (2001). Item-based collaborative filtering recommendation algorithms. Proceedings of the 10th international conference on World Wide Web, WWW '01 (p. 285–295). New York, NY, USA: ACM. doi:10.1145/371920.372071

Su, X., & Khoshgoftaar, T. M. (2009). A survey of collaborative filtering techniques. Advances in Artificial Intelligence, 2009, 4:2–4:2. doi:http://dx.doi.org/10.1155/2009/421425

Tollmar, K., Yeh, T. and Darrel, T. (2004), IDeixis – Image-based Deixis for Finding Location-Based Information, Proc. Mobile HCI 2004.

Chapter 5

Dynamic Visualization of Geospatial Data on Small Screen Mobile Devices

Fangli Ying, Peter Mooney, Padraig Corcoran and Adam C. Winstanley

Department of Computer Science, National University of Ireland Maynooth, Co. Kildare, Ireland
{fangli.ying, peter.mooney, padraig.corcoran, adam.winstanley}@nuim.ie

Abstract

Vector-based geographic datasets are widely used as a source of data for Location-based Services (LBS) applications. However these datasets are often very large and it is a challenge for mobile devices to access this spatial data remotely especially with poor Internet connectivity. In these applications when users perform spatial operations on the small screen it is very resource intensive to pre-load large scale map representations for processing and visualization using the limited computation capability of the mobile device. In this paper we discuss the development of a flexible clipping method to help reduce the amount of unnecessary data for querying of large scale spatial data on mobile devices in LBS. A multi-resolution data structure is then described associated with a progressive transmission strategy. Clipped levels of detail can be efficiently delivered for quick processing and rendering on the small display. A discussion of this approach is provided which shows the potential usability of this clipping approach in spatial queries for LBS.

Keywords: Progressive transmission, Mobile devices, Clipping

5.1 Introduction

The increasing use of Location based services (LBS) on mobile devices corresponds to the development of mobile GIS technologies. These technologies aim to enhance user experience when interacting with large collections of geospatial information

G. Gartner, F. Ortag (Eds.), *Advances in Location-Based Services*, Lecture Notes in Geoinformation and Cartography, DOI 10.1007/ 978-3-642-24198-7_5, © Springer-Verlag Berlin Heidelberg 2012

and data. When users view digital maps on the mobile screen, the most popular type of spatial query performed would includes search and selection in a specific spatial range. It is very difficult for traditional raster based mapping technologies to handle such a very wide range of user queries. Meanwhile, adaptive visualization necessary to access the geospatial information in real time is also a problem. On the other hand, vector-based spatial data can potentially be used for adaptive rendering on mobile device screens after data generalization has been performed in the background. However one must be careful with which data generalization approaches are used. The traditional GIS technologies for desktop applications can not be used directly with LBS orientated spatial data since there are many limitations of the mobile device that must be considered before processing. These limitations will now be outlined as follows:

1. The small screen and resolution issues of the display on mobile devices strongly restrict the data representation and visualisation on the mobile device. Generalization is necessary to reduce information content in the spatial data. Then suitable levels of detail are delivered to the mobile device for the users to perform specific tasks under the knowledge that some data transformations have taken place.

2. The Internet communication for mobile devices is often slow (e.g., 170 kbps in GPRS), expensive, and unreliable (e.g., due to frequent disconnections) according to the work of Ilarri et. al. (2010). The authors comment that the latency of network communication heavily constrains the set of geospatial data that can be delivered to the end user. They also argue that there is an urgent need for the efficient transmission of spatial information to be improved.

3. The available memory, processing, and storage specifications of most mobile devices restrict the amount of spatial data that can be processed locally on mobile devices. This also places restrictions on the types of algorithms can be used to perform the processing. Consequently some computationally expensive algorithms are not acceptable on a mobile device.

4. The types of user interaction with the interfaces of applications on mobile devices can be complex. In the case of web-based mapping applications on mobile devices there are a significant number of zoom in/out and pan operations. As Reichenbacher (2003) states these are "tedious and cognitively complicated due to global context loss". Many users expect "instant information access" using these interfaces. It is therefore necessary to deliver understandable and flexible geospatial information to mobile user for instant access.

Despite these issues that the power of mobile device graphics hardware is improving quickly and this makes it possible to provide users on the move with more flexible visualizations (Chittaro, L. 2006). The development of these LBS applications can allow the users to access geospatial information in real time, anywhere and anytime, in a dynamic and flexible way.

Our paper will be organised as follows. We will introduce the significance of progressive transmission of vector based geospatial data and survey research that investigated the visualization of geographic data on mobile devices. Many of the functions that users require for map-based applications on mobile devices (spatial exploring, panning, range selection, etc) depends on where users are located. Progressive transmission takes the advantage of efficient transmission of spatial data in under potentially limited bandwidth situations. In this paper we will take account of the specific region of interest of the map from the user actions. Using a dynamic clipping approach for the vector data to provide quick generalisation of several levels of detail based on the priority of the parts in the map are computed. Based on a proposed dynamic data structure for this approach we reduce the amount of insignificant spatial information transmitted to the mobile device. To finish we show potential usage in a wide range of LBS.

5.2 Overview of Related Work

In this section we provide an overview of the advantages of progressive transmission strategies for vector data in LBS. This has been the subject of a number of papers written on our previous work. For example (Ying et al 2010) used shape-complexity as a metric to decide the most important aspects of a given dataset for transmission to a mobile device. We also review some literature on the area of adaptive cartography and on-demand mapping. This provides us with the opportunity to introduce the concept of "clipping". Clipping texture is well a known technique in the area of Computer Graphics and we describe it here to provide a linkage to our work on spatial data transmission and visualisation.

5.2.1 The Advantage of Progressive Transmission

Interactive mapping with LBS offers opportunities for high user interaction with digital mapping applications. In comparison to purely raster-based approaches vector-based geospatial data is more suitable for adaptive geo visualization and efficient data delivery for LBS applications requiring dynamic visualisation. Vector data offers flexibility for real time rendering and the adaptability for generalisation. The Volunteered Geographic Information (VGI) data can potentially provide the vector-based geospatial data source for the current information demand (Goodchild et. Al 2007) of LBS. One of the most promising VGI projects in recent years is OpenStreetMap (OSM) (Haklay et. al 2008). There is, of course, a warning to LBS developers before using OSM or VGI. The representation quality of spatial data obtained using such a crowd-sourcing model can vary significantly in terms of geometry, accuracy, and metadata richness and correctness. This is mainly due to differences in skills, equipment and information technology tools used by those who contribute data (Metzger et. al 2007). Similar problem can be seen in other survey datasets (Mooney and Corcoran 2010). However authoritative datasets usually document their data collection and analysis techniques allowing developers to understand the limitations of their data. In our work we use OSM data but the approaches outlined are applicable to any vector-based spatial data resource. The spatial query module of the work can be configured to work with different databases and spatial data models.

For vector-based spatial data, progressive transmission has many advantages including: the transmission of smaller data sizes, quick response time, and possibly the transmission of only relevant details (Bertolotto et. al 2007). Extensive research has been carried out on vector data progressive transmission and has been documented in the literature over the last decade. Some examples of these include Buttenfield (2002), Bertolotto et. al (2001), and Haunert et. al (2009). Their approaches provide coarser Levels of Detail (LOD) for real time rendering initially. This map is then iteratively refined through the transmission the integration of further data. The user can immediately perform analysis using the coarser map and terminate further downloads when the map approaches their desired LOD. Many generalization techniques are investigated to generate varying LOD to adapt to the user particular usage requirements (van der Poorten 2002). Methods from the domain of computer vision may be used to perform map simplification in an evolutional manner which preserves a contours overall shape across levels of detail (Latecki et. al 1999). However such a local criteria for a single map feature cannot be used to determine a suitable evolution (through several LOD) for a collection of map features. For maps containing many polygons and lines a methodology for determining a globally suitable generalization is required. The display capabilities of mobile devices are more restrictive than the traditional desktop machine. Sester et. al (2004) proposed an approach for generalization of vector data tailored specifically for mobile devices with limited screen resolution. Kjeldskov et. al (2003) states that more research is required to understand the usability of LBS and desktop applications using progressive transmission approaches. This research should be focused from a user-perspective and not a purely theoretical viewpoint. Previous research performed by the authors of this paper investigated the implementation of a selective progressive transmission based on the similarity between LOD and original datasets. We found that initially transmitting the most significant data can effectively reduce the differences, in future LOD, with the original, full resolution dataset. In Ying et. al (2011) we show the potential of a selective progressive transmission scheme. The visualisation of a set of geographical features is controlled and ordered by comparing important characteristics of the features (for polygons for example: area, number of points, overall complexity). However, there appears to be a critical point when it becomes more efficient to rebuild the original map dataset rather than simply reverse all the processes of the generalization. The key advantage of these adaptive and dynamic data transmission approaches is that they can enhance the user experience of the map interaction functionality when accessing and browsing geospatial information while offering the possibility of significant data reduction.

5.2.2 Adaptive Cartography and On-Demand Mapping

The visualisation of spatial data in a small screen mobile context raises usability issues. The adaptation for many user requirements is related to on-demand mapping (Crampton, 1999). If on-demand mapping happens in real-time, it relies on on-the-fly generalisation which Crampton defines as "the creation of a cartographic product upon a user request appropriate to its scale and purpose". This point is reinforced by the work of Reichenbacher, (2004). Some on the fly generalization such as the work by Lehto et. al (2005) uses the XSLT transform for XML-based vector data. The authors claim that this allows emphasis of user requirements and works well for multi-scale data. The Cartogen project (Boulos, M. et. al 2010) provides dynamic mapping with the canvas element in HTML5. Cartogen allows the user to customise the vector data style for display. Agrawala (2001) gives a good example for generalized route maps to improve the map usability in specific user requirements for personal navigation tasks. However, the latency of network communication and I/O performance limitations of mobile devices can not allow the transmission of large quantities of large scale data. As Argrawala pointed out the cartographic data is not only adapted to the usability requirements of the user-interface in the mobile context but also adapted to the limited nature of mobile device in terms of its technological environment.

5.2.3 Clipping in Computer Graphic

In the area of computer graphics most adaptive visualisation technologies use multi-resolution techniques to deal with the interaction with large datasets. Typically a user starts by exploring the lowest resolution version, which is much smaller than the original dataset then progressively transforms to a higher resolution. In this scheme there is more time to render a higher resolution version of the entire dataset. Tanner et al. (1998) present the "clipmap" which is based on the texture representation to support arbitrarily large texture sizes to be visualized in real time. The mipmap (Shreiner et. Al 2006) stores a pyramid of resolutions for a texture. Each texture level is usually a lower resolution representation of the previous level with level zero being the highest resolution. The clipmap is used to determine which area should be clipped to a specified maximum size which is usually the screen resolution. Clipped levels represent only a region of the entire surface and levels of detail. Those areas which are not clipped represent lower resolutions of the entire surface. Within this algorithm the clip area can be moved dynamically in response to user interactions. As the level number increases the textures in different non-clipped levels in a mipmap cover the same surface at decreasing resolution whereas the textures in different clipped levels in a clipmap cover concentric regions of increasing size and decreasing resolution. The performance also shows good flexibility for low cache memory devices.

To summarise, all the technologies mentioned in this section take account of the limitations of mobile devices for the transmission and dynamic visualisation of spatial data. The aim of the research work is to enhance the usability of visualisations of spatial data for mobile users accessing LBS. In particular we focus on LBS where there is a requirement to download and visualise spatial data on small mobile devices in dynamic environments – for example forest fires, humanitarian situations. The investigation of progressive transmission for efficient data delivery of geosptatial data through limited bandwidth shows that there is promise in using these approaches. We also see advantages for adaptive cartographic techniques to generalise maps with adaptive levels of details to meet users' requirements while working with the constraint of the resolution of the mobile screen. This paper proposes a clipping strategy for the visualization of vector based data. Clipping will be used to clip the levels of detail for flexible management of geospatial data regarding the small size cache and computation capability of mobile devices.

5.3 Disadvantages of Current Approaches for Progressive Transmission Approaches

The system needs to perform a number of operations for displaying a seamless map. As a user zooms out on a map, the scale of the map decreases allowing the user to make sense of globe context. To display a map with a very small scale, most of the geometric objects in a map are retrieved and displayed. However, transmitting of too many objects will reduce the user's effectiveness in viewing the relevant information. Some retrieved geometric objects may become too small to be displayed meaningfully on the screen. In some cases large objects, such as national road, highway and national forest with several hundred segments, may occupy only a single pixel on the display, but previous approach transmit all these segments from the server side and then draw them onto the client screen. Moreover, most users only require a section of the object and therefore transmitting the complete object is a waste of CPU and memory resources. This problem becomes more severe for applications where the data set is large and the map is viewed in real-time. Thus, it is important to customize (or clip) a specific region and scale for retrieval, transmission, and subsequent display of the necessary vector data at different levels of detail.

Some approaches we mentioned previously try to store the same map in different scales with multiple copies in a database. Each copy is assigned a different level of detail. This method not only introduces a high degree of data redundancy, but is also not suitable for quick updating. Even if this method is adopted, one needs to know how many scales are required. Another alternative is to store spatial data in their entirety, and apply a generalisation algorithm on the fly to obtain the desired level of detail. This is obviously a more efficient method than the previous approach, as it needs only one copy of the geometric object to store in the server disk. However, this approach still retrieves more data than necessary.

Therefore, we slightly change the model in previous work to better accomplish the clipping methods, which will be demonstrated in the next section. Considering the large amount of OSM XML data, we store the OSM vector data on the server hard disk for more efficient query performance. However the processing technique we use is still in real time without pre-processing and the OSM database can be Spatial-Temple database allowing frequent update. Generalization of the OSM vector data will be performed on the fly when user requests a specific geographic area. Simplification of polygons can adversely affect the overall shape of the polygon contour structure. Therefore only insignificant vertices can be considered for removal during simplification. As shown in *Figure 5.1,* the significant nodes will be promoted to a higher level while the unnecessary nodes will be demoted to a lower level for transmission later. The process continues until all the nodes are

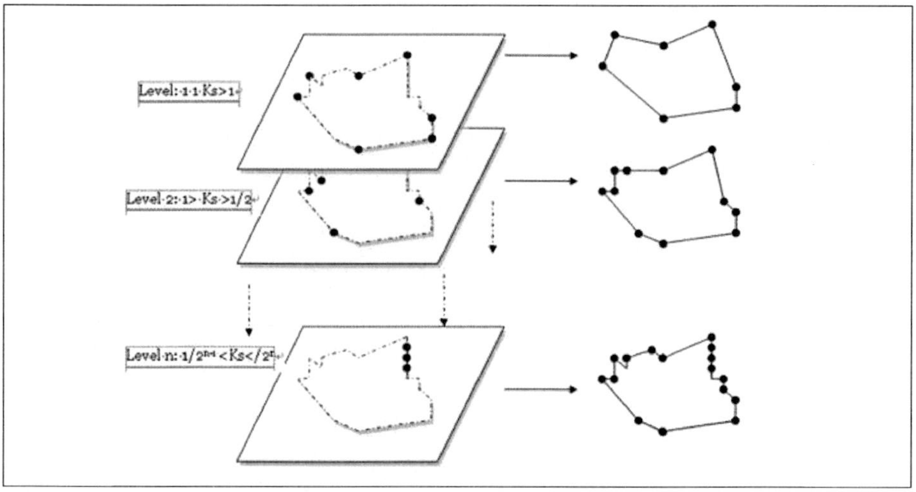

Fig. 5.1. The levels of detail of geospatial data in a specific order

stored in the special data structure ranked by levels of details. The highest levels of detail level 0 which keep the lowest representation will be delivered immediately for quick rendering and visualization in mobile devices. The lower levels of detail which correspond to higher resolution will be sent incrementally then to refine the previous levels of detail. However, when the user moves the screen area, the clipping algorithm will accelerate this refinement process much more only in user specific region to enhance the visualization quality rather than non-clipping region.

5.4 Continuous Clipping

In this section we attempt to describe the continuous clipping algorithm to clip the specific region corresponding to a viewing area for subsequent geospatial data transmission. A coarser version of the map will allow the user to make sense of global context when performing panning and zooming. The clipping region will be considered a priority for additional levels of detail. *Figure 5.2* shows a simple example for the clipping the spatial objects when user moving the region. When the user moves the region being viewed, we only update the changed region with additional levels of detail; other parts of the region with low resolution are still stored in the main memory without being reloaded.

The well known Liang-Barsky algorithm (Wenjun Huang, 2009) is suitable for determining the intersection between the rectangle bounding box area of the mobile screen and polygonal objects in spatial datasets. Therefore, the server side requires a data structure to represent the polygonal objects with levels of detail as well as clipping parts. As shown in Table1, feature vertices maintain the coordinates

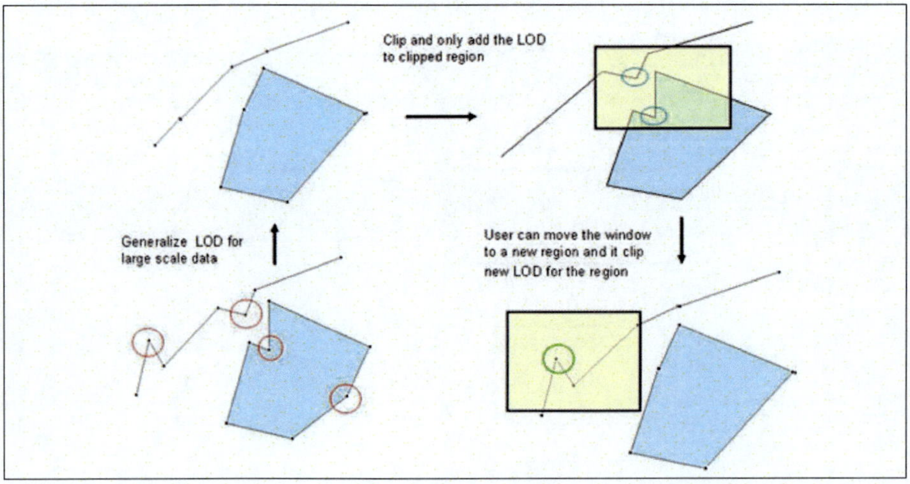

Fig. 5.2. A simple example to demonstrate the work-flow of continuous clipping

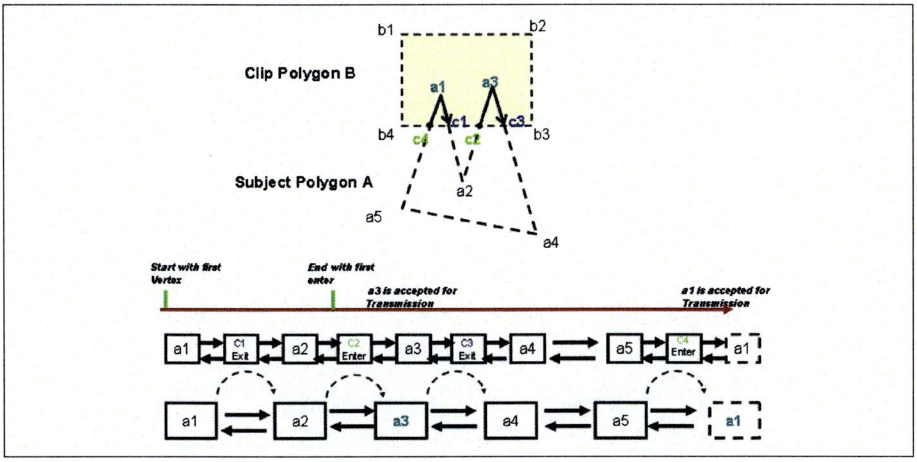

Fig. 5.3. Diagram of the data structure of the clipping algorithm and the work-flow of interaction clipping

information and levels of detail in a doubly linked list. Each vertex will initially have the Intersection attribute set to default value -"False", which means there's no intersection with this vertex. After this we must iterate through the vertices in clock-wise order for all polygons with the bounding box, then we mark the Intersection value – "Enter_Exit" for Vertex and "inside" values for polygons if there are any intersections. If there is intersection between two vertices, a new Vertex will be constructed and inserted into the Vertices (VertexList) and marked as Intersect to "TRUE" *(Figure 5.3)* and determine the Entry (Entry_exit = TRUE) or not(Entry_exit = False). If one of the Vertices lies on the bounding box it will be marked as "Intersect" value to TRUE and also used to determine the Entry boolean value.

When the user moves the viewing window, the bounding box request will be sent to get the clipped segments. We retrieve all the polygons which lie inside the window, and trace once for all vertices in one polygon, the clipped vertex should be started with ENTER which means "entry_exit" value is TRUE. The polygon should be closed and consequently end with the first vertex ("entry_exit" value is False) or with no end for polyline. Using the "CompSim" value to compare the similarity of the changing polygons with original ones we then selectively send the incremental levels of detail with a set of clipped vertices.

The existing algorithms for geometry clipping are not suitable for multi-levels of detail and we optimize the algorithm. We perform the K-levels of detail for the vertices list. Therefore, we describe a flexible data structure to represent the polygonal objects with levels of details within the clipping region. As shown in *Table 5.1,* Vertex has the attribute of LOD (levels of detail) in the data structure. All of the Vertex will be set default "Intersect" to False, which means no intersection with this vertex. After a full iteration of the algorithm in clock-wise order for all polygons

Table 5.1. Data structure for processing the geometry objects

Vertex

Data Name	Data Type	Description
lat,lon	geocoordinates	Location information
Next,prev	vID	Previous and next vertices
Entry_exit	boolean	Record the entry or exit
Interesect	boolean	If the vertex is on the intersection point
Clipped	boolean	Accepted as a candidate
K_level	int	Record the levels of detail
VetexOrder	vID	The vertex ID

Polygon

Data Name	Data Type	Description
PolygonID	pID	Polygon ID
Vetice	VertexList	List of vertices
CompSim	double	Similarity score compare with original score
Inside	boolean	If the intersection existed in the polygon

within the bounding box, we then mark the "Intersection" and "Enter_Exit" for Vertex and inside values for polygons if all of the vertex are on the right side of the clip polygon region and there are no any intersections – all the vertices are marked as accepted as candidates for immediate transmission. Obviously, if all the vertices are in the left side of segment of the clip frame and this polygon is not involved and all vertices are rejected. When user moves the view region very frequently and fast, the mobile device will get less detail since the user continuously updates the bounding box for clipping the spatial objects, but user still can get the global context in the low level of detail to recognize the whole scale of geospatial datasets. This can significantly reduce the volume of data to be transmitted and the delay time for visualization. If user starts to focus on a specific area, the clipping region will then be fixed and the LOD can accumulate in the clipping region quickly. User can get the highest detail in the screen region immediately. After finishing this step, we then can transmit the LOD to non-clipping region for the related spatial objects, which can help to enhance the user visual experience when scrolling the rest part of map later as shown in *Figure 5.4*.

For tile-based mapping systems the spatial data is usually stored in a very large data based in the background. When the data in this database changes the set of tiles must be regenerated. For large areas such as countries or cities with multiple levels of resolution this can result in millions of tiles. For small areas there can be several hundred tiles. This approach is not suitable for dynamic datasets like VGI which changes very quickly over the time particularly in situations of natural disasters. Efficiency is also an issue as the retrieval of such data for users' spatial queries is somewhat constrained by the fact that the map tiles are gird-based. Moreover, after users have downloaded the tile-based data when the semantic objects are

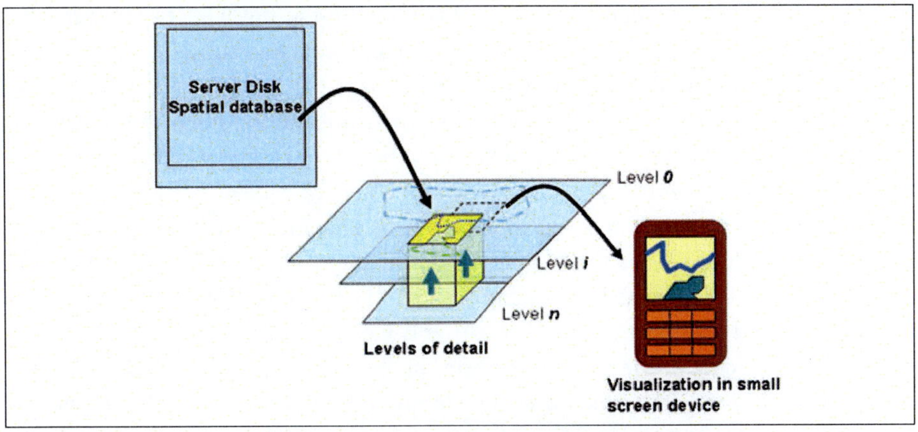

Fig. 5.4. Example when user browses the clipped map on mobile device

often reduced to an incoherent bunch of pixels if there are a rich set of attributes corresponding to the spatial data. In this case of spatial information overload it is very difficult for users to select some specific spatial objects or classes of objects within these datasets. For some progressive transmission approaches, the approach outlined here also can provide the advantage of providing a customized range of queries by using a flexible data structure. The levels of detail in a specific area of the small screen region can be progressively pushed to the client much faster than other parts at coarser levels of detail (level 0 in *Figure 5.4)*. It also facilitates the reasonably flexible spontaneous movement of the mobile device users. The user will not have to wait for a large map download to complete before they decide to scroll the map or move to another location. In comparison to the exchange and download of compressed data to a mobile device the progressive transmission scheme is more robust in the environment of internet disconnects or latency occurring on the mobile device. The user can move quickly and freely while not required to wait for compressed data to download to their device and be subsequently visualisation on screen. Our approach adapts to low bandwidth situations by sending the data in most important region, based on the user's current location. As the user moves around, this strategy can always drain most high resolution available to user mobile for real-time visualization with least cost of data delivery and process for mobile. One of the key issues for the future work in this area is a full evaluation of the performance of the data retrieval and real-time efficiency of this approach.

5.5 Conclusion and Discussion

In this paper, we attempt to design a framework for efficient spatial query and adaptive transmission of spatial data and information to mobile devices. Software on the mobile device then visualizes this data on-the-fly. The paper has also addressed some limitations of the delivery of spatial data and subsequent visualisation and query on small mobile devices. The dynamic framework we describe can potentially be used in location based services in the future since the geospatial datasets for user centred services should serve the spatial content in suitable levels of detail. Currently only some initial experiments have been carried out which show the query results in the Android platform. The amount of data delivered is significantly reduced since the clipped region is allocated by users. In the future, our work will focus on the following issues: (1) Enhance the efficiency of the clipping method, the continuous clipping will get the least overlap of data reload and will be smooth for the user to perform quick panning operations. (2) Evaluate the usability of the whole system in the solving of real world tasks. Users will be recruited to perform tasks in fields like tourism, surveying, or way-finding. These tests could help to inform us about the usability and potential adaptation of this approach on mobile devices. (3) We will extend the basic set of Cartographical tags which will be rendered on the map allowing the improved visualisation of the real map with our strategy. Finally, (4) it is important that this approach is evaluated amongst the more widely accepted strategies of tile-based delivery of maps and data compression exchange. At present this type of analysis is beyond the scope of the work presented here. However quantitative comparisons will be made between the three approaches and shall be presented in future papers. The overall goal of the work is to enhance the user experience when using mobile mapping and location based services on small screen devices.

References

Iarri, S., & Mena, E. (2010). Location-Dependent Query Processing : Where We Are and Where We Are Heading. Computing, 42(3). doi: 10.1145/1670679.1670682.

Reichenbacher T. (2003) Adaptive methods for mobile cartography, Proceedings of the 21st International Cartographic Conference ICC: Cartographic Renaissance, 10–16th August 2003, Durban, South Africa., pp. 311–132.

Goodchild, M. (2007). Citizens as sensors: the world of volunteered geography. GEOJOURNAL. 69(4): p. 211–221.

Haklay, M.M. and A. P. Weber (2008). OpenStreetMap: User-Generated Street Maps. IEEE Pervasive Computing, October–December.

Metzger, A.J.F.a.M.J. (2008). The credibility of volunteered geographic information. GEOJOURNAL, 2008. Volume 72, Numbers 3–4, (Volunteered geographic information: key questions, concepts and methods to guide emerging research and practice): p. 137–148.

Mooney, P. and Corcoran, P. (2010) Can we look at your Polygons? Presentation at the OpenStreetMap State of the Map (SOTM) 2010 conference in Girona, Spain. July 2010.

Chittaro, L. (2006). "Visualizing information on mobile devices," Computer, vol.39, no.3, pp. 40- 45, March 2006 doi: 10.1109/MC.2006.109

Bertolotto, M. (2007). Progressive Techniques for Efficient Vector Map Data Transmission:An Overview, in Spatial data on the Web: modeling and management A.C. Belussi, B.; Clementini, E.; Ferrari, E. (Eds.), Springer. p. 65–85.

Buttenfield, P. (2002). Transmitting vector geospatial data across the Internet. In: Egenhofer, M.J., Mark, D.M. (Eds.), Lecture Notes in Computer Science, vol. 27, no. 48, pp. 51–64.

Bertolotto, M. and M.J. Egenhofer (2001). Progressive Transmission of Vector Map Data over the World Wide Web. GeoInformatica, 5(4): p. 345–373.

Haunert, J.-H., A. Dilo, and P. van Oosterom (2009) Constrained set-up of the tGAP structure for progressive vector data transfer. Computers & Geosciences, 35(11): p. 2191–2203.

P. M. van der Poorten, S.Z.a.C.B.J. (2002). Topologically-Consistent Map Generalisation Procedures and Multi-scale Spatial Databases.

Latecki, L.J. & R. Lakämper (1999) Convexity Rule for Shape Decomposition Based on Discrete Contour Evolution. Computer Vision and Image Understanding, 73(3): p. 441–454.

Sester, M., Brenner, C. (2004). Continuous generalization for visualization on small mobile devices. In: Fisher, P. (Ed.), Developments in Spatial Data Handling – 11th International Symposium on Spatial Data Handling. Springer, Berlin, pp. 469–480.

Kjeldskov and Graham, 2003 Kjeldskov, J., & Graham, C. (2003). A review of mobile HCI research methods. In Proceedings of Mobile HCI 2003 (pp. 317–335), Udine, Italy.

Ying, F. and Mooney, P. and Corcoran, P. and Winstanley, A.C. (2011). How little is enough? Evaluation of user satisfaction with maps generated by a progressive transmission scheme for geospatial data In proceedings of the 14th AGILE International Conference on Geographic Information Science, Utrecht, The Netherlands, April 18–21, 2011

Ying, F. and Mooney, P. and Corcoran, P. and Winstanley, AC (2011). Selective progressive transmission of vector data, In Proceedings of the 11th International Conference on GeoComputation, UCL, London, England, July 2011

Ying, F. and Mooney, P. and Corcoran, P. and Winstanley, A.C (2010). A model for progressive transmission of spatial data based on shape complexity ACM SIGSPATIAL GIS 2010: Appears in The SIGSPATIAL Special Volume 2, Number 3, November 2010.

Crampton, JW, (1999). Online Mapping: Theoretical Context and Practical Applications. In Cartwright W, Peterson MP, and Gartner G (eds) Multimedia Cartography. Springer-Verlag, Berlin Heidelberg New York.

Reichenbacher, T. (2004). Mobile Cartography – Adaptive Visualisation of Geographic Information on Mobile Devices. Methods. Technical University, Munich.

Lehto, L., & Sarjakoski, L. T. (2005). Real-time generalization of XML-encoded spatial data for the Web and mobile devices. International Journal of Geographical Information Science, 19(8–9), 957–973. doi: 10.1080/13658810500161260.

Boulos, M., Warren, J., Gong, J. and Yue, P. (2010). Web GIS in practice VIII: HTML5 and the canvas element for interactive online mapping, International Journal of Health Geographics 9(1): 14.

Agrawala, M. and Stolte, C., (2001). Rendering Effective Route Maps: Improving Usability Through Generalization. Proceedings of SIGGRAPH01, pp. 241–249.

Christopher C. Tanner, Christopher J. Migdal, and Michael T. Jones. (1998). The Clipmap: A Virtual Mipmap. In Computer Graphics and Interactive Techniques (SIGGRAPH), pages 151–158, New York, NY, USA, 1998. ACM.

Dave Shreiner, Mason Woo, Jachie Neider, and Tom Davis. (2006). OpenGL Programming Guide, Fifth Edition, chapter 9. Addison Wesley, 2006.

Latecki, L.J. and R. Lakämper (1999) Convexity Rule for Shape Decomposition Based on Discrete Contour Evolution. Computer Vision and Image Understanding, 73(3): p. 441–454.

Wenjun Huang; Wangyong;, "A Novel Algorithm for Line Clipping," Computational Intelligence and Software Engineering, 2009. CiSE 2009. International Conference on , vol., no., pp.1–5, 11–13 Dec. 2009

Chapter 6

What's up that Street? Exploring Streets Using a Haptic GeoWand

Ricky Jacob, Peter Mooney and Adam Winstanley

Department of Computer Science, National University of Ireland Maynooth, Co. Kildare, Ireland
{ricky.jacob, peter.mooney, adam.winstanley}@nuim.ie

Abstract

In this paper we describe a Location-based Service (LBS) 'point to query' system called Haptic GeoWand where we use the orientation of the user to retrieve information for that user along a given street. Haptic GeoWand provides the user with location, orientation, and distance information using varying vibration patterns. This helps to reduce the visual mobile interaction required from the user. With the textual description of query results we use the vibration alarm to provide haptic feedback to the user about the availability and distance to their desired points of interests (POI) along a particular path or street. To test the usability of Haptic GeoWand we developed two mobile prototypes – whereHaptics and whichHaptics. The distance to Point of Interest (POI) information is provided by the *whereHaptics* which uses vibration feedback of varying frequency to represent distance to the POI. The *whichHaptics* uses vibration alarms of varying frequency to deliver information about the density of features along a particular street. With our system we demonstrate that the user can point and query the availability of POI along the street which are not directly visible to them but are along the path he/she intends to take. After selecting a POI, the user can follow this route without increased cognitive burden of interaction with the mobile interface. The user is not required to constantly look at the mobile screen. Haptics has proven to be powerful notification technique used to convey quantitative information. To test the system we integrated this with OpenStreetMap as the spatial data source.

Keywords: haptics, orientation, pointing, gesture, geowand

G. Gartner, F. Ortag (Eds.), *Advances in Location-Based Services*, Lecture Notes in Geoinformation and Cartography, DOI 10.1007/ 978-3-642-24198-7_6,

6.1 Introduction

There has been tremendous growth in the use of smartphones that are equipped with GPS receivers with affordable Internet access available on these devices. This has lead to a rise in location based services. With the development of micro sensor technologies most smartphones these days are equipped with other sensors such as orientation (magnetometer), tilt sensors (accelerometer), and of course GPS receivers. Apart from the location information about the device we now also have the orientation of the phone. This provides application developers with the ability to understand which direction the user is facing. It is now possible to integrate the output of these sensors to provide a more location and orientation-aware interaction instead of the standard location (only)-based interaction on mobile devices. Over the last decade research work has been carried out on integrating the orientation component with the location information. In 1999 Egenhofer (1999) predicted that various 'Spatial Appliances' will appear in the future that could be used with gyroscopes and GPS receivers to use both location and orientation information for knowledge retrieval. In early 2000 some interesting applications appeared that included orientation information integrated with other types of sensors such as the AudioGPS (Strachan et al. 2005) and M3I (Wasinger 2003). Haptic technology, or haptics, is a tactile feedback technology that takes advantage of our sense of touch by applying forces, vibrations, and/or motions to the user. Several important pieces of research work has been undertaken by a number of authors using haptic feedback as an alternative mode for providing user's with location and/or orientation information for use in navigational assistance. These works include: Erp et al (2005), Elliott et al (2010), Jacob et al (2010), Amemiya & Sugiyama (2008), and Robinson et al (2010). All of these authors agree that the key advantage of haptic feedback over other map or audio based assistance is that it does not require the user's constant attention and alertness.

In this paper we describe the development of a point to query system for specific queries along a street/path. We integrate haptic feedback with the point to query system using the GeoWand concept. Two haptic prototypes are developed to represent two different kinds of information that can be conveyed to the user. The 'distance to features' is conveyed using the whereHaptics module while the 'density of features' is provided by the whichHaptics module. Vibration alarms with varying frequency and patterns are used to provide the user with haptic-feedback for information such as -distance, density, location, and orientation.

Our paper is organised as follows. In *Section 6.2* we provide an overview of related literature. This focuses on orientation-aware devices and applications. This also helps us motivate haptic-feedback as an interface modality. In *Section 6.3* we provide the system model and also the implementation of two haptic based prototypes. *Section 6.4* discusses the experimental setup for the tasks to test a haptic based system. The results of user trials of the system prototypes are discussed in *Section 6.5.* In the final section we provide a review of the advantages offered by our approach and discuss some issues for future work.

6.2 Background

In this section we review literature where orientation aware interaction systems have been developed or used. Researchers have focused on retrieving information about a feature the user can see or is within his/her line of sight. Our literature overview shows that researchers are now beginning to integrate haptic feedback as a modality for navigation assistance in way finding and route planning.

6.2.1 Orientation-Aware Interaction

One of the most interesting applications listed by Egenhofer (1999) was the GeoWand. As described by Egenhofer a GeoWand is an intelligent geographic pointer, which allows users to identify remote geographic objects by pointing to them (Egenhofer 1999). Several researchers like Carswell et al (2010), Simon et al (2007), and Robinson et al (2008) have used this concept in their work. In Gardiner et al (2003), their approach restricted the search space to a user's field-of-view (FoV) using the concept of an observer's two dimensional query frustum to determine what the user can actually see from their location in the environment. The framework used by Simon et al (2006) relied on a 2.5D environment block model. For performing the query they use a basic scan-line algorithm to compute a 360-degrees line of sight. Similar work has also been carried out by Gardiner (2009) for their system called EgoViz which is a mobile based spatial interaction system. Carswell et al (2010) focuses on directional querying and the development of algorithms for mobile 2D and 3D spatial data interaction. Simon et al (2007) developed a system around the concept of 'GeoWand' which implements a Point-to-Discover block model using a mobile device. Here the user points the mobile device at a building and based on the position and direction the address and other information about the building is obtained. Here with the help of location direction and orientation information directional queries in the real world are performed on a mobile device. The 'quality of the sensor data' determines the quality of results returned from the query and thus becomes the most important aspect of directional queries. Simon et al (2008) has investigated GPS inaccuracy and compass error. They commented that these factors can be the deciding factor for systems returning accurate query results. Strachan and Murray-Smith (2009) have demonstrated an audio based system for using gestures and user orientation.

Robinson et al. (2008) demonstrates how a user can record interesting locations along the path by pointing and tilting a mobile device. Points of Interests (POIs) along the path of the user that were recorded can be downloaded and viewed and even blogged about. The authors use lightweight approaches for gathering location-orientated material while the user is mobile; the sensor data is used here both to collect and to provide content; and integrating map visualisation is used as the basis of the journey record. Costanza et al (2006) introduced a novel notification display embedded in eyeglass frames. The device delivers peripheral visual cues to the wearer without disruption to the immediate environment. The display is designed to take advantage of the visual field narrowing phenomenon, so that it tends to become less noticeable under high workload. Wasinger (2003) developed a location based interaction system where the user could ask 'What is that?' by pointing the device in that direction. The platform combines 2D/3Dgraphics and synthesized route descriptions, with combined

6.2.2 Haptics as a Modality

Research from Erp et al (2005) and Robinson et al (2010) have shown how haptics can assist where users would not want their visual and/or sound channel interrupted by interaction with mobile. Haptic feedback has been integrated into orientation aware navigation systems such as: Erp et al (2005), Elliott et al (2010), Jacob et al (2010), Amemiya & Sugiyama (2008), and Robinson et al (2010). Elliott et al (2010) validated the use of haptic navigation devices and found promising results. Elliott et al developed a device that was a torso-based eight-tactor belt which was used to provide direction information. The authors find that the system is viable and found it effective in extremely demanding conditions. Tactile displays enabled the user to more easily make a detour from the straight routes when terrain circumstances made this advantageous. Jacob et al. (2011) describes a system that integrates haptic feedback into pedestrian navigation applications that uses the OpenStreetMap database with the routing service provided by the Cloudmade API (Cloudmade API). In this system the authors use orientation information along with location for navigation assistance. Haptic feedback has been used in various other systems including alerting passengers using public transport of the arrival at their bus stop using the GPS sensor on the user's mobile device where feedback is provided using the vibration alarm in the phone (Jacob et al. 2011a).

This overview of the literature has shown that haptics has been demonstrated as an effective way of providing feedback to users in a more subtle way. It provides an alternative to the traditional "neck-down" approach required by the user of most mobile devices. We have reviewed work where gesture based queries are used and have also seen where haptic feedback has been used to provide spatial information.

Our work combines haptic feedback with pointing gestures with haptic feedback (using the vibration alarm) using distinctly encoded spatial information (orientation, distance, density).

Our paper argues that assuming that an application commands the full attention of the user on his/her mobile device while moving can be impractical at various times. In a busy city, with large numbers of pedestrians, it is very difficult to stand at a particular spot trying to make decisions (based on feedback provided on the mobile device) about which way one needs to move from the current location or somewhere nearby. A familiar sight in many cities is one of tourists standing at street junctions and intersections with tourist-orientated paper maps attempting to orientate themselves with their current location. In all of these cases a haptic-based system with a visual interface can help users make decisions more quickly and without feeling under pressure from the busy streets around them. A point and query system as described in this paper, with haptic feedback as a feedback modality, could be very useful in these real-world situations. The literature on haptics has shown that using haptic feedback can reduce the overhead of looking into the screen at all times while walking and allows users more time to interact with the real world around them. In the next section we describe the design of the *Haptic GeoWand* system.

6.3 System Design

The *Haptic GeoWand* is a system that allows users query geo-tagged data along a street or streets by using a point-and-scan technique with their mobile device. The feedback is provided using haptic feedback to reduce the cognitive burden on the user by delivering quantitative information about the distance/density to interested feature type using variations in vibration frequencies.

6.3.1 The Haptic GeoWand Model

The *Haptic GeoWand* is a system that helps the user take quick decisions at road intersections regarding which street to take in order to visit their POI of choice. There is often information overload in the typical visual map interface with overlays showing all the features of interest to the user. *Figure 6.1* represents the *Haptic GeoWand* model. The input to the system is via a visual interface. The user selects the maximum distance he/she wishes to walk in any direction and the type of feature(s) they wish to visit within that distance. *Figure 6.2* shows how the query region is reduced based on the pointing gesture performed by the user where the user holds the phone parallel to the ground and '*scans*' the area. Scanning is performed by moving

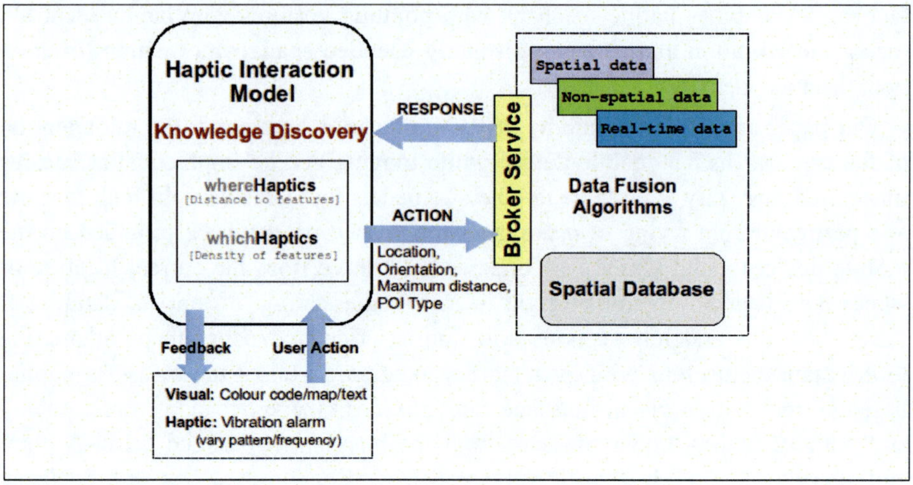

Fig. 6.1. The Haptic GeoWand Interaction Model

the phone parallel to the ground along each street from the user's current location. The user's current location and orientation along with the selected maximum distance and type(s) of POIs are sent to the server. The broker service receives these input values and queries the spatial database for matching features. The broker service then sends responses back to the interaction model. This response is provided to the user by various feedback mechanisms. Textual description of the *n* nearest features with name and distance along a street is provided. Haptic feedback is used for two different types of information requested by the user – a) distance to features and b) density of features.

Information regarding the distance(s) to features is provided by the *whereHaptics* module of the system where varying pattern and frequency of the vibration alarm is used as shown in *Figure 6.3*. Haptic feedback is used to inform the user about the zone (near, far, or very far) that the features are available along any particular street. Vibration frequency increases from *near* to *very far* with distinctly varying patterns ensuring the user is able to understand the feedback without having to look into the mobile. Information regarding density of feature(s) along a street is provided by the *whichHaptics* module of the system. Varying frequency and patterns of vibration are used to inform the user about the density of the number of features/POIs along any given path. This enables the user to check which street he/she should take when they have multiple choices of a particular feature(s) available to them.

6.3.2 Implementation

To implement the *Haptic GeoWand* prototypes we used a smartphone running on the Android operating system. The smartphone must have an inbuilt GPS receiver, digital compass, and an accelerometer. The Android Software Development Kit (SDK) enables us to customise the vibration alarm pattern and frequency. Thus by varying the frequency and vibration pattern, we can provide meaningful information to the user. The geographic data used to test the system was OpenStreetMap. A version of the Ireland OpenStreetMap database was stored locally on our server in a PostGIS database. OpenStreetMap provides us with a rich spatial dataset to test the application. However, we have designed the application with flexibility to access other sources of vector spatial data if available. In our system we log a number of important variables: the location, the orientation of the user's phone, and the accelerometer reading. These log files are retained and used for further analysis. The ability to integrate all the sensors into an off-the-shelf smartphone has allowed quick development and implementation of the system.

The three methods providing control of the vibration of the device are

- public void cancel()
- public void vibrate (long[] pattern, int repeat)
- public void vibrate (long milliseconds)

6.3.3 Prototypes

The *Haptic GeoWand* has two prototypes that can be a useful to provide mainly two kinds of information. The *whereHaptics* is a prototype used to provide information (via haptic feedback) about distance and *whichHaptics* is the prototype that is used to provide information about density of POIs. Thus these prototypes are used to ensure the user is quickly alerted of the ideal street he/she could possibly takes based on choice of POIs and its availability and distance. The user manually inputs the maximum walking distance and POI. Unlike other usual visual interfaces based POI searches where the query range is a circular region around the user of a specific radius; here we see in *Figure 6.2* that the 'query region' is narrowed down based on the direction the user is pointing the device.

The *whereHaptics* prototype is used to provide users with haptic feedback to inform the user about the distance to the feature by providing vibration patterns of different frequencies. *Figure 6.3* represents the vibration pattern and frequency used by the *whereHaptics* prototype to convey distance information to the user. If the feature is *very far away* from the user along a particular street, the user is provided information by high frequency vibration feedback with a distinct pattern of short duration (300ms each). Vibration of very low frequency is provided if there are POIs which are *very close* to the user and a different vibration pattern of medium frequency is used to represent distances from the user which is not very close.

The *whichHaptics* prototype is used to provide users with haptic feedback to inform the user about the density of features along a street by varying vibration

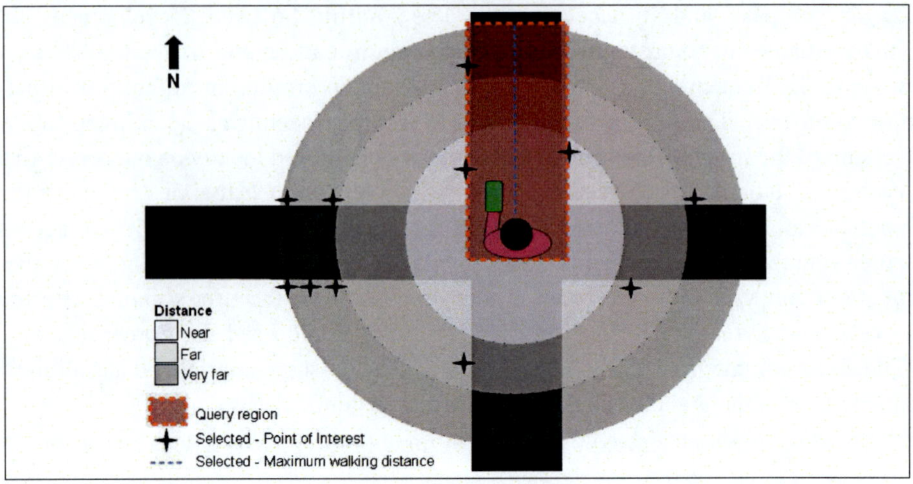

Fig. 6.2. The Haptic interaction model query description.

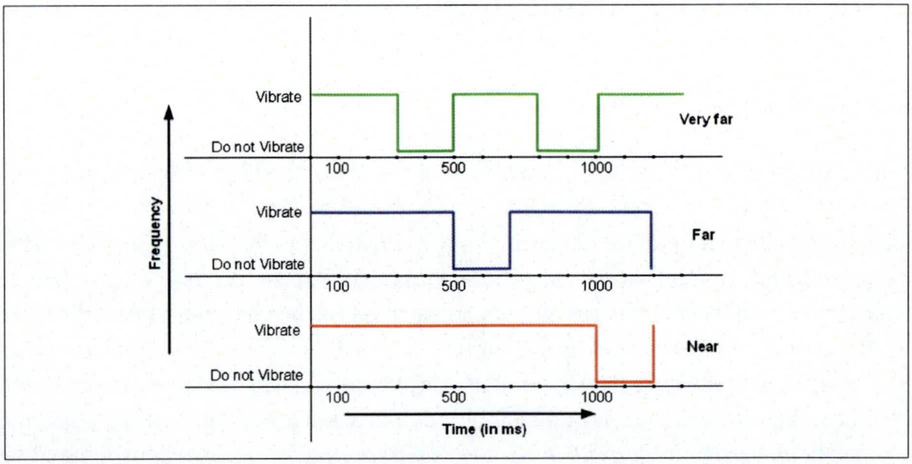

Fig. 6.3. The haptic feedback provided by whereHaptics for 'distance to features' information.

patterns of different frequencies. *Figure 6.4* represents the vibration pattern and frequency used by the *whichHaptics* prototype to convey the density of features information to the user. Based on the density of features along a particular street the user is provided with information by high frequency vibration feedback with a distinct pattern of short duration (300ms each) when there is just one feature within the query region. Vibration of low frequency and a clear distinct pattern (500ms vibrations with 200 ms interval between each) is used to represent density of features where a small number of features are available. To represent very high density of POIs vibrations of very low frequency (1000ms long) are used and are shown in *Figure 6.4*.

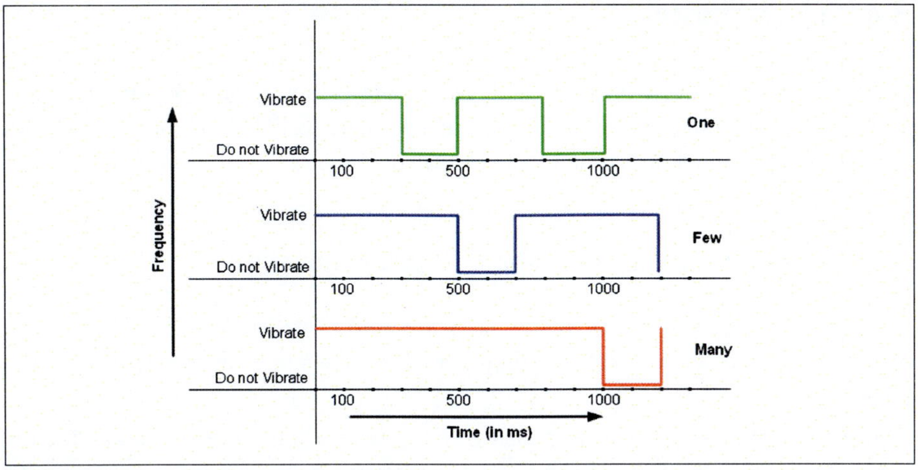

Fig. 6.4. The haptic feedback provided by whichHaptics for 'density of features' information.

To understand the working of the prototype, let us consider a real-world example of the query with the distribution of POIs as shown in *Figure 6.2.* We will obtain the following feedback for the two prototypes. If the user is using *whereHaptics* and thus looking along streets for POIs that are nearest to him the low frequency feedback (1000ms long each) will be presented when they are pointing along the north. The user will obtain feedback of medium frequency (500ms vibration with 200ms intervals) when the user points towards east or south and will obtain very high frequency feedback when they point towards the west. In this case all of the POIs fall in the outermost zone and thus are farthest away from the user. If the user chooses whichHaptics the user will get low frequency feedback (1000ms long each) when the user points towards west as it is along east where there is the highest density of POIs within the specified 'maximum walking distance'. The user will get a varying pattern (300ms vibrations with 200ms intervals between them) of high frequency when they point towards the south direction as there is only one POI along that street. In both the *whereHaptics* and *whichHaptics* prototypes the visual feedback provides names and distances of the *n* nearest POIs along a particular direction. The user may or may not choose to use the visual interface and instead chooses to walk in the direction based on the haptic feedback received. As they walk in the chosen direction the names and distances are updated on the visual interface.

6.4 Experiments

Tests were carried out within the NUI Maynooth campuses and Maynooth town. Ten participants took part in the test where the haptic feedback at intersections and navigation task based on haptic feedback was performed.

6.4.1 Test: Can Users Understand Haptic Feedback?

A test was carried out to test if the users are able to understand haptic feedback at an actual intersection with environmental noises and disturbances and other pedestrians.

Method: The task was to establish if the user was able to understand what the haptic feedback meant along each of the directions from a given intersection. To test we chose an intersection with 4 roads originating. The users in the test thus had to identify the street(s) which provided haptic feedback to represent that POI was 'very near' and also identify the road where POI was 'very far'. A test was also carried out to evaluate the *whichHaptics* prototype where the user must determine which street had the highest density of the selected features based on the feedback provided and which street had only one or no POI.

6.4.2 Test: Can Users Navigate Using Haptic Feedback?

A pilot study was carried out to test if users were able to travel from their current location to a given destination without the help of any map interface. The user relied fully on haptic feedback delivered on the mobile device when the user pointed in the direction of the destination. Based on the distance to destination at any given point the haptic feedback will vary as the user gets closer to the destination.

Method: The users in the test are familiar with the area where the test is carried out and so the destination where the user must reach is not revealed before hand. The users have to navigate using haptic feedback towards the destination. The visual display with the distance to the destination at any given point was provided for added assistance in the case where the user was in doubt at any point in the path. The total time taken and distance traveled were recorded. The location, orientation, and accelerometer values were recorded and analysed to understand parts of the path where the user paused or stopped unexpectedly.

One of the parameters which are logged for further analysis is orientation of the user and their scanning patterns with the mobile device. In *Figure 6.5* we can visualise the orientation pattern of the user by viewing their trip log. The polar graph shows periods of scanning through a wide angular range. The horizontal axes represent time from the beginning of the journey to the destination. The times when the user was heading in a straight line is shown by the concentration of orientation values between 0 and 45 degrees. Using the accelerometer data we can perform analysis on where (and for how long) the user paused or was stationary as they following the directions from our applications. On comparison of this data with a map interface, we can try to understand as to what could be the possible reasons for pauses/confusions and need for assistance at such regions in the path. This is illustrated in *Figure 6.6*.

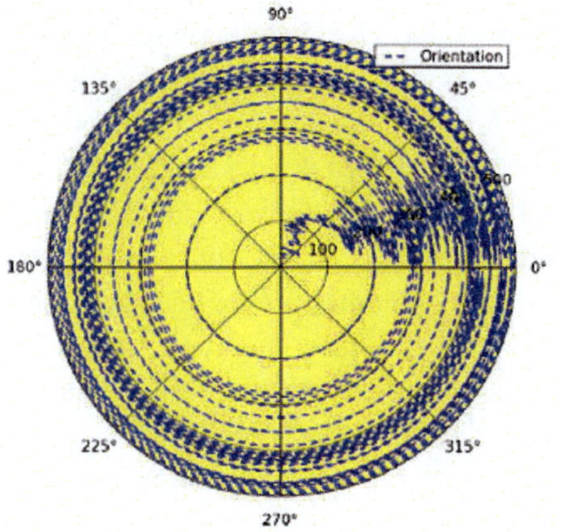

Fig. 6.5. Orientation data of a user during the travel is used to analyse the query angle regions.

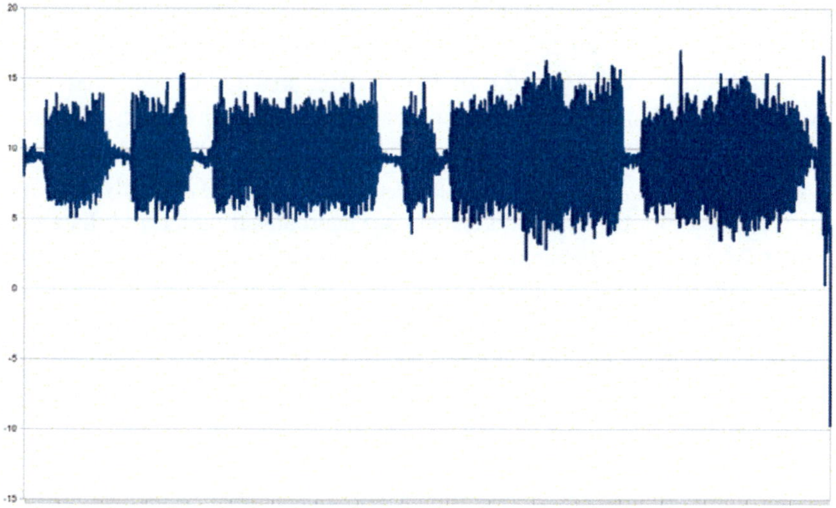

Fig. 6.6. Accelerometer data of a user during the travel is used to analyse the regions in the path when the user was standing still and where he was moving faster.

6.5 Results and Discussions

The *whereHaptics* and the *whichHaptics* tests were carried out with the task to find if the user was able to understand haptic feedback along each of the direction from a given intersection meant. Of the ten users all except one user were able to understand the feedback and thus were able to identify the street(s) where the POI was 'very near' and where POI was 'very far'. All the users were able to accurately find which street had the highest density of the selected POI and which street had just one or no POI. In the second task the users were able to reach the destination from the start point following the haptic feedback without the help of any map interface. The users were able to reach the destination in about the same time that a visual map interface using Cloudmade API (Cloudmade 2011) suggests as travel time. The users were able to effectively travel to the destination without having to look down at the mobile screen and thus ensuring travel without continuously interacting with the mobile device. The navigation time had not improved greatly as compared to a visual interface. The positive outcomes are that the user did not having to continuously interact with the visual interface and could thus continue talking with the friend who was walking alongside as the test user traversed the route. On getting direction to destination feedback, the user could 'cut across corners' and walk over open area (grass) as they walked towards the destination and not necessarily stick to paved walkways.

One of the drawbacks of the query functionality in the current implementation is that the system returned POIs in adjacent/parallel streets which were not necessarily accessible from the street along the user path. Some further work is necessary to investigate if this is an issue that can be resolved by development of more complex spatial queries or if it is an artifact of how OpenStreetMap mapping is performed. Building outlines are often traced from poor resolution aerial imagery and uploaded to the OpenStreetMap database. This approach could allow poorly constructed and represented polygon shapes be uploaded to OpenStreetMap which are not very similar to the real-world building footprint of the building. Currently our application is restricted to look ahead at adjacent streets/roads at the current location and the street "ahead" of the user. In the future development to this system we will provide a deeper look-ahead function allowing streets/roads which branch off the current street "ahead" to be listed in the results. This could potentially introduce additional learning for the user of the application. It could also require the user be provided with shortest route information. These issues will be carefully considered before this functionality is integrated into the prototypes.

6.6 Conclusion and Future Work

As outlined above in the literature review section of this paper haptic-feedback based Location-based service applications are beginning to appear. From our tests we found that users are able to understand the haptic feedback using the pointing gestures and thus has great potential as an effective communication tool in situations when a visual interface is not appropriate to use. A multimodal system for user interaction brings the added advantage of choosing one modality over the other when either one is not-appropriate. In our *Haptic GeoWand,* the user can choose either haptic-feedback or the more traditional text-based feedback. We have purposely avoided using maps in this implementation but are carrying out research into the best way to integrate mapping without reducing the opportunities for the user to choose haptic-feedback. We believe that there is now a trend which looks beyond the location-aware system to more orientation and context aware interactive applications. *Haptic GeoWand* integrates haptic feedback to provide a 'point to query' scanning application which reduces the visual interaction required with the mobile device. Scanning the immediate area around the current location is a very simple physical task. There is a reduction in the overall time the user must be attentive to the mobile screen for information. They are only required to "look down" into the screen when there is some relevant, additional, information displayed. The user in the meantime can be attentive to their actual physical environment. Sjölinder (1998) summerises spatial abilities as the cognitive functions that enable people to deal effectively with spatial relations, visual spatial tasks and orientation of objects in space. And with *Haptic Geowand*, we see that the cognitive burden on the user is reduced the interaction with the visual interface in minimal here. The users are quickly and easily able to orient themselves in the direction of the destination and there by improving their spatial abilities. Immediate future work will be undertaken to provide additional user testing of our application in busy urban environments. The application relies heavily on GPS signal accuracy and continuity along with uninterrupted internet connectivity. In our future work we wish to investigate how to reduce the use of GPS and haptic feedback to a minimal amount so as to conserve the battery charge whilst still ensuring effective communication of spatial information to the user.

Acknowledgements

Research in this paper is carried out as part of the Strategic Research Cluster grant (07/SRC/I1168) funded by Science Foundation Ireland under the National Development Plan. Dr. Peter Mooney is a research fellow at the Department of Computer Science and he is funded by the Irish Environmental Protection Agency

STRIVE programme (grant 2008-FS-DM-14-S4). The authors gratefully acknowledge this support.

References

Amemiya T, and Sugiyama H (2008) Design of a Haptic Direction Indicator for Visually Impaired People in Emergency Situations. Computers Helping People with Special Needs 5105:1141–1144

Costanza E, Inverso S, Pavlov E, Allen R, and Maes P (2006) eye-q: eyeglass peripheral display for subtle intimate notifications. Proceedings of the 8th conference on Human-computer interaction with mobile devices and services, 211–218.

Carswell James D, and Gardiner K, and Yin J (2010) Mobile Visibility Querying for LBS. Transactions in GIS, 14(6):794–809

Cloudmade (2011), Cloudmade API http://cloudmade.com/products/web-maps-api, last accessed September 2011

Egenhofer M(1999) Spatial Information Appliances: A Next Generation of Geographic Information Systems. Proceedings of the First Brazilian Workshop on GeoInformatics, Campinas, Brazil.

Elliott L, Erp J, Redden E, and Duistermaat M (2010) Field-Based Validation of a Tactile Navigation Device. IEEE Transactions on Haptics. 3(2)78:87

Erp J, Veen H, Jansen C, and Dobbins T (2005) Waypoint navigation with a vibrotactile waist belt. ACM Trans. Appl. Percept. 2:106–117

Gardiner K, and Carswell James D (2003) Viewer-based directional querying for mobile applications. Proceedings of the Fourth international conference on Web information systems engineering workshops, Roma, Italy. pp 83–91

Heikkinen J, Olsson T, and Vaananen-Vainio-Mattila K (2009) Expectations for user experience in haptic communication with mobile devices. Proceedings of the 11th conference on Human-computer interaction with mobile devices and services, 28: 1–10.

Jacob R, Mooney P, Corcoran P, and Winstanley A.C (2010) Haptic-GIS: Exploring the possibilities. ACM SIGSPATIAL Special, November 2010. 2(3):16–21

Jacob R, Mooney P, Corcoran P, and Winstanley A.C (2011) Integrating haptic feedback to pedestrian navigation applications. In Proceedings of the GIS Research UK 19th Annual Conference, Portsmouth, England

Jacob R, Salaik B, Winstanley A.C and Mooney P (2011a) Haptic feedback for passengers using public transport, H. Cherifi, J.M. Zain, and E. El-Qawasmeh (Eds.), DICTAP 2011, Part I, CCIS,166:24–32. Springer Heidelberg.

Robinson S, Eslambolchilar P, and Jones M (2008) Point-to-GeoBlog: gestures and sensors to support user generated content creation. Proceedings of the 10th international conference on Human computer interaction with mobile devices and services, 197–206.

Robinson S, Jones M, Eslambolchilar P, and Smith R, and Lindborg M (2010) I did it my way: moving away from the tyranny of turn-by-turn pedestrian navigation. Proceedings of the 12th international conference on Human computer interaction with mobile devices and services, 341–344.

Simon R, Fröhlich P, and Anegg H (2006) Beyond Location Based – The Spatially Aware Mobile Phone. Web and Wireless Geographical Information Systems, 4295:12–21. Springer Berlin/Heidelberg.

Simon R, Kunczier H, and Anegg H (2007) Towards Orientation-Aware Location Based Mobile Services. Location Based Services and TeleCartography, 279–290. Springer Berlin/Heidelberg.

Simon R, Fröhlich P, and Grechenig T (2008) GeoPointing: Evaluating the Performance of Orientation-Aware Location-Based Interaction Under Real-World Conditions. Journal of Location Based Services, 2 (1): 24–40.

Sjölinder, Marie (1998) Spatial Cognition and Environmental Descriptions, In Navigation in Electronic Spaces, Swedish Institute of Computer Science (SICS).

Strachan S, Eslambolchilar P, Murray-Smith R, Hughes S, O'Modhrain S (2005) GpsTunes: controlling navigation via audio feedback. Proceedings of the 7th International Symposium on Human-Computer Interaction with Mobile Devices and Service. Salzburg, Austria

Strachan S, and Murray-Smith R (2009) Bearing-based selection in mobile spatial interaction. Persuasive Ubiquitous Computing, 13:265–280.

Wasinger R, Stahl C, Krüger A (2003) M3I in a Pedestrian Navigation & Exploration System. Proceedings of the 5th International Symposium on Human-Computer Interaction with Mobile Devices and Service. Udine, Italy

Section II

Modeling and Computing

Chapter 7

Tracking Movements with Mobile Phone Billing Data: A Case Study with Publicly-Available Data

Matthew Kwan, William Cartwright and Colin Arrowsmith

RMIT University, Melbourne, Australia
matthew.kwan@student.rmit.edu.au,
{william.cartwright, colin.arrowsmith}@rmit.edu.au

Abstract

This paper analyzes six months of one individual's Deutsche Telekom mobile phone billing records to evaluate its effectiveness for tracking movements. As well as describing the contents of the records, the paper suggests a technique for improving the spatial accuracy of the data, and explores the relationship between the frequency of billing records and the accuracy with which a handset can be located.

Keywords: mobile phone, billing data, movement tracking

7.1 Introduction

Whenever a mobile phone communicates with a network it reveals its approximate location through the ID of the fixed antenna that received its signal. Given the location of the antenna and the direction it points, its coverage area, or *cell*, can be estimated, providing an indication of where the handset is located. Additionally, when the communication consists of a *billable event* – such as a phone call, an SMS, or internet access – details of the event, including the ID of the cell where the transmission was received, are stored in a billing record that can later be used to generate an invoice.

Storing cell IDs in billing records gives carriers the ability to charge distance-based tariffs, since it is simple to calculate the distance between two antennae. However, the records also allow the movements of a handset to be tracked when-

G. Gartner, F. Ortag (Eds.), *Advances in Location-Based Services*, Lecture Notes in Geoinformation and Cartography, DOI 10.1007/ 978-3-642-24198-7_7, © Springer-Verlag Berlin Heidelberg 2012

ever it is used, accurate to the nearest cell. The accuracy of cells ranges from a few hundred metres in a city centre to a kilometre or two in suburbs, and up to 100km or more in rural areas (Trevisani & Vitaletti 2004).

Due to privacy laws and commercial sensitivity, mobile phone billing records are not generally available for analysis. However, in 2010 Malte Spitz, a German MP from the Green Party, sued Deutsche Telekom to release all billing data relating to his account (Biermann 2011). After redacting some fields to protect the privacy of others, he chose to make this data publicly available on the internet (linked from Biermann's Zeit Online story (Biermann 2011)).

Although a number of other researchers have had private access to mobile billing data (Ahas *et al.* 2007, Calabrese *et al.* 2010, González, Hidalgo & Barabási 2008, Yuan & Raubal 2010), this appears to be the first dataset to be made publicly available, making it a valuable resource for researchers. Not only does it show exactly what information is available in billing records, it also provides sufficient data to estimate how accurately billing data can track handset movements.

In this paper we go through the records in detail, describing the contents of each field and summarizing the information they contain. We then apply a new technique to improve the spatial accuracy of the data, using estimates of cell centroids rather than antenna locations. Finally, we deliberately discard records from the data to investigate how changes to the sample rate of billing data affects accuracy.

7.2 Data Analysis

The spreadsheet of Spitz's billing data covers a six month period from 31 August 2009 to 27 February 2010, containing 35,830 records, or just under 200 per day. As an indication of the data provided, the first ten lines of the file are as follows

```
Beginn,Ende,Dienst,ein/ausgehend,Laenge,Breite,Richtung,Cell-Id_A,Cell-Id_B,Hint
ergrundinfo: http://www.zeit.de/vorratsdaten
8/31/09 7:57,8/31/09 8:09,GPRS,ausgehend,13.39611111,52.52944444,30,45830,XXXXX
X XXXX
8/31/09 8:09,8/31/09 8:09,GPRS,ausgehend,13.38361111,52.53,240,59015,XXXXXXXXXX
8/31/09 8:09,8/31/09 8:15,GPRS,ausgehend,13.37472222,52.53027778,120,1845,XXXXX
X XXXX
8/31/09 8:15,8/31/09 8:39,GPRS,ausgehend,13.37472222,52.53027778,120,1845,XXXXX
X XXXX
8/31/09 8:20,,,ausgehend,,,,,XXXXXXXXXX
8/31/09 8:20,,SMS,ausgehend,13.38361111,52.53,240,9215,XXXXXXXXXX
8/31/09 8:39,8/31/09 9:09,GPRS,ausgehend,13.37472222,52.53027778,120,1845,XXXXX
X XXXX
8/31/09 9:09,8/31/09 9:39,GPRS,ausgehend,13.37472222,52.53027778,120,1845,XXXXX
X XXXX
8/31/09 9:12,8/31/09 9:12,Telefonie,ausgehend,13.37472222,52.53027778,120,1845,XXX
XXXXXXX
```

The fields are fairly self-explanatory (although a working knowledge of German helps). All the records contain a beginning date and time, and, when the service is not instantaneous (e.g. SMS), an end date and time as well. The third column describes the type of service (e.g. GPRS internet access, SMS, or voice telephony), and the fourth indicates whether it was inbound or outbound.

Columns five and six contain the longitude and latitude, respectively, of the antenna covering the cell used by the phone. If the antenna is directional, column seven contains its direction, as a bearing in degrees clockwise from north, otherwise the value is **null**.

Column eight contains the cell's ID, and column nine contains the redacted cell ID of the other party to the service. Cell IDs range from 1 to 65523, suggesting that they are stored as 16-bit values. Note that the cell IDs are not all unique, since there are 2,830 different cells but only 2,797 different IDs. For example, in the records below cell ID 2334 refers to two different cells because they have different latitudes and longitudes.

```
9/9/09 14:56,9/9/09 15:04,GPRS,ausgehend,9.650833333,50.40638889,0,2334,XXXXXXX
X XX
9/18/09 19:36,9/18/09 19:36,Telefonie,ausgehend,13.48472222,52.50194444,30,2334,XX
XXXXXXXX
```

In a mobile network, cells are grouped into Location Areas, each with its own Location Area Identifier (LAI) (Lin & Chlamtac 2001). It appears that the two records above refer to cells in different Location Areas, but since there are no LAI columns it isn't possible to uniquely identify cells by their cell ID. However, since cells can be distinguished by their latitude, longitude, and direction, this doesn't affect analysis.

The majority of the billing records are related to GPRS internet access, with SMS records making up another quarter, as shown in *Figure 7.1*.

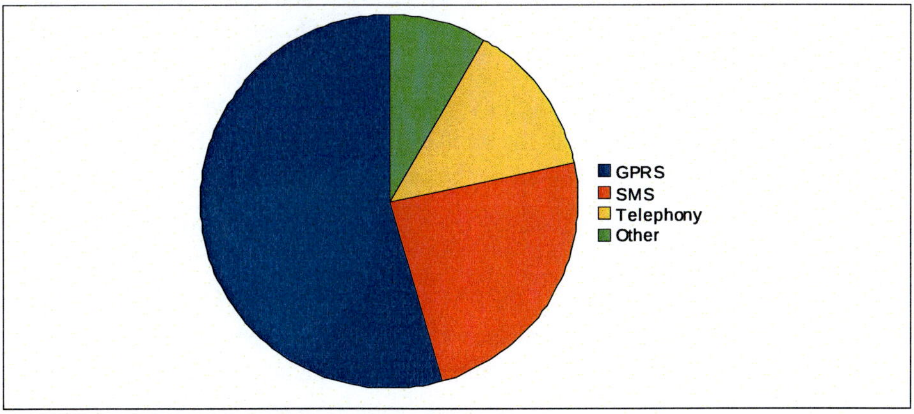

Fig. 7.1. Breakdown of service types in the billing data.

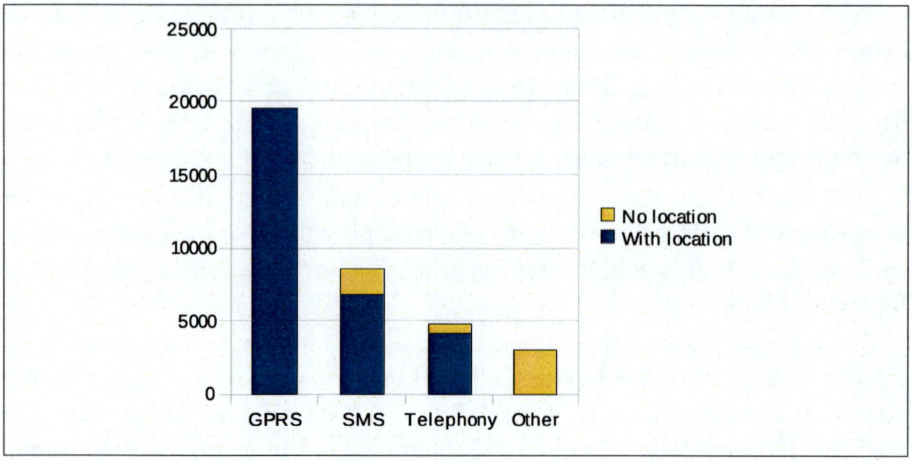

Fig. 7.2. Breakdown of records containing location data.

Of the 35,830 records, 30,374 (84.8 percent) contain a latitude and longitude, averaging 168 per day. *Figure 7.2* shows the numbers of these records, broken down by service type. According to the Zeit Online story (Biermann 2011) many of the SMS and voice call records that don't provide a location occurred when the handset was overseas, roaming on a different network.

7.3 Location Data

The latitude and longitude are expressed to eight decimal places, but all appear to be decimal conversions of degrees-minutes-seconds values. When converted to that form, the seconds component is always within 0.00002 of an integer value, so the locations are presumably accurate only to the nearest second. This implies a north-south accuracy of 31 metres, and, at a latitude of 52°N, an east-west accuracy of 19 metres.

Since many records provide both a start and end time, it was initially hypothesized that the handset remains in the same cell for the duration. However, there are numerous records where this cannot be the case, for example

```
2/27/10 17:30,2/27/10 17:35,GPRS,,13.3725,52.5575,null,47124,XXXXXXXXXX
2/27/10 17:33,,SMS,,13.41861111,52.49916667,null,52924,XXXXXXXXXX
```

The first record indicates that GPRS is being used from 17:30 to 17:35 in cell 47124. However, the second record shows an SMS being sent or received at 17:33 from cell 52924, indicating that the handset was *not* in cell 47124 at 17:33. Since there are 1,782 such conflicts in the data, the hypothesis appears to be incorrect. As a result, it is assumed that the location is only valid at the start time of a record.

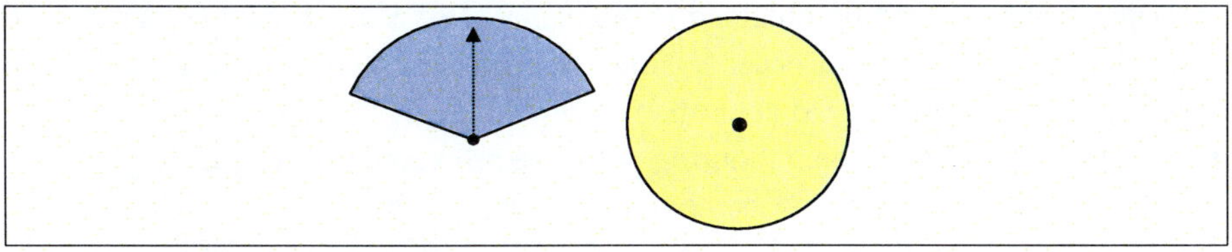

Fig. 7.3. Coverage areas of directional (left) and omni-directional antennae (right).

7.4 Location Accuracy

It should be noted that the locations in the dataset are actually those of the *antennae* covering the cells, which are not necessarily the best values for estimating handset locations. Ideally the values would be the centroid of each cell's coverage area, or, even better, the average position of all handsets ever to communicate with that cell.

In practice, there are two types of antenna – directional and omni-directional, as shown in *Figure 7.3*. Directional antennae tend to cover an arc between 120 and 180 degrees, and are usually mounted three to a tower. Omni-directional antennae, on the other hand, transmit in all directions and are usually alone on their tower.

Cells tend to be laid out in a hexagonal pattern, so the effective coverage area of omni-directional antennae is usually hexagonal, while for directional antennae it is often diamond-shaped. In the dataset provided, most of the cells had directional antennae – of the 30,374 cells, only 1,111 (3.7 percent) were omni-directional.

7.5 Centroid Estimation

When it comes to estimating the centroid of a cell, the location of an omni-directional antenna is a reasonable approximation. It broadcasts with equal power in all directions, and so long as its neighbouring antennae are arranged roughly symmetrically it is likely to be located near the centre of its coverage area.

However, this is not the case for directional antennae, which are located at a corner of their coverage area. This becomes clear in cases where a handset moves between two cells covered by antennae on the same tower, as shown below.

```
8/31/09 13:35,,SMS,ausgehend,12.67444444,51.86833333,60,19583,XXXXXXXXXX
8/31/09 13:37,8/31/09 13:38,GPRS,ausgehend,12.67444444,51.86833333,290,59260,XXX
XXXXXXX
```

Although the handset has clearly moved to a new cell – from cell 19583 with an antenna direction of 60° to cell 59260 with a direction of 290° – the location in both records is the same.

Better results would be obtained if the centroid of a directional antenna's cell could be estimated. But in order to find the centroid of a cell with a directional antenna it is first necessary to determine its cell boundary. In theory this can be estimated if the positions of neighbouring antennae are known, and assuming roughly equal transmitting power and homogeneous terrain. The Deutsche Telekom data provides the locations of 2,830 antennae on 1,642 different towers, so there may be enough information to do those calculations in some regions.

7.6 A Simpler Method

However, a simpler method is to assume that an antenna transmits symmetrically on either side of the direction it points in, so the cell centroid will lie somewhere along its directional vector. Then it simply becomes a case of estimating the distance along that vector.

One approach to calculating that vector distance is to assume it's the same for all cells, and to find the value that minimizes the average positional error when locating handsets. However, without knowing the *actual* position of the handset at all times, a proxy for positional error is needed. In this case the proxy used was overall distance travelled.

If we assume that a handset in the real world generally travels in straight lines, taking the shortest path between destinations, then cell centroid estimates that minimize total distance travelled should be the most accurate, since they generate the smallest amount of "zigzagging".

In F*igure 7.4,* the estimated total distance travelled is plotted against cell centroid distances from the directional antennae. As can be seen, that total distance is minimized when the centroid is roughly 320 metres in front of the antenna – a distance

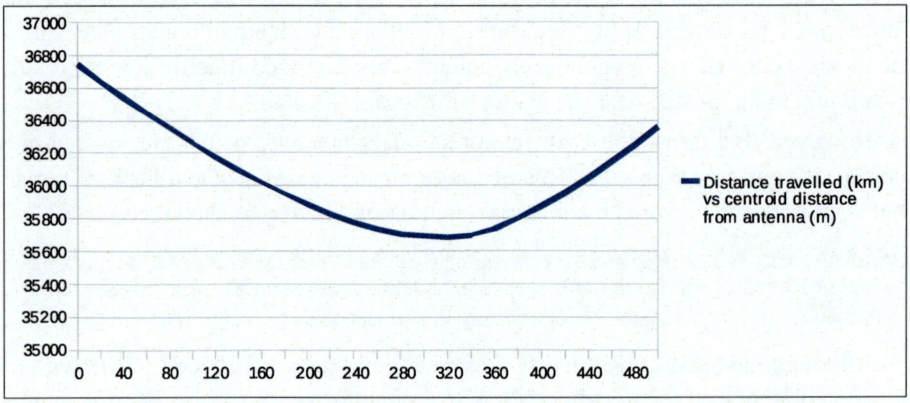

Fig. 7.4. Graph of total distance travelled vs cell centroid distance from antenna.

of 35,687km, or nearly 200km per day. This leads to the conclusion that, on the Deutsche Telekom network at least, a more accurate estimate of a handset's location is 320 metres in front of a directional antenna, and not the antenna itself.

7.7 Sample Rate

As mentioned earlier, the data released by Malte Spitz contains just under 200 records per day, of which 168 provide location data. However, previous studies of mobile phone billing data have indicated a significantly lower sample rates than this. Due to his job as a full-time politician, Spitz may be an outlier in terms of mobile phone use.

Data from almost a million subscribers in the Boston area over the period 30 July to 12 September 2009 (45 days) consisted of 130 million records (Calabrese *et al.* 2010). This implies a sampling rate of only 2.9 records per subscriber per day, barely one sixtieth as many as Spitz. However, it should be noted that this data was "aggregated and anonymous" so it is not clear if it contains all the original data.

But assuming that most mobile phone users generate fewer than 168 records per day, it is useful to know how much positional accuracy is lost due to lower sample rates. This can be estimated by comparing the original "high resolution" dataset with artificially-degraded "low resolution" versions taken at lower sample rates.

The high resolution dataset is used to create a "reference" set of locations, using the positions of omni-directional antennae and 320 metres in front of directional antennae. During the period between samples the handset is assumed to be at the last known location.

A low resolution dataset is created the same way, except that samples are discarded based on the new sample rate. For example, to simulate a rate of eight samples per day, only every 21st record from the original dataset would be used. The accuracy of this dataset can then be evaluated by calculating the distance between its estimated handset position and that of the high resolution data, averaged across the six-month sample period.

The results are shown in *Figure 7.5.* Note that the X axis uses the mean time between samples rather than samples per day. The original high resolution data has a mean sample period of 8.58 minutes, while, for comparison, the Boston data from 2009 had a mean period of 500 minutes.

It should be noted that in the data provided by Malte Spitz the average distance travelled per day was nearly 200km, which may also be an outlier among the general population. As a result, the average location error should be scaled down in line with the average distance travelled by people in the region of interest.

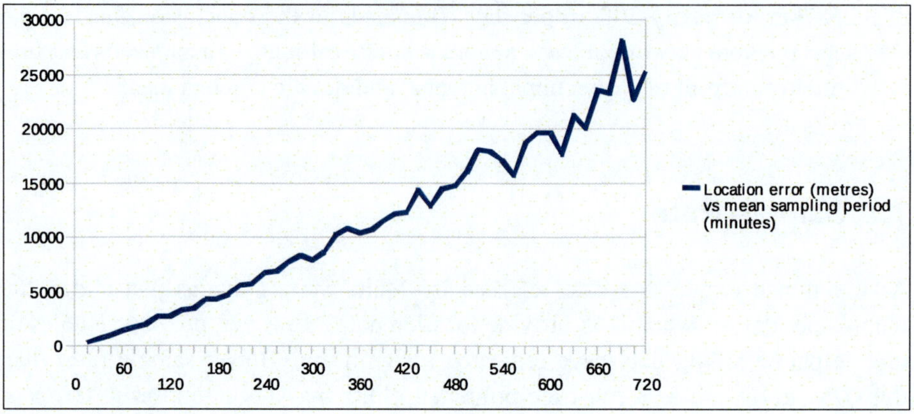

Fig. 7.5. Average location error vs mean sampling period.

But regardless of the scale, the data shows a strong linear correlation between the mean sampling period and the average distance error, with a correlation coefficient of 0.9872. In fact the correlation is so strong it should be possible to estimate the average error of the original dataset by comparing it to an "ideal" dataset with a sampling period of zero.

The line of best fit through the points has a gradient of 35.6 metres average location error per minute of mean sampling period. Given an original sampling period of 8.58 minutes, that implies an error of 305 metres due to the sampling rate. Note that this error is in addition to errors caused by the use of cell centroids to estimate the handset position.

7.8 Conclusions

Although it is difficult to generalize based on the billing data of a single individual – an individual whose movements and phone usage may not be typical – the release of Malte Spitz's records has provided valuable insights into what information is collected by carriers, and how it can be used. The data showed that, assuming people take the shortest path between locations, using a point 320 metres in front of a directional antenna is a better estimate of a handset's location than the antenna itself (at least on the Deutsche Telekom network). Using this knowledge, applications making use of billing data to track movements can produce more accurate results.

The data also showed that the frequency of billing records directly impacts the accuracy with which a handset can be located at any point in time. Knowing this, applications using billing data can more accurately estimate the errors involved and take appropriate action.

Finally, because the data is publicly available, other researchers can perform their own analysis and generate their own results. That allows the work in this paper to be reproduced or invalidated, something that is not possible with privately-held datasets. Moving forward, it would be desirable to have more billing data to analyse, from a diverse range of individuals, to see if the results apply generally.

References

Ahas R, Aasa A, Mark Ü et al. (2007), 'Seasonal tourism spaces in Estonia: Case study with mobile positioning data', Tourism Management, vol. 28, no. 3, pp. 898–910.

Biermann K (2011), Betrayed by our own data, viewed 26 March 2011, http://www.zeit.de/digital/datenschutz/2011-03/data-protection-malte-spitz.

Calabrese F, Pereira FC, Di Lorenzo G et al. (2010), 'The geography of taste: analyzing cell-phone mobility and social events', International Conference on Pervasive Computing.

González MC, Hidalgo CA & Barabási A (2008), 'Understanding individual human mobility patterns', Nature, vol. 453, no. 7196, pp. 779–782.

Lin Y & Chlamtac I (2001), Wireless and Mobile Network Architectures, Wiley & Sons.

Trevisani E & Vitaletti A (2004), 'Cell-ID location technique, limits and benefits: an experimental study', IEEE Workshop on Mobile Computing Systems and Applications, December 2004.

Yuan Y & Raubal M (2010), 'Spatio-temporal knowledge discovery from georeferenced mobile phone data', Proceedings of the Workshop on Movement Pattern Analysis 2010, Zurich, Switzerland, 14 September 2010.

Chapter 8

Dynamic Simulation of Vehicle Routing Using Timed Colored Petri Nets

Yang Cao[1] and Pengyan Xing[2]

[1] Spatial Information Research Center, South China Normal University, Guangzhou, China
yangcao@scnu.edu.cn
[2] School of Geosciences and Info-physics, Central South University, Changsha, China
xpycy@21cn.com

Abstract

Routing is one of the most common services for vehicle navigation applications. Since the traffic conditions are variable and the location of the vehicle changes continuously, how to efficiently find a real-time optimal route based on the traffic level on the roads poses significant design challenges. In this paper, we present a dynamic modeling approach to simulate vehicle routing and actual traffic information using Timed Colored Petri Nets (TCNPs). The dynamic simulation capabilities of TCNPs overcome the static nature of maps or directed graphs. And the simulation results will provide timesaving route guidance reflecting real-time varying traffic conditions.

Keywords: Vehicle Routing, Timed Colored Petri Nets, Dynamic Simulation

8.1 Introduction

Vehicle navigation system is currently the most widespread and successful application of GPS and GIS. Route planning is one of its main functions, which is a process of calculating an optimal driving route from the origin to the destination. However,

G. Gartner, F. Ortag (Eds.), *Advances in Location-Based Services*, Lecture Notes in Geoinformation and Cartography, DOI 10.1007/ 978-3-642-24198-7_8, © Springer-Verlag Berlin Heidelberg 2012

although vehicle route planning problems are dynamic and real-time traffic information is one of the most important and essential criteria for drivers during route selection, the traditional vehicle navigation systems are isolated (Song Ying & Yu Yang 2008), which perform routing based on static data instead of real-time or dynamic traffic information.

Recent developments in information and communication technologies and their impact on transportation researches lead to intelligent transportation systems (ITS) (Saeed Nadi & Mahmoud Reza Delavar1 2010). ITS tries to provide drivers with efficient instructions, in order to avoid congested roads and reach their destination sooner. The traffic management center processes the real-time traffic data and sends them to the drivers. A series of projects for the ITS practice have being carried in many countries, such as Japan's VICS, America's TravTek and Europe's Trafficmaster (Chen L et al. 2007). The common solution is just to offer traffic information by RDS TMC (Radio-Data-System – Traffic Message Channel) or via the Mobile Internet. In recent years, some researches about navigation services enhanced by real-time dynamic traffic information have been published (Lu Lin 2010, Saeed Nadi & Mahmoud Reza Delavar1 2010, Song Ying & Yu Yang 2008, John Garofalakis et al. 2007, Qiang Li 2004). They studied the dynamic vehicle routing problem and proposed some navigation models based on real-time traffic information. However these models always are developed based on static maps or directed graphs, which is difficult to develop analytical models and calculate dynamic traffic parameters efficiently.

Petri nets are graphical and mathematical modeling tools applicable to many systems. They are promising tools for describing and studying information processing systems that are characterized as being concurrent, asynchronous, distributed, parallel, nondeterministic and/or stochastic (T. Murata 1989). The advantages of Petri nets – intuitive graphical representation and formal semantics – are retained and supplemented with a uniform way to model mobility and mobile systems (Köhler M et al. 2003). Therefore, Petri nets have become an attractive option for traffic modeling and control (Shuming Tang et al. 2006). More precisely, Petri nets are employed both for describing traffic signals controlling urban signalized areas, as well as for modeling concurrent activities that are typical of traffic networks (Jorge Júlvez et al. 2010, M. Dotoli et al. 2008, M. Fanti et al. 2005).

In this paper we present a dynamic modeling approach to simulate vehicle routing using Timed Colored Petri Nets (TCNPs). The detailed description and simulation capabilities of TCNPs overcome the limitations of static maps and graphs. And the simulation results will provide route guidance reflecting real-time varying traffic conditions. This model can be used in vehicle route guidance system and provide a location based service which provide decision support by finding and visualizing the best path regarding the context such as the position of the user and current traffic condition.

The rest of the paper is organized as follows: *Section 8.2* introduces the theory of TCNPs. In *Section 8.3,* the method of modeling road network is presented. *Section 8.4* describes how to model dynamic traffic condition in our TCPNs model. The vehicle routing simulation proceeds are described in *Section 8.5*. *Section 8.6* provides a discussion about the proposed model. Finally conclusions are drawn in *Section 8.7.*

8.2 Timed Coloured Petri Nets

The Petri net was first proposed by Carl Adam Petri (1962). Formally a Petri net can be defined as follows (D. Azzopardi, S. Lloyd 1994):

$PN = (P, T, I, O, M)$

where

P is a set of *places*, graphically represented by circles;

T is a set of *transitions*, graphically represented by bars, with $P \cup T \neq \Phi$ and $P \cap T = \Phi$;

I: $P \times T \rightarrow \{0,1\}$ is the *input function* that specifies the arcs directed from places to transitions;

O: $P \times T \rightarrow \{0,1\}$ is the *output function* that specifies the arcs directed from transitions to places;

M: $P \rightarrow N$ is a *marking* whose ith component is the number of tokens in the ith place, graphically represented by dots. M_0 is an *initial marking*.

Enabling and Firing rules. A transition $t \in T$ is *enabled* iff $M(p)>0$ when $I(p, t)=1$, $\forall p \in P$.

An enabled transition t may *fire* at marking M', yielding the new marking $M(p_i) = M'(p_i)+O(p_i, t)-I(p_i, t)$ for $i =1, 2, ..., |P|$. The marking M is said to be *reachable* from M'.

Based on the original Petri net, extensions have also been made over the years in many directions (Gao Shang & Dew Robert 2007). TCPNs is an extension to Petri nets which combines Petri nets with high-level programming language (Aalst W. & Hee K. 2002).

The **Colour Extension** of TCPNs (Aalst W. & Hee K. 2002) solves the problem that arises in traditional Petri net of distinguishing tokens belonging to different cases. In traditional Petri net, it is not possible to separate two tokens in the same net. A coloured token has one or more attributes associated with it.

By adding "colour" to tokens, it is possible to distinguish them. The Colour Extension also gives the opportunity to add preconditions to transitions restricting it to fire until a certain condition is met by checking an attribute associated with the coloured tokens.

The **Time Extension** of TCPNs (Abdul-Kader et al. 2010) handles the modeling of the time aspect. This is done by associating a time delay variable with the transition, which represents the execution time of that transition. Time delays inscribed on the transition apply to all output tokens created by that transition. When the transition is enabled, tokens remain on the input places of the transition for a time, which at least equals to the time delay of the transition, before their removal by firing this transition.

Furthermore, it can be used to indicate how long an operation takes and how long it should wait before a retransmission is made.

The formal definition of TCPNs can be found in (K. Jensen & L.M. Kristensen 2009, Safiye Kızmaz & Mürvet Kırcı. 2011)

Furthermore, in order to model road network, we introduce the concept of *immediate transitions* and *timed transition* (Alexandre Zenie 1985) in our TCPNs model. Immediate transitions fire in zero time once they are enabled. Timed transitions fire after a certain time. Immediate transitions are drawn as thin bars, and timed transitions are drawn as thick bars.

8.3 Road Network Modeling

In our proposed TCPNs model, the road is spatially discretized, i.e., the road is divided into several sections due to different traffic characteristics. As shown in *Figure 8.1,* each uni-directional road section is modeled separately by a timed transition. A bi-directional road section is represented with two timed transitions. A link cell is modeled by a place. A crossroad is represented by the combination of immediate transitions, timed transitions and places. A directed arc always connects a place to a transition or a transition to a place, which indicates the particular direction of traffic flow. It is illegal to have an arc between two nodes of the same kind, i.e., between two transitions or two places. Places and transitions are called nodes. Together with the directed arcs they constitute the road network structure. In our navigation system, the main structure of the TCPNs model will be built according to the road map in advance.

For example, *Figure 8.2* shows a simplified road map. There are two kinds of road: the blue thick lines represent the highways; the black thin lines represent the urban roads. They have different traffic characteristics, such as the speed of the vehicles. All of the roads are bi-directional. The crossing point of the road is represented by the brown point p_i (i=1,…, 8). The TCPNs model for this road network is shown in *Figure 8.3.* This model contains 20 timed transitions, of which T_{ij} (i,

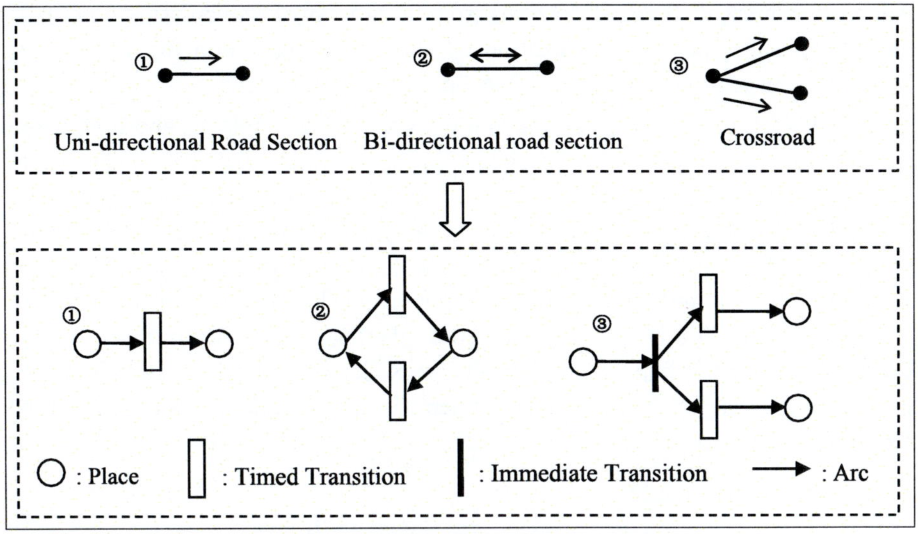

Fig. 8.1. The fundamental components of road network modeling

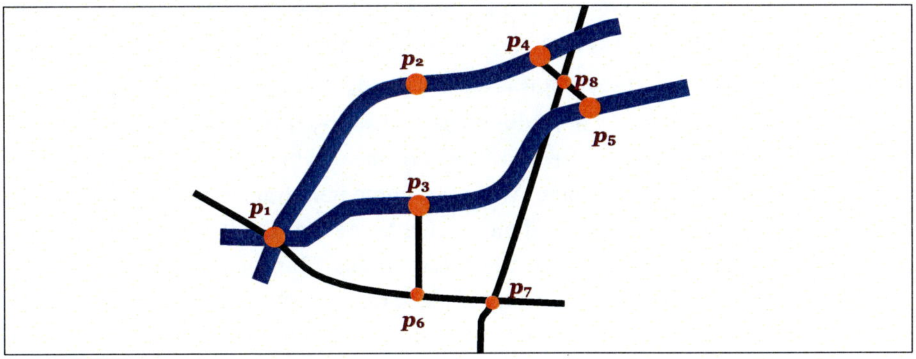

Fig. 8.2. The simplified road map example.

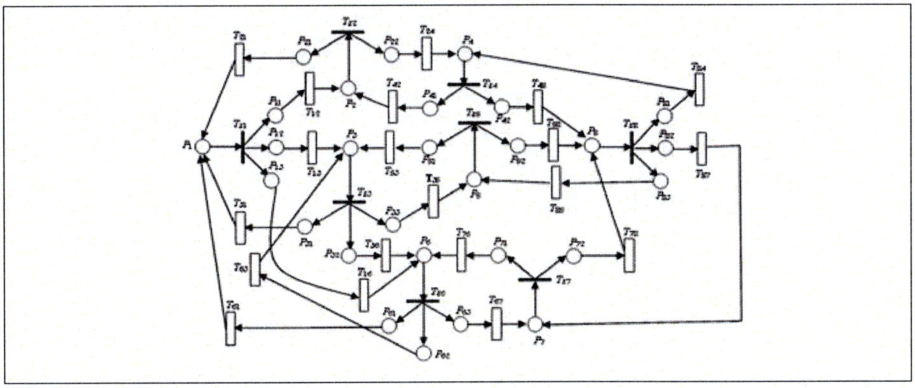

Fig. 8.3. The TCPNs model of *Figure 8.2*.

j=1,..., 8) represents the road sections from point p_i to p_j. The crossing point p_i is model by place P_i, immediate transition T_{Si}, and several accessorial places P_{ik}. (k=1, 2, ...). A number of directed arcs connect places and transitions.

In TCPNs, Each place can be marked with one or more tokens, and each token has a data value attached to it. This data value is called the token colour. The set of possible token colours is specified by means of a type (as known from programming languages), and it is called the colour set of the place (K. Jensen et al. 2007). In our proposed TCPNs model, a coloured token in a place simulates a vehicle. The colour value of each token indicates the path that the vehicle has traveled starting from the original position. This path is represented by the list of the place number that the vehicle has passed. Using CPN ML language (K. Jensen & L.M. Kristensen 2009), the set of possible token colours is defined as a **list** colour set, which is named **Routing** declared as follows:

colset **Routing** = list INT;

The **Routing** is a variable-length sequence of integer values $[r_1, r_2, \ldots, r_i]$, where i is an integer and $1<=i<=n$, n is the number of the places that represent the certain positions in the road network, such as link cells and crossing points. In the example of *Figure 8.3, n* is 8. r_x ($1<=x<=i$) is the subscript of a place (i.e. the serial number of the place in *Figure 8.3)* that represents a crossing point in *Figure 8.2*.

In the initial marking M_0, only the place that represents the current position of the vehicle contains a token. The **Routing** sequence of the token contains only the serial number of the current place. When an immediate transition fires, the tokens in the input place will be replicated and added to the output place. The new tokens represent the vehicles that choose other new routes. During the execution of the Petri net, once the token passes through a place that represents a certain position of the road network, the serial number of the place is added in the **Routing** sequence. At the end of simulation, the **Routing** value of each token is the sequence of the place numbers that the token has passed through.

In order to avoid loop in the route, our model add a fire restriction rule about the transitions: if the **Routing** sequence of a token in the input places includes the serial number of the output places, the transition can not fire.

8.4 Traffic Condition Modeling

The traffic condition is always characterized by the average vehicle speed on the road. For example, Google maps (http://maps.google.co.uk/) uses the colour codes to provide an indication of traffic speed compared to normal free-flowing traffic conditions, such as green=50mph or more, yellow=25–50mph, red=less than 25mph and Red/Black= stop-and-go traffic.

The NAVIGATOR (http://www.georgianavigator.com), a website implemented by the Georgia Department of Transportation, provides an online service that offers information about traffic conditions, road conditions and accidents. Its colour codes and the corresponding vehicle speeds are similar to Google map.

In this paper, traffic conditions are also represented by the average vehicle speeds of the various road sections. Let x denotes the serial number of the road section. And in the road section x, v_x denotes the average vehicle speed and l_x denotes the length of the road section. Then the time of the vehicle traveling through the road section x is calculated by the formula $t_x=l_x/v_x$. In our TCPNs model, t_x is used as the execution time of the transition that represents the road sections x.

The time-delay variable of transition can be either a deterministic variable or a random variable. If the average vehicle speed is expressed as an interval just as the Google maps, the time-delay variable of transition can be calculated using the function **discrete**, shown as follow:

Int t_x = **discrete** (a, b);

The function **discrete** is a predefined function in CPN ML language to provide a discrete uniform distribution over the closed interval specified by its arguments. This means that a call discrete (a, b) returns an integer from the interval [a, b] and that all numbers in the interval have the same probability of being chosen. Intuitively, this represents that the travel time of vehicle may vary between a and b time units, e.g., due to the load on the road network.

Moreover, our model is able to react to any traffic events by adapting the parameters according to the corresponding traffic state. The feasible events are mainly related to the occurrence of traffic congestion and the corresponding restoring of regular traffic conditions. Typically, if an event occurs in section i, at least both section i-1 and i+1 will be affected by the resulting changes in the traffic flow behavior. After tuning the traffic parameters according to the type of event occurred, the new values of the transition time delay will be determined. Furthermore, if the traffic conditions in the vehicle route are changed, a new route should be calculated according to the location of the vehicle and the real-time traffic information. In order to calculate a new route, the simulation of TCPNs model will be restarted.

8.5 Vehicle Routing Simulation

Based on the proposed TCPNs model, the vehicle routing simulation proceeds as follows:

Step 1: Initialization.

(1) The structure of the TCPNs model is built according to the road network.

(2) The time-delay variables of all timed transitions are calculated according to the current traffic information.

(3) The place P_r and P_d, which represent the origin and destination respectively, are found in the TCPNs model.

(4) Set j=0, M_0 (P_r)=1, and M_o (P_i)=0, where $i \neq r$.

(5) Set **Routing** of $token_1$= [r].

Step 2: Begin the simulation.

The main loop of the simulation runs:

(1) If the M_j (P_d)⩾1, then the simulation terminates.

(2) Check whether any transitions are enabled. If any, the corresponding transitions fire according to some specific firing rules noted in *Section 8.3,* shown as follows:

- Restricting the transitions to fire if the Routing sequence of a token in the input places includes the serial number of the output places.
- If an immediate transition is enabled, it fires in zero time and the tokens in the input place will be replicated and added to the output place. The serial number of the output place is added in the **Routing** sequence of the moving tokens.
- If a timed transition is enabled, it fires after a certain time delay. The serial number of the output place is added in the **Routing** sequences of the tokens that move through the transition.

If no transition is enabled, then the simulation terminates.

(3) Check if there is any traffic condition changing. If any, go to Step 3.

(4) Set j=j+1, and go back to the beginning of the loop.

Step 3: Adjust traffic parameters.

(1) Adjust the parameters according to the current traffic state.

(2) Assign P_r to the place that represents the current position of the vehicle. And go back to step 1 (4).

After the simulation terminates, check if M_j (P_d)⩾1. If so, the **Routing** sequences of the tokens in place P_d record the optimal routes. There may be more than one token in place P_d and corresponding optimal routes based on the real-time traffic information and the road network data. The minimum travel time is the sum of the time spent by the token to move from P_r to P_d. If M_j (P_d)=0, it means that there is no route to destination.

8.6 Discussion

In this paper, we present a dynamic modeling approach to simulate vehicle routing and actual traffic information using TCNPs. This model can be used in vehicle route guidance system and provide a location based service helps driver to find the best route regarding real-time situation of the traffic network. When the vehicle driver requests for navigational guidance from his current position to his destination, the model will simulate the vehicles driving on all possible routes. The travel time of the vehicles indicate the different travel costs of the different routes. The output of the simulation will provide decision support to the driver to find the optimal path regarding the current traffic condition.

The advantages of the method are summarized in the following. First, the dynamic simulation capabilities of TCNPs overcome the static nature of maps or directed graphs. It can simulate the vehicle's travel behavior and to estimate their travel times under various roadway conditions and traffic situations. It can integrate and incorporates the traffic simulation modeling and dynamic routing capabilities into a dynamic vehicle routing system.

Additionally, the simulation results will provide timesaving route guidance taking into account not only the length of the road sections that constitute the final route but also the real-time traffic condition. The implemented model within the vehicle navigation system will enhances the existing location based navigation services.

Finally, because all possible paths will be traveled by the tokens in the model, the output of the simulation may include not only one shortest-time route. Moreover, if do not stop the simulation at step 2 (1) and terminate it until there no transition is enabled, all possible routes and their travel time will be found. The driver can select his route based on his practical knowledge and experiments.

Besides the advantages mentioned above, there are also a number of limitations with this research that should be addressed for future assessments. First, the simulation of the model needs many computer resources. It should be implemented in the application servers, and mobile navigation system can just be used for visualizing results to drivers. Furthermore, as there are route planning criteria other than the travel time such as number of stop signs, travel distance and number of right and left turns, the model should be extended to incorporate such multiple dynamic and static criteria.

8.7 Conclusion

This paper introduces a dynamic modeling approach to simulate vehicle routing and actual traffic information using TCNPs. The result of simulation will determine the optimal route based on traffic information collected in real time as well as the static road network data. It can help drivers to avoid congestion and choose timesaving and safe route.

Further research may study in more depth the variables involved in real time traffic information. In addition, more work is needed to investigate usability and usefulness of the system by conducting tests addressing both the user interface design and the suggested approach.

References

Aalst W., Hee K (2002) Workflow Management: Models, Methods, and Systems (Cooperative Information Systems). MIT Press

Abdul-Kader, Hatem, El-Kholy, Warda (2010) Modelling of updating moving object database using timed Petri net model. Int. Arab J. Inf. Technol. 7(1), 79–85

Alexandre Zenie (1985) Colored Stochastic Petri Nets. International Workshop on Timed Petri Nets

Chen L, Jiang C, Li J (2007) VGITS: ITS based on intervehicle communication networks and grid technology. Journal of Network and Compute, r 31(3): 285–302

D. Azzopardi, S. Lloyd (1994) Scheduling and simulation of multi-product batch process plant through Petri net modelling. 4th International Conference on Advanced Factory Automation: 273–279

Gao, Shang, Dew, Robert (2007) Enhancing web-based adaptive learning with colored timed petri net. Lecture Notes in Computer Science, vol. 4798: 177–185

John Garofalakis, Nakou Polyxeni, Plessas Athanasios (2007) Vehicle Routing and Road Traffic Simulation: A Smart Navigation System. 11th Panhellenic Conference on Informatics (PCI 2007), Patras, Greece, Proc. 611–623.

Jorge Júlvez, René K. Boel (2010) A continuous Petri net approach for model predictive control of traffic systems. IEEE Transactions on Systems 40(4):686–697

Köhler M, Moldt D, Rölke H (2003) Modelling Mobility and Mobile Agents using Nets within Nets. International Conference on Application and Theory of Petri Nets 2003 (W. v. d. Aalst, E. Best, Eds.), 2679, Springer-Verlag

K. Jensen and L.M. Kristensen (2009) Coloured Petri Nets – Modelling and Validation of Concurrent Systems, Springer, ISBN: 978-3-642-00283-0

K. Jensen, L.M. Kristensen, L. Wells (2007) Coloured Petri Nets and CPN Tools for Modelling and Validation of Concurrent Systems. International Journal on Software Tools for Technology Transfer, Springer Verlag, 213–254

M. Dotoli, M. P. Fanti, G. Iacobellis (2008) An urban traffic network model by first order hybrid Petri nets. Proc. IEEE Int. Canf. SMC, Singapore, Oct., 1929–1934.

M. Dotoli, M. P. Fanti (2006) An urban traffic network model via coloured timed Petri nets. Control Eng. Pract. 14(10): 1213–1229

M. Fanti, M.P. Iacobellis (2005) Validation of an Urban Traffic Network Model using Colored Timed Petri Nets. IEEE International Conference on Systems, Man and Cybernetics: 1347–1352

Lu Lin (2010) Dynamic vehicle routing problem and its simulation. 2010 International Conference on Computer Design And Appliations

Petri Carl Adam (1962). Kommunikation mit Automaten (Communication with Automata). Bonn: Institut für Instrumentelle Mathematik, Ph.D. diss

Qiang Li (2004) Integration of Dynamic Vehicle Routing and Microscopic Traffic Simulation. 2004 IEEE Intelligent Transportation Systems Conference

Saeed Nadi, Mahmoud Reza Delavar (2010) Location-based Service for In-vehicle Route Guidance with Real Time Traffic Information. The 12th World Conference on Transport Research

Safiye Kızmaz, Mürvet Kırcı (2011) Verification of Session Initiation Protocol Using Timed Colored Petri Net. Int'l J. of Communications, Network and System Sciences 4(3): 170–179

Shuming Tang, Fei-Yue Wang Qinghai Miao (2006) ITSC 05: Current Issues and Research Trends. IEEE Intelligent Systems 21(2): 96–102

Song Ying, Yu Yang (2008) Study on Vehicle Navigation System with Real-Time Traffic Information. CSSE (4): 1079–1082

T. Murata (1989) Petri Nets: Properties, Analysis and Applications. Proceedings of the IEEE 77(4): 541–580

Chapter 9

Using Context-Aware Collaborative Filtering for POI Recommendations in Mobile Guides

Haosheng Huang and Georg Gartner

Institute of Geoinformation and Cartography,
Vienna University of Technology, Austria
{haosheng.huang, georg.gartner}@tuwien.ac.at

Abstract

Mobile guide is one of the most popular Location Based Services. Currently, providing context-aware services/information is still very challenging in mobile guides. Collaborative filtering (CF), known as "Amazon-like recommendations", is a promising solution for providing context-aware recommendations. The paper investigates how context-aware CF (CaCF) can be introduced into mobile guides. Specifically, we focus on applying CaCF methods on the highly available GPS trajectories to enhance visitors with context-aware POI (Point of Interest) recommendations.

After analysing the key issues of CaCF, we present a methodology for addressing them. Firstly, a two-stage method is proposed to identify context parameters which are relevant and thus needed to be modelled in a CaCF application. After identifying relevant context parameters, we explore a statistic-based approach (SBA) to measure similarity between different contexts (situations). In considering two different ways of incorporating context information into the CF process, two CaCF methods are designed: SBA_CP_CaCF (using SBA and contextual pre-filtering), and SBA_CM_CaCF (using SBA and contextual modelling). With these CaCF methods, smart services like "in similar contexts, other people similar to you often …" can be provided.

Finally, the proposed methods are evaluated with some real GPS trajectories collected from Vienna Zoo (Austria). The results of the experiments show that the proposed CaCF methods are feasible and useful for providing context-aware POI recommendations in mobile guides. More importantly, we show that

G. Gartner, F. Ortag (Eds.), *Advances in Location-Based Services*, Lecture Notes in Geoinformation and Cartography, DOI 10.1007/ 978-3-642-24198-7_9, © Springer-Verlag Berlin Heidelberg 2012

including context information in the CF process can improve the recommendation performance.

Keywords: Location Based Services, mobile guides, context-aware recommendations, collaborative filtering, GPS trajectory

9.1 Introduction

Recent years have seen an increasing interest in Location Based Services (LBS) with the continual evolution of mobile devices and communication technology. Mobile guide is the largest group of LBS applications (Raper et al. 2007). One of the key goals of mobile guides is to provide users with relevant information/services for satisfying their needs, e.g. recommending some Points of Interest (POIs) to visit.

Providing context-aware POI recommendations in mobile guides is still very challenging. Currently, in order to provide context-awareness, mobile guides mostly rely on an adaptation engine to determine the appropriateness of POIs for satisfying users' needs and context. However, building the adaptation engine has to undergo a long process of knowledge acquisition, which is very time-consuming and impractical for many mobile guides.

Additionally, the increasing ubiquity of GPS-enabled devices has led to the accumulation of large spatio-temporal datasets, such as GPS trajectories. These trajectories may reflect the perspectives and experiences of other people who solve their spatial tasks (e.g. choosing which POI to visit next) in this situation. It is obvious in daily life that experiences from past users (especially similar users) in similar contexts can help current users solve their own problems efficiently (Wexelblat 1999). Therefore, by aggregating the trajectories, mobile guides can provide users with smart services, such as providing "social advice" for making decisions. However, little research has addressed these considerations.

Collaborative filtering (CF, "Amazon-like recommendations") is a promising solution for the above problems. It uses "opinions" of similar users in similar contexts to help the current user efficiently identify information of interest (Resnick & Varian 1997). As a result, by incorporating CF into mobile guides, relevant information (e.g. POIs) matching the user's current situation can be identified (by aggregating opinions from similar users in similar contexts).

The goal of this paper is to investigate methods of introducing context-aware CF (CaCF) into mobile guides to provide context-aware recommendations. Specifically, we aim at applying CaCF methods on the highly available GPS trajectories to enhance visitors with context-aware POI recommendations. With CaCF, smart services like "in similar contexts, other people similar to you often visited POI A" can be provided in mobile guides.

The rest of this paper is structured as follows. In *Section 9.2,* we outline related work. *Section 9.3* identifies the key issues of incorporating CaCF into mobile guides. In *Section 9.4,* a methodology for addressing these issues is presented. In *Section 9.5,* the proposed methods are implemented and evaluated with GPS trajectories collected from Vienna Zoo (Austria). Some major results are also discussed in *Section 9.5.* Finally, we draw conclusions and present future work in *Section 9.6.*

9.2 Related Work

The research concerns how CaCF can be incorporated into mobile guides for providing context-aware POI recommendations. It integrates several mainstream trends and concepts, such as POI recommendations in mobile guides, CF in LBS applications and trajectory mining. From these aspects, we summarize the related work.

9.2.1 POI Recommendations in Mobile Guides

To provide context-aware POI recommendations, current mobile guides mainly rely on knowledge about POIs (domain model, DM), knowledge about the user (user model, UM) and her/his context (context model, CM). An adaptation engine is also employed: It measures the appropriateness of objects (DM) for satisfying a user's needs and context (UM and CM), and returns the relevant objects.

UM includes information about users' interests, preferences, and needs. It can be static (Baus et al. 2001, Kray 2003) and dynamic (Cheverst et al. 2000, Wiesenhofer et al. 2007). For building DM and the adaptation engine, a long underlying learning (knowledge acquisition) process has to be carried out in the field, which is often very time-consuming and impractical for many mobile guides. More importantly, current mobile guides are unable to effectively provide users with context-aware services in unseen situations or situations with little previous knowledge. However, real-world applications often need to operate in these kinds of situations.

In contrast to the above approaches, we aim to use CF to providing context-aware recommendations. In the following, we analyse CF's potential in solving the problems of current context-aware recommendation systems (i.e. a long process of knowledge acquisition in providing context-awareness).

One of the key goals of context-awareness in mobile guides is to provide a user with relevant information according to her/his context. This goal matches with the aim of CF, especially CaCF, which aggregates opinions of similar users in similar contexts to help individuals efficiently identify interesting information (Resnick & Varian 1997). Opinions from other users may reflect their perceptions on the fitness/

appropriateness of a particular item (information) for their contexts. If similar users in similar contexts like that item, this item can be considered as a matching item for the current user in the current context. As a result, CaCF can be viewed as a real-time underlying learning process of building DM and UM, and an automatic engine for identifying relevant information. Moreover, as CF solely relies on user feedback (i.e. trajectories in this paper) and requires no previous domain knowledge, mobile guides employing CF will be able to effectively provide context-aware services in dynamic environments and unseen situations. Therefore, CaCF can be a novel method of providing context-aware recommendations in mobile guides.

9.2.2 CF in LBS

CF is often applied in Web-based applications, such as movie recommendations, and product recommendations (see Amazon.com). Currently, context-aware CF in Web-based applications is still very challenging (Adomavicius & Tuzhilin 2010).

There are some tries on applying CF in LBS, such as restaurant recommendations (Horozov et al. 2006), event recommendations (de Spindler et al. 2006, Li et al. 2009), shop recommendations (Takeuchi & Sugimoto 2006), recommendations in museums (Bohnert et al. 2008), and POI recommendations for tourism (van Setten et al. 2004). However, most of the researches only employed location as contextual factor, and did not consider other contextual factors which are also relevant for generating recommendations, e.g. weather and companion (with whom).

Context-awareness is a key element when introducing CF into LBS. For example, recommending a place for the same person to visit may vary according to different weather conditions and different companion conditions (alone or with children). "There is more to context than location" (Schmidt et al. 1999). When incorporating more context information into CF for LBS, the problem becomes very challenging. A comprehensive investigation of how context can be incorporated into LBS-based CF is urgently needed. It is also important to note that none of the research focused on experimentally studying whether including context information in a CF for LBS can improve the recommendation performance.

9.2.3 Trajectory Mining

With the increasing ubiquity of GPS-enabled devices, more and more people start to record their travel/sport experiences with GPS loggers. Therefore, large spatio-temporal datasets (e.g. trajectories) are created every day, or even every hour. Recently, mining these kinds of user-generated GPS data is receiving considerable attention.

There are researches focusing on mining personal location history based on individual trajectories, e.g. detecting significant locations of a user, predicting a user's behaviour among these locations, identifying a user's spatio-temporal behaviour patterns, and recognizing a user's activities on each location (Li et al. 2008). In the meantime, many other researches mined multiple users' trajectories to understand mobility-related phenomena, e.g. Giannotti et al. (2007) aggregated many users' trajectories to identify spatio-temporal behaviour patterns. Zheng et al. (2009a) mined interesting locations and travel sequences from multiple users' trajectories. Li et al. (2008) proposed a user similarity measure based on different users' trajectories; however, they did not incorporate the similarity measure into the CF process.

There are also some researches focusing on using trajectories for recommendations. Takeuchi & Sugimoto (2006) recommended shops to users based on their individual preferences and needs, estimated by analysing their location history (i.e. GPS trajectories). Bohnert et al. (2008) developed a system for exhibit recommendations based on users' trajectories in a museum. However, it is necessary to note that context information (except location) was not considered in these researches.

In contrast to the above researches, this paper aims at designing context-aware CF methods to utilize trajectories for context-aware POI recommendations. Incorporating context information into the CF process for LBS applications is the main research focus.

9.3 Key Issues of Context-Aware CF

CaCF aggregates what similar users chose in similar contexts for recommendations. Several key issues have to be considered when providing CaCF in LBS: annotating user profiles with context, measuring similarities between contexts and similarities between users, and incorporating context information into the CF process.

1) Annotating user profiles with context: In CaCF, user profiles (e.g. trajectories[1]) should be annotated with context. A context (situation) can be characterized by a set of context parameters. Not all context parameters are relevant for generating recommendations. In order to annotate user profiles with context, a main question has to be answered: Which context parameters are relevant and thus needed to be modelled. Many researches chose some features of the world as context parameters from their own views (e.g. Panniello et al. (2009), and Adomavicius et al. (2005)). What is missing, however, is a method of identifying relevant context parameters for CaCF.
2) Measuring context similarity and user similarity: The goal is to determine which ratings (or user profiles, i.e. trajectories in this paper) are more relevant for the

[1] Raw trajectories can be processed and enriched with some semantic information to generate meaningful user profiles (e.g. places where users visit). Alvares et al. (2007) proposed a method on this aspect.

current user in the current context (i.e. more useful for making recommendations for the current context).

3) Incorporating context information into the CF process: Adomavicius & Tuzhilin (2010) proposed three approaches to incorporate context information into CF: 1) contextual pre-filtering: filter out irrelevant ratings (i.e. trajectories in our case) before using non-contextual CF method (i.e. the traditional method); 2) contextual post-filtering: use the traditional CF method, and then filter the results with context information; 3) contextual modelling: use context information directly inside the recommendation process. Currently, the approaches have not been applied to provide CaCF in LBS. How these approaches can be combined with the other key issues to provide CaCF in LBS should be carefully investigated.

9.4 Methodology

In this section, we explore some methods to address the key issues identified above.

9.4.1 Identifying Relevant Context Parameters

As mentioned before, context-dependent user profiles are important for context-aware recommendations. For annotating user profiles with context, we have to answer the question: Which context parameters are relevant and thus needed to be modelled.

We adopt an "interactional perspective" on context (Dourish 2004). Something is context (parameter) only if users' decision-making (e.g. choosing which POIs to visit), interaction with the system, or the behaviour of the system depends on it, otherwise it is just a feature of the world (Winograd 2001, Huang & Gartner 2009). For example, the temperature of the room is a relevant context parameter only if the adaptation of the interaction between human and the current system depends on it (or the behaviour of the system depends on it, e.g. when the temperature is higher than 30°C, start the air-conditioner), but otherwise it is just a feature of the world.

Based on this understanding, a two-stage method to identify relevant context parameters is designed:

1) A preliminary set of context parameters can be identified from literature or brainstorming. Data from users are then collected in different situations characterized by the preliminary set of context parameters.
2) The final set of context parameters can be created by refining the preliminary set according to the collected data. The basic strategy of refining is to analyse how some key aspects (e.g. the number of visited POIs, length of visit, and duration

of visit) of users' trajectories differ with different values of each context parameter in the preliminary set. If context parameter *c1* has *n* values, and the differences of the key aspects of visits are significant among these *n* values, then the current context parameter is relevant and thus needed to be modelled, otherwise it is irrelevant. T-test or analysis of variance (ANOVA) can be employed to test the significance. For example, if context parameter *"weather"* has two values *"sunny"* and *"rainy"*, and the difference between the key aspects of visits (e.g. the number of visited POIs) in *"sunny"* and the key aspects of visits in *"rainy"* is significant (i.e. "people behave differently in different weather condition"), then *"weather"* is relevant and thus needed to be modelled for CaCF, otherwise it is irrelevant.

It is necessary to note that we do not need to consider location as a relevant context parameter when annotating user profiles (trajectories) with context. The reason is that users' current location and location history are already stored in their trajectories. A trajectory includes a series of different stops (i.e. POIs). Every POI has a location. When recommending a POI for the current user, his/her current location (POI) is used to select relevant POIs which are *"close"* to the user (see step 1 of SBA_CP_CaCF in *Section 9.4.4)*.

9.4.2 Measuring User Similarity

For each user, a series of stops[2] (POIs) visited by him/her can be identified from his/her trajectory, e.g. using the SMoT method developed by Alvares et al. (2007). Therefore, a simple user similarity measure is adopted. It compares the POIs visited by the two users. It is obvious that two users accessed a POI visited by a few people might be more correlated than others who share a POI history accessed by many people (Zheng et al. 2009b). As a result, the visited popularity of a POI is considered when measuring similarity between users. Following is the proposed user similarity measure:

$$SIM_{user}(a,b) = \frac{\sum_{p \in POIS_{a,b}} \frac{1}{F_p}}{\sqrt{(\sum_{p \in POIS_a} \frac{1}{F_p}) * (\sum_{p \in POIS_b} \frac{1}{F_p})}}$$

where $POIS_a$ and $POIS_b$ are the set of visited POIs of user *a* and user *b*. $POIS_{a,b}$ is the set of POIs which are visited both by user *a* and user *b*. F_p is the visited popularity of POI *p* considering all the trajectories.

[2] When a visitor has stayed in a certain distance threshold over a time period, we consider him/her has a stop. Visitors tend to have a longer stay at a POI if they are interested in it. Therefore, in this paper, the time threhold is set big enough to eliminate POIs which were passed and not liked by the visitor.

9.4.3 Measuring Context Similarity

The similarity between the context (situation) in which the visit (trajectory) was made and the current context of the active user (who asks for recommendations) determines the usefulness of the trajectory in recommending POIs for him/her. In the following, we explore a statistic-based approach (SBA), which adopts a machine-learning technique.

With the method proposed in *Section 9.4.1,* relevant context parameters can be identified. By varying values for each parameter, all different situations can be identified. In the following, we propose an approach to measure the similarity between any two situations.

We assume that if visits in a situation (e.g. A) are similar to visits in another situation (e.g. B); these two situations can be considered as similar. Therefore, similarity between different contexts (situations) can be measured as some statistical metrics.

1) Measuring the distance of visits in situation A and visits in situation B:

$$Dist(A,B) = \sqrt{\frac{\sum_{p \in \bar{P}} \frac{1}{F_p} * (A_p - B_p)^2}{\sum_{p \in \bar{P}} \frac{1}{F_p}}}$$

$\bar{P}$ is the set of all POIs. A_p and B_p are the visit frequencies of POI p in situation A and B. F_p is the visited popularity of POI p considering all the trajectories.

2) Translating the distance measure into a similarity measure:

Shepard (1987) proposes that distance and perceived similarity are related via an exponential function. As a result, the following context similarity measure is designed:

$$SIM_{conx}(A,B) = e^{-Dist(A,B)}$$

With these two steps, similarity between any two situations can be calculated.

9.4.4 Making Recommendations

As mentioned in *Section 9.3,* context information can be incorporated into CF by contextual pre-filtering, contextual post-filtering, and contextual modelling. In this paper, we mainly focus on contextual pre-filtering and contextual modelling. As a result, two kinds of CaCF methods are designed. The steps of each method are as follows. We assume that the current user u is finishing the current POI p, and asking "which POI to visit next".

SBA_CP_CaCF: Using SBA and contextual pre-filtering

1) Identifying users whose next POI after visiting *p* has not been visited by the current user.
2) Filtering users whose context similarities with the current user do not exceed a threshold δ. Context similarity is measured by the SBA method.
3) For the results of step 2, identify the N most similar users. The user similarity measure proposed in *Section 9.4.2* is employed.
4) For the N most similar users, aggregating every similar user's next POI after visiting *p* (considering user similarity values).
5) Selecting the POI with the highest predicted value, and recommending it to the current user.

SBA_CM_CaCF: Using SBA and contextual modelling

1) The same as step 1 in SBA_CP_CaCF.
2) For the results of step 1, identify the N most useful users. The usefulness is measured by considering both context similarity and user similarity.

   $$Utility(a,b) = l * SIM_{user}(a,b) + (1 - l)SIM_{conx}(C_a, C_b)$$

   where C_a and C_b are the contexts of user *a* and *b*. SIM_{conx} (C_a, C_b) is calculated using the SBA method in *Section 9.4.3*.
3) For the N most useful users, aggregating every useful user's next POI after visiting *p* (considering usefulness values).
4) The same as the step 5 in SBA_CP_CaCF.

With the above CaCF methods, context-aware recommendations can be provided in mobile guides.

9.5 Evaluation and Discussion

In this section, we discuss some experimental evaluations. The data collection and processing are discussed in *Section 9.5.1*. *Section 9.5.2* employs the proposed method in *Section 9.4.1* to identify relevant context parameters for the CaCF methods. We describe the experiment setting in *Section 9.5.3*. The evaluation and results are presented and discussed in *Section 9.5.4,* and summarized in *Section 9.5.5*.

9.5.1 Data Collection and Processing

Thanks to a cooperation with Vienna Zoo (Tiergarten Schönbrunn), we collected trajectories in the zoo in the first half of 2010. We encouraged visitors to carry GPS loggers with them while walking through the zoo. Before they start, we recorded some additional information (e.g. context information) about them and their visits, i.e. weather condition (sunny or rainy), age (>=45 or <45), time limit (Yes or No), annual ticket (Yes or No), first time in the zoo (Yes or No), and companion with small children (Yes or No). In total, we collected 53 valid trajectories of all kinds of visitors in different situations. For every trajectory, we extracted the following information: visited POIs and their orders, duration of visit, and length of visit.

We employed the SMoT method developed by Alvares et al. (2007) to identify stops from every trajectory. To simplify the process of identifying the visited POIs from every trajectory, 36 POIs (candidate stops) were defined in the zoo by considering the layout of the zoo and GPS accuracies. When a user's stop is within a defined POI, the user is considered to have been visited the POI. With this, a series of POIs which users visited can be identified. Therefore, for every trajectory, the following information was modelled:

<ID, visited POIs and their orders, the number of visited POIs, length of visit (km), duration of visit (hour), age, first time in the zoo, companion with small children, time limit, annual ticket, weather>

We only considered trajectories with at least 6 POIs for the experimental evaluations. In total, we had 41 valid trajectories. The average number of POIs visited was 13.9 (ranging from 6 to 23), with a standard deviation of 4.

9.5.2 Identifying Relevant Context Parameters

The recorded context information (i.e. < "age", "first time in the zoo", "companion with small children", "time limit", "annual ticket", "weather">) was the preliminary set of context parameters. We applied the proposed method in *Section 9.4.1* to identify relevant context parameters from this preliminary set.

We mainly compared the follow key aspects of visit among different situations: the number of visited POIs, length of visit, and duration of visit. In order to test whether the differences among different conditions for each context parameter were significant, we employed the independent group two-tailed t-test. Due to the small size of the dataset, $p < 0.2$ was used to denote statistical significance. *Table 9.1* shows the results of the comparison. Each data cell in *Table 9.1* contains the following information: p-value of t-test, mean of condition1, and mean of condition2.

The t-tests indicated that the numbers of visited POIs among different "age" conditions were significantly different. Similar results can be found for length of

Table 9.1. How visits differed among different conditions for each context parameter.

	The number of visited POIs	Length of visit (km)	Duration of visit (hour)
Age (>=45, <45)	**p=0.18 (15.45 vs. 13.23)**	**p=0.19 (4.56 vs. 3.09)**	**p=0.16 (2.88 vs. 2.01)**
First Visit (Yes, No)	p=0.52 (14.46 vs. 13.56)	p=0.26 (4.19 vs. 3.18)	p=0.30 (2.58 vs. 2.08)
Annual Ticket (Yes, No)	p=0.63 (13.50 vs. 14.77)	p=0.79 (3.66 vs. 3.41)	p=0.28 (2.62 vs. 2.05)
Companion (Yes, No)	p=0.93 (13.88 vs. 14.00)	p=0.71 (3.39 vs. 3.89)	p=0.74 (2.30 vs. 2.03)
Time Limit (Yes, No)	p=0.29 (13.00 vs. 14:32)	p=0.31 (2.98 vs. 3.74)	p=0.60 (2.10 vs. 2.32)
Weather (Sunny, Rainy)	**p=0.01 (15.07 vs. 11.64)**	**p=0.04 (4.01 vs. 2.52)**	**p=0.01 (2.62 vs. 1.52)**

visit in different "age" conditions, and duration of visit in different "age" conditions. In other words, people at different "age" groups visited the zoo differently. Similarly, for different "weather" conditions, people also behaved differently (see the bold parts in *Table 9.1). Therefore*, "age" and "weather" were considered as relevant, and taken as the final set of context parameters. In *Section 9.5.4,* we provide an evaluation to test the correctness of this decision.

9.5.3 Experiment Setting

We used the dataset in *Section 9.5.1* to evaluate the recommendation performance of SBA_CP_CaCF and SBA_CM_CaCF. In order to experimentally study whether including context information in a CF can improve the recommendation performance, we also implemented a non-contextual CF method (nonCa_CF, i.e. SBA_CP_CaCF ignoring step 2) as a benchmark.

Due to the small size of the dataset, we used a leave-one-out validation. We trained the recommendation models on 40 of the 41 visitors (trajectories) in the dataset, and tested them on the remaining visitor (the active user). We used accuracy to evaluate the performance of the CaCF methods, and accuracy was defined as the ratio of the number of correct recommendations (i.e. the recommended POI was actually viewed immediately by the active user) and the number of recommendation processes (i.e. 41 in our dataset).

In order to identify optimized values for different parameters in the proposed CaCF methods, we evaluated several thousand parametrisations (i.e. varying the threshold in SBA_CP_CaCF, and the important weights in SBA_CM_CaCF), and used the best-performing one for the final experiments.

Two evaluations were performed. The first evaluation studied whether using the proposed set of context parameters *"<age, weather>"* can achieve the best recommendation performance. As we had 6 preliminary context parameters (candidates), in total we had another 62 $(=C_6^1+C_6^2+C_6^3+C_6^4+C_6^5+C_6^6-1)$ possible sets of context parameters. The proposed CaCF methods using different sets of context parameters were compared when making recommendations for the last 5 POIs of every trajectory (we used the average accuracy of the 5 POIs). This evaluation is very useful for testing the effectiveness of the method proposed in *Section 9.4.1* and *Section 9.5.2* (i.e. identifying relevant context parameters).

The second evaluation focused on how the recommendation performances of the proposed CaCF methods differed when making recommendations for different places of a visit (i.e. from the 1st last to the 5th last). *"<age, weather>"* was used as the set of relevant context parameters. nonCa_CF was implemented as a benchmark. This evaluation can help us answer the following questions: 1) Does including context information in a CF for LBS improve the CF recommendation performance (context-aware CF vs. non-contextual CF)? 2) How do the recommendation performances of the proposed methods change when making recommendations for different places of a visit?

9.5.4 Results and Discussion

9.5.4.1 Making Recommendations with Different Sets of Context Parameters

Figure 9.1 shows how the recommendation performances of the proposed CaCF methods changed when using different sets of context parameters.

Using different sets of context parameters: *Figure 9.1* shows that among all the possible sets of context parameters, both CaCF methods using the proposed set "<age, weather>" achevied the best recommendation performances. Therefore, the proposed method (*Section 9.4.1* and *9.5.2)* to identify relevant context parameters is feasible and useful.

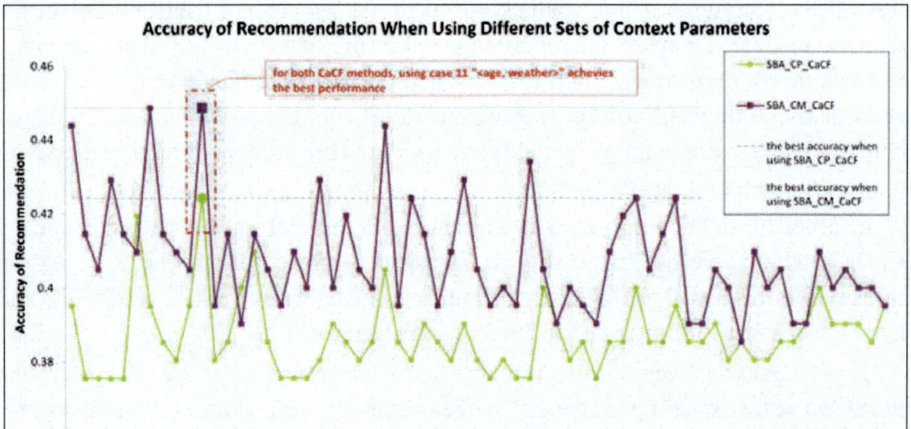

Fig. 9.1. The recommendation performances of the proposed CaCF methods changed when using different sets of context parameters (when predicting for the last 5 POIs of every trajectory).

In the meanwhile, It is also important to note that incorporating more context parameters into the CF process did not mean an improvement of performance. This can be explained by the increasing difficulty of developing an accurate context similarity measure when using more context parameters, and the increasing demand of data.

Contextual modelling vs. contextual pre-filtering when using different sets of context parameters: For most different sets of context parameters, the performance of SBA_CM_CaCF was at least as good as the performance of SBA_CP_CaCF. The two-tailed t-test showed that the difference between performance of SBA_CM_CaCF and that of SBA_CP_CaCF was statistically significant ($p=2.08E-15 << 0.01$). An explanation for this would be: The latter suffered from the problems of sparsity as lots of trajectories were filtered out, while in contextual modelling, more users (trajectories) were involved in making recommendations.

9.5.4.2 Making Recommendations for Different Places of a Visit

Figure 9.2 shows the results of how the recommendation performances of the CaCF methods changed when making recommendations for different places of a visit (i.e. from the 1st last to the 5th last). The proposed set of context parameters *("<age, weather>")* was employed.

non-contextual CF vs. context-aware CF: when making recommendations for different places of a visit, the performances of CaCF methods (i.e. SBA_CP_CaCF, and SBA_CM_CaCF) were considerably better than the performances of non-contextual CF method (i.e. nonCa_CF). This is consistent with what we expected: as CaCF methods were aware of the context (suitation) the user was in, they might generate more appropriate recommendations.

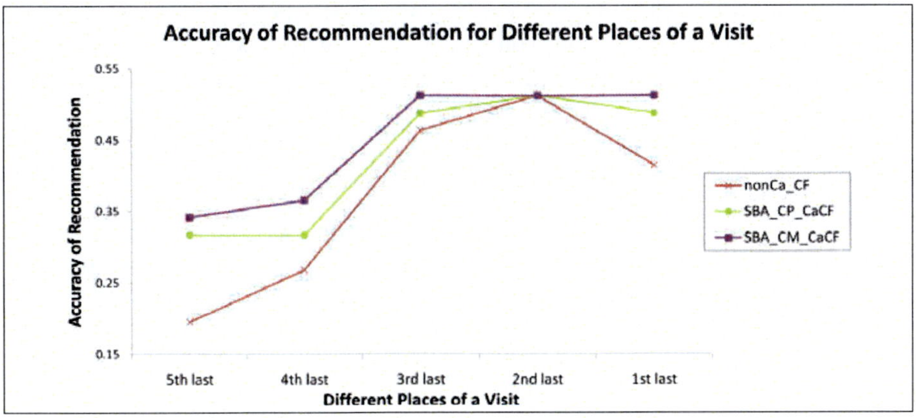

Fig. 9.2. The recommendation performances of the proposed CaCF methods changed when making recommendation for different places of a visit (using "<age, weather>" as relevant context parameters).

Making recommendations at different places of a visit: Different places (positions) reflect the different amount of information available about a visitor. *Figure 9.2* shows an upwards trend for the accuracy of all CaCF methods and the nonCa_CF when the positions of the predicted POI increased. This is consistent with our expectation: with the increase of the positions of the predicted POI, more information about a visitor was available for the recommendation methods, and therefore the recommendation performances were improving.

Contextual modelling vs. contextual pre-filtering when making recommendations for different places of a visit: When making recommendations for different places of a visit, the performance of contextual modelling approach (i.e. SBA_CM_CaCF) was at least as good as the performance of contextual pre-filtering approach (i.e. SBA_CP_CaCF). Similarly, this might be explained by the sparsity problem as lots of trajectories were filtered out in the contextual pre-filtering approach.

9.5.5 Summary

In summary, the main findings of the experiments are as follows:

1) When including context information in the CF process, choosing a suitable set of relevant context parameters is very important and may affect the recommendation performance.
2) The proposed method to identify relevant context parameters is feasible and useful, and using the proposed "<age, weather>" can achieve a higher recommendation accuracy for all the designed CaCF methods.
3) The recommendation performance of contextual modelling approach is at least as good as the performance of contextual pre-filtering approach.
4) Most importantly, the proposed CaCF methods can provide better recommendation performance than non-contextual CF, which means including context information in a CF for mobile guides can improve the recommendation performance.

9.6 Conclusions and Future Work

Currently, providing context-aware services/information is still very challenging in mobile guides. In this paper, methods of introducing CaCF into mobile guides were proposed. To be more specific, CaCF methods were applied on the highly available GPS trajectories to enhance visitors with context-aware POI recommendations in mobile guides.

The main contributions are as follows:

1) Key issues of incorporating CaCF into LBS applications (e.g. mobile guides) were identified.
2) A two-stage method was designed to identify relevant context parameters for CaCF.
3) A statistic-based approach (SBA) was proposed to measure context similarity.
4) In considering two different ways (i.e. contextual pre-filtering, and contextual modelling) of incorporating context information into the CF process, two CaCF methods were designed for LBS applications: SBA_CP_CaCF, and SBA_CM_CaCF.
5) Experimental studies were designed to evaluate the proposed methods. The results of the experiments show that the proposed CaCF methods are feasible and useful for providing context-aware recommendations in mobile guides.

From the experiments, following conclusion can be drawn: including context information in the CF process can provide more appropriate recommendations to users.

Our next step is to collect more trajectory data in both outdoor and indoor to evaluate the methods. We propose that with more trajectories available, the recommendation performances of CaCF methods will be improved and will have a significant difference with that of non-contextual CF methods. We are also interested in exploring more complex CaCF methods in considering different types of context information.

Acknowledgement

Haosheng Huang is a recipient of the Austrian Marshall Plan Scholarship. Special thanks to Prof. Martin Raubal at the ETH Zurich for his supervision on Huang and his useful comments to the draft of the paper. We also gratefully acknowledge the contributions of students from the LBS class (VU 126.099, SS2010) for collecting the data in Vienna Zoo.

References

Adomavicius G, Sankaranarayanan R, Sen S, Tuzhilin A (2005) Incorporating contextual information in recommender systems using a multidimensional approach. ACM Transactions on Information Systems, 23 (1): 103–145.

Adomavicius G, Tuzhilin A (2010): Context-aware recommender systems. In: Ricci F, Rokach L, Shapira B, Kantor P (Eds), Recommender Systems Handbook. Springer, 217–253.

Alvares L, Bogorny V, Kuijpers B, Macedo J, Moelans B, Vaisman A (2007) A Model for Enriching Trajectories with Semantic Geographical Information. In Proc. of ACM-GIS 2007.

Baus J, Kray C, Krüger A (2001) Visualization of route descriptions in a resource-adaptive navigation aid. Cognitive Processing, 2 (2–3):323–345.

Bohnert F, Zukerman I, Berkovsky S, Baldwin T, Sonenberg L (2008) Using collaborative models to adaptively predict visitor locations in museums. In Proc. of the Fifth Intl. Conf. on Adaptive Hypermedia and Adaptive Web-Based Systems (AH-08).

Cheverst K, Davies N, Mitchell K, Friday A, Efstratiou C (2000) Developing a context-aware electronic tour guide: Some issues and experiences. In Proc. of the Conference on Human Factors in Computing Systems (CHI 2000), 17–24.

de Spindler A, Norrie M, Grossniklaus M, Signer B (2006) Spatio-temporal proximity as a basis for collaborative filtering in mobile environments. In Proc. of UMICS'06, Luxemburg, June 5–9, 2006.

Dourish P (2004) What we talk about when we talk about context. Personal and Ubiquitous Computing, 8(1): 19–30.

Giannotti F, Nanni M, Pedreschi D, Pinelli F (2007) Trajectory pattern mining. In Proc. of KDD'07, ACM Press, 330–339.

Horozov T, Narasimhan N, Vasudevan V (2006) Using location for personalized POI recommendations in mobile environments. In Proc. of the International Symposium on Applications on Internet 2006, 124–129.

Huang H, Gartner G (2009) Using Activity Theory to Identify Relevant Context Parameters. In Gartner G, Rehrl K (Eds) Location Based Services and TeleCartography II – From Sensor Fusion to Context Models, Springer Verlag Berlin Heidelberg.

Li Q, Zheng Y, Xie X, Chen Y, Liu W, Ma W (2008) Mining user similarity based on location history. In Proc. of GIS'08 (Santa Ana, CA, Nov.2008). ACM Press, 298–307

Li L, Lee F, Chen Y, Cheng C (2009) A multi-stage collaborative filtering approach for mobile recommendation. In Proc. of the 3rd International Conference on Ubiquitous Information Management and Communication, January 15–16, 2009, Suwon, Korea.

Kray C (2003) Situated interaction on spatial topics. PhD thesis, Saarland University, Germany.

Panniello U, Gorgoglione M, Palmisano C (2009) Comparing pre-filtering and post-filtering approach in a collaborative contextual recommender system: An application to E-Commerce. In Noia T, Buccafurri F (Eds), Proc. of EC-Web 2009, Springer LNCS 5692, 348–359.

Raper J, Gartner G, Karimi H, Rizos C (2007) Applications of location-based services: a selected review. Journal of Location Based Services, 1(2): 89–111.

Resnick P, Varian H (1997) Recommender systems. Communications of the ACM, 40(3):50–58.

Schmidt A, Beigl M, Gellersen H (1999) There is more to context than location. Computers and Graphics, 23(6): 893–901.

Shepard R (1987) Toward a universal law of generalization for psychological science. Science, 237:1317–1323.

Takeuchi Y, Sugimoto M (2006) CityVoyager: An outdoor recommendation system based on user location history. In: Ma J. et al. (Eds), UIC 2006. Springer Berlin/Heidelberg, 625–636.

van Setten M, Pokraev S, Koolwaaij J (2004) Context-aware recommendations in the mobile visitor application COMPASS. In Nejdl W, de Bra P (Eds), Proc. of AH 2004, Springer LNCS 3137.

Wexelblat A (1999) Footprints: Interaction history for digital objects. PhD thesis, MIT Program in Media Arts & Sciences.

Wiesenhofer S, Feiertag H, Ray M, Paletta L, Luley P, Alexander A, Mathias S, Josef R, Beyer P (2007) Mobile city explorer: An innovative GPS and camera phone based travel assistant for city visitors. In Location Based Services and TeleCartography, Springer, 557–573.

Winograd T (2001) Architectures for context. Human-Computer Interaction, 16(2/4): 401–419.

Zheng Y, Chen Y, Xie X, Ma W (2009a) GeoLife2.0: A Location-Based Social Networking Service. In Proc. of the 2009 Tenth International Conference on Mobile Data Management: Systems, Services and Middleware, p.357–358, May 18–20, 2009

Zheng Y, Zhang L, Xie X, Ma W (2009b) Mining interesting locations and travel sequences from GPS trajectories. In Proc. of WWW 2009.

Chapter 10

Spatial Data Computations in a Toolkit to Improve Accessibility for Mobile Applications

Janne Kovanen[1], Paul D. Kelly[2], Stuart Ferguson[2], L. Tiina Sarjakoski[1] and Tapani Sarjakoski[1]

[1] Department of Geoinformatics and Cartography,
Finnish Geodetic Institute, Masala, Finland
{janne.kovanen, tiina.sarjakoski, tapani.sarjakoski}@fgi.fi

[2] Queen's University Belfast, Belfast, Northern Ireland
{paul.kelly, r.ferguson}@ee.qub.ac.uk

Abstract

The HaptiMap toolkit is an open-source library enabling location-based mobile application developers to include features advancing accessibility in their applications. This paper presents the structure of the geospatial data access of the HaptiMap toolkit. In addition, an evaluation is carried out on the processing performance of the toolkit. The operation performances are compared with the corresponding operations of a leading open-source library, which supports computational geometry calculations. The results indicate that implementing custom-optimized algorithms for the toolkit's data model is a feasible approach when the algorithms are specific or simple enough, while complex algorithms should be accessed through wrapped functions.

Keywords: Geographic information, computational geometry, accessibility, open-source, audio, tactile

10.1 Introduction

The HaptiMap toolkit is a cross-platform, open-source library that enhances accessibility for mobile location-based applications. The toolkit is built as part of the

G. Gartner, F. Ortag (Eds.), *Advances in Location-Based Services*, Lecture Notes in Geoinformation and Cartography, DOI 10.1007/ 978-3-642-24198-7_10,

research in the European Commission-funded HaptiMap project (Magnusson et al., 2009), the overall objective of which is to make maps and location-based applications more accessible by utilizing touch, hearing and vision senses.

The target group of the library is applications that aim to advance the perceptualization for visually impaired users. The toolkit focuses especially on applications for pedestrian navigation. It has been built to solve the problem of taking people with disabilities into account in applications developed mainly by experts other than human-computer interaction (HCI) specialists. To reach a wide audience, the toolkit can be used on a variety of popular mobile platforms, including Android, iPhone, Windows Mobile, Meego and Symbian.

The toolkit is a collection of application programming interfaces (APIs). Some of the interfaces are only available internally to support others, and the interfaces have different levels of complexity and functionality. Hence, general API development rules and principles, such as robustness, durability, modularity and ease of use, apply to the toolkit. The toolkit design concentrates especially on addressing the usability of the toolkit by its users. The users of the toolkit can be categorised in two distinctive groups: HCI designers and map-based mobile application developers. The internal API is mainly for the HCI designers, who want to share their knowledge on how to make mobile applications more accessible for navigation. The HCI designers extend the toolkit, and take the end users into account in the toolkit's functionality. The public API is for application developers who might only have a moderate understanding of geographic information, or how to channel this information to visually impaired end users. Thus, to succeed, the toolkit interfaces should follow the ideal functionality described by Henning (2007): *the right call for a particular job is available at just the right time, can be found and memorized easily, is well documented, has an interface that is intuitive to use, and deals correctly with boundary conditions*.

The toolkit aims to support several modalities and types of HCI in order to help the end user to perceive the spatial information. The applications could use visual feedback for visually impaired people or the elderly through common visualization rules. Tactile feedback for blind or visually overloaded users, such as bikers, can be based on vibrations. Vibrations may come from the in-built vibrators of smart phones. The toolkit core supports access to the vibrators. Vibrations may also come from external devices; for instance, Anastassova & Roselier (2010) defined preferred vibration patterns for a handheld haptic device called VIFLEX (Roselier & Hafez, 2006), which can be accessed through the toolkit. Similarly, vibrations may be given by a tactile waist belt (Erp. et al. 2005; Pielot & Boll, 2010) or wristbands (Bosman et al. 2003). For toolkit support, these actuator devices would need their own output drivers in the toolkit core.

Audio-based guidance can be based on spoken instructions, as has been presented by several systems, like Chittaro & Burigat (2005). The toolkit supports text-to-speech, but auditory feedback is not restricted to speech because non-verbal descriptions of the surroundings may also be given through other means, such as earcons (Blattner et al. 1989), spearcons (Walker et al., 2006), auditory icons (Gaver 1986) or soundscapes (Schafer 1977). Similarly, directions and distances may be given with audio patterns or metaphors, such as the Geiger counter (Holland et al. 2002), Hearcons (Klante et al. 2004), gpsTunes (Strachan et al. 2005), Soundcrumbs (Magnusson et al. 2009) and AudioBubbles (McCookin 2009). These functionalities are not part of the toolkit, but similar and broader ones may be implemented in the future as part of the toolkit.

Several toolkit modules support the HCI modules. One of the support modules handles computational geometry tasks, such as calculating the distance to the next obstacle on a path. The module is made available for internal and external use. The demand for such a module was incontrovertible; however, whether or not to implement the necessary algorithms, especially for the toolkit, or use existing open source geometry engines was not self-evident. Hence, our research question was formed: What is the benefit of implementing an engine of our own in terms of performance and usability?

Our solution was to implement the functions that had mostly been applied from the beginning under the licensing model of the toolkit, and to empirically validate our approach. In addition, we decided to make the rest of the algorithms available through wrapping them inside functions that perform the necessary data and geometry type conversions. The fact that not all of the required functionalities were available in computational geometry engines had a substantial impact on the decision. This paper presents an assessment of the solution, and compares it with the option of only using wrapped functions.

10.2 The Architecture of the HaptiMap Toolkit

10.2.1 Core, Mantle, Crust

The toolkit is composed of three principal modular layers: core, mantle and crust. All modules can access the public interfaces of the same layer as well as the lower layers. Furthermore, a set of plug-ins is used to read geospatial data into the internal data model of the toolkit. The core layer has two main functions. Its first task is to connect to external sensor hardware, such as an accelerometer, for input or a tactile vibrator belt for output. To make use of platform-specific operating system interfaces, some parts of the core are also platform specific. The second task of the toolkit core is handling and caching geographical data in a simplified data model. Geographical data is stored within a custom database based on memory-mapped disk files. Memory-mapped disk files reduce processing requirements by presenting disk files as virtual memory blocks (Scheiber 1997). Our main alternative for memory-mapped files was SpatiaLite (2011), which is based on SQLite. However, we did not want to have additional external dependencies, and we were curious to see whether memory-mapped files could be used – as it turns out they can be.

The mantle layer is situated on top of the core. This platform-independent layer acts as a middle layer between the core and the toolkit utilizing software components – the crust. The modules of the mantle layer are building blocks that involve analysis and processing functionalities. Consequently, the mantle layer modules contain the primary accessibility-enhancing logic. For instance, a module might employ the text-to-speech functionality of the core together with access to the vibrations to signal the direction to a target location. The text-to-speech functionality would be provided using the eSpeak speech synthesis engine (eSpeak 2011) and the vibrations through the in-built vibrator of the mobile device.

The modules of the mantle are written in pure ANSI-C. Both the core and mantle layers are published under the Lesser General Public License (LGPL). The crust components may just as well be called 'convenience-modules', because they present alternative ways to a mobile application user for how to realize the functionality of the core and mantle. For instance, the crust can include view activities and fragments for the Android platform, while for the iPhone platform the crust can contain view controllers. A view controller could, for example, visualize how passable a route is for a physically restricted person, by providing the starting and end points, or a reference to the route. The crust functionality might also simply be used to modify the mantle behaviour, such as to provide settings for vibration patterns. The crust layer can contain modules in any suitable license. Similarly, the functionality in the crust is written in a platform-specific language.

Figures 10.1 and *10.2* represents two of the demonstrators built to use the toolkit functionality. *Figure 10.1* represents the PocketNavigator that is built for the Android platform (Pielot et al. 2011), whereas *Figure 10.2* contains a screenshot of the Terrain Navigator that is run on the iOS and is an extension of the prototype presented by Kovanen et al. (2009). These two demonstrators are meant for different environments, they utilise different data sources and interfaces for the background maps and they use different crust functionalities of the toolkit.

10.2.2 The Data Model and Internal Coordinate Reference System

The abstraction level of the toolkit was a compromise between the degree of functionality versus its level of simplicity. The solution was to keep the abstraction high, because most of the accessibility-enhancing software implementations do not need a complex geographical data model and it is anticipated that their developers will have only meagre experience with geographical data; thus, an overly complex

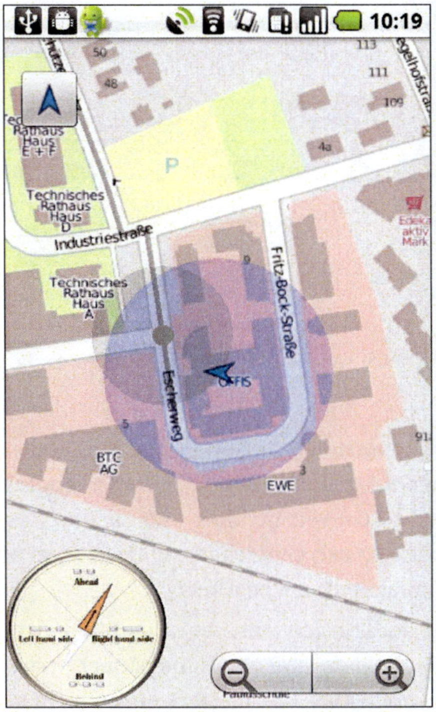

Fig. 10.1. PocketNavigator represents an Android application that makes use of the HaptiMap toolkit through a TactileCompass (Pielot et al. 2011).

Fig. 10.2. The Terrain Navigator application. The application is run on the iOS and a part of its functionality utilizes the toolkit.

model would be unpropitious (Kelly et al. 2011). Consequently, the toolkit may be easy to learn, easy to use and hard to misuse; however, it does not allow for enhancing accessibility using a comprehensive geometry model.

The geometry model of the toolkit is composed of two principal geometrical primitives: points and linestrings. In addition, a linestring may be used to form a planar surface, a polygon, without any interior holes. Formalising a linestring as a polygon is performed through the geographical data storage API. A comparison of this geometry model with a model of the ISO Simple feature access standard (ISO 2004) is presented in *Figure 10.3*. The information model of the toolkit allows the geometries to be linked with any number of attributes. The attributes are key-value encoded. The keys are integer values whose meaning is defined in resource files. The value of an attribute may be an integer, integer array or string.

We performed another simplification in order to increase efficiency on mobile devices. We made the decision to use centimetres as the internal unit of measure and store all co-ordinate values as 32-bit integer values. The whole numbers reduce round-off errors (i.e., they increase robustness) and slightly increase perform-

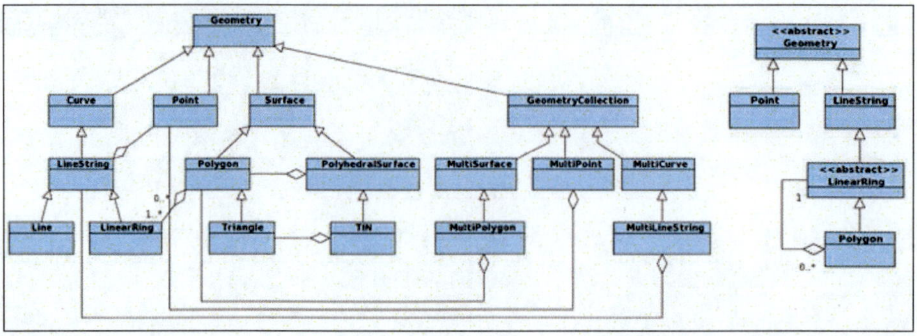

Fig. 10.3. The ISO Simple feature access geometry model (left) and HaptiMap toolkit geometry model (right).

ance; however, during spatial operations, a conversion to floating point numbers is commonly required to avoid integer over- and underflows. Similarly, while reading data into the internal data model, we needed to convert the data into centimetres. As the unit of measure for angles, we decided to use degrees in public interfaces because most platform APIs use them for digital compass bearings.

The coordinate reference system of the toolkit can only be a compound or projected coordinate reference system. In both cases, the two-dimensional part is confined so that it has a metric-map projection; thus, for instance, Pseudo Plate Carrée would be unsuitable as a projection. The second part of the compound system is a gravity-related coordinate reference system; in other words, height values are normal or orthometric. The coordinate reference system is based on a leading data source; however, for most operations, it is irrelevant what the actual system is. It is enough to have coherent units and to know that even the height axis is almost perpendicular to the two-dimensional plane.

10.2.3 Geospatial Data Work-Flow

The toolkit is built to handle only geographical vector data, because it takes far too much time to perform analysis on raster data on a mobile device while in a hurry. The data is loaded into the toolkit from external data sources using logically separate plug-ins for different types of data sources. The relevant servers are connected directly by the plug-ins. For example, in case of the open and free OpenStreetMap API, first a HTTP request is emulated, next the plug-in connects to a XAPI server, and finally the plug-in downloads and parses an XML file containing the data. Similarly, NAVTEQ's commercial MapTP API is accessed using the C API provided by NAVTEQ,

A plug-in is required to map the feature type and attributes from the downloaded data into the toolkit information model. For instance, the input model may be the

OGC simple feature model. Additionally, the plug-in may have to generalize the input geometries to make them suitable for the toolkit, which may involve operations such as removing polygon holes and splitting a multi-geometry into individual geometries. Separating the plug-ins outside the core and mantle allows them to be written under some other license than LGPL. Besides, the plug-ins' data may be added to the internal data storage through the public interface of the core layer. The application making use of the library may decide to only temporarily add data to the internal storage of the toolkit, while maintaining its own database. This scenario is especially suitable for applications where the toolkit library is included at a late stage of development, or as an optional extension. Applications also use own data sources, and the applications may use any internal data format. Consequently, an application can, for example, let the toolkit store all required vector data while it stores raster data for background map visualization. Similarly, the toolkit does not restrict an application from exchanging data with the server-side, but the toolkit itself does not perform writing operations through server interfaces, because the output of the HCI modules is only directed towards the current end user – a human.

10.2.4 Geospatial Operations

The requirements for the toolkit computational geometry functions are based on the requirements gathered from their intended users, the HCI developers. HCI developers are, for the most part, not interested in the absolute positions of features in their applications; instead, relative measures between objects play an important role. For instance, pointing or scanning (or sweeping) types of applications, such as a geiger counter, hearcons or soundcrumbs, could be implemented by knowing only relative measures to the closest points. Hence, the distances and directions to nearby objects measured from the location of the user are of greatest importance for building the mental model of the surroundings of the user and guiding her or him to the final destination.

Operations such as distance and bearing are required to guide the user. For instance, an application could give, while using multiple modalities, the direction and distance to a visual or otherwise notable landmark. These operations, together with length and area, can also be used to describe the environment. To obtain the values, it is necessary to offer supporting operations for fundamental static computational geometry problems, like geometry intersection or convex hull calculation. Similarly, support for search problems, spatial predicates and buffering is required to solve queries made by accessibility enhancing functionality. For instance, the point-line intersection function may be essential for determining the distance to the centre of a footpath, buffering may be applied to validate that the user is inside a passable safe zone, and ray-tracing can be used to verify how far it is to the next obstacle or attention point. The requirements for algorithms depend on the application at hand. At the moment, we can only implement our own applications and predict some of the problems that others might need to solve, because the library will evolve in the future and more human-computer interaction modules with new needs will emerge.

A single module in the mantle layer primarily performs geometry-related computations. A client-server solution for the computations was not suitable, because our usage scenarios require the temporary unavailability of a network connection, which is particularly true in an outdoors environment. The module is made available for the HCI modules in the mantle layer and all modules in the crust layer. The input

Table 10.1. Some operation interfaces of the HaptiMap toolkit concerning geospatial operations.

```
/* Metric methods */
HM_RESULT hm_geom_area(hm_t *hm, int lid, double *area);
HM_RESULT hm_geom_bearing(hm_t *hm, int pid1, int pid2, double *angle);
HM_RESULT hm_geom_distance(hm_t *hm, enum HM_GEOMETRY_TYPE gtype1, int fid1,
          enum HM_GEOMETRY_TYPE gtype2, int fid2, double *dist);
HM_RESULT hm_geom_distance_hausdorff(hm_t *hm, enum HM_GEOMETRY_TYPE gtype1,
          int fid1,enum HM_GEOMETRY_TYPE gtype2, int fid2,double *dist);
HM_RESULT hm_geom_length(hm_t *hm, int lid, double *length);
/* Spatial predicates */
HM_RESULT hm_geom_contains(hm_t *hm, int polyfid,
          enum HM_GEOMETRY_TYPE gtype, int fid, int *r);
HM_RESULT hm_geom_within(hm_t *hm, int fid,int polyfid,
          enum HM_GEOMETRY_TYPE gtype, int *r);
HM_RESULT hm_geom_intersects(hm_t *hm, enum HM_GEOMETRY_TYPE gtype1,
          int fid1, enum HM_GEOMETRY_TYPE gtype2, int fid2,int *r);
/* Overlay methods */
HM_RESULT hm_geom_intersection(hm_t *hm,
          enum HM_GEOMETRY_TYPE gtype1, int fid1,
          enum HM_GEOMETRY_TYPE gtype2, int fid2,
          enum HM_GEOMETRY_TYPE *r_type, int *r);
/* Buffering */
HM_RESULT hm_geom_buffer(hm_t *hm, int fid, enum HM_GEOMETRY_TYPE gtype,
          double buffer_width, int *r);
/* Generalisation etc */
HM_RESULT hm_geom_simplify(hm_t *hm, int lid, double tolerance, int *r);
HM_RESULT hm_geom_centroid(hm_t *hm, int lid, int *r);
HM_RESULT hm_geom_interior_point(hm_t *hm, int lid, int *r);
HM_RESULT hm_geom_convex_hull(hm_t *hm, int lid, int *r);
HM_RESULT hm_geom_mbr(hm_t *hm, int lid, int *mbr_id);
HM_RESULT hm_geom_ray_intersection(hm_t *hm, int lid, int pid,
          double angle, double *distance);
```

geometries are directly read from the public core interface, and the calculations take into account the characteristics of the toolkit. The algorithms support common spatial operations for two-dimensional geometries. The implemented algorithms are not all robust. Some algorithms are affected by round-off errors, but typically the targeted types of HCI applications do not require the robustness. For instance, an application wants to inform the user how far the next obstacle is located. To calculate the intersection point a ray is cast from the user's location to a certain direction. If the obstacle is a point or parallel line then due to floating point arithmetic the ray might pass the geometry even if it should intersect in reality.

The interfaces support only centimetres for distances and degrees for angles. *Table 10.1* presents interfaces for some of the operations. All operations return a status code indicating whether they were successful or not. If the geometry type is not implicit in the operation, the geometries are given as references together with the type of geometry; for instance, length is not computed for a point feature. Possible output geometries are directly written to the internal data storage and references to the identifiers of the new geometries are returned.

10.3 Assessment of the Operational Performance of Geometry Handling

We tested the spatial operations of the toolkit against one of the oldest, most widespread and stable open source geometry engines available, the GEOS (Geometry Engine – Open Source), which is a part of the Java Topology Suite (JTS). GEOS was originally written in C++, but has also been wrapped in the C-language. We tested against GEOS version 3.2.2. We also designed the geometry engine to be wrapped for the most complex functions required by the toolkit.

10.3.1 Fit for Use and Ready for Testing

We conducted unit tests for every implemented spatial function. The test cases are comprehensive, independent and can be run continuously while extending the interfaces or upgrading the functionality behind the interfaces. The unit tests also provide examples for HCI developers on how to use the API, that is to say, the test cases also act as documentation. We validated the unit testing values by comparing them to other geometry libraries; thereby, during performance comparisons, it was no longer necessary to assess the correctness of the library by making separate result comparisons.

The functions for testing were not randomly selected. We only tested and validated functions that were based on the same algorithms in GEOS and the toolkit module. We performed the validation by going through the open source code of GEOS. For instance, both the library and GEOS implement convex hull calculations using Graham Scan algorithm (Graham 1972) with slightly different syntaxes.

10.3.2 Operational processing power comparison

We compared the operation performances with the corresponding operations of GEOS. The operations included the most common functions, such as distance calculation, but also more sophisticated algorithms, such as convex hull and intersection calculation. We performed the benchmarking on an iPhone mobile phone simulator, iPhone 4 device, and iPad 2 tablet. The simulator was used to find defects in the benchmarks before running them on the real devices.

We ran each operation several times (100–1000) in order to obtain realistic benchmark numbers and remove noise. In addition, each benchmark was repeated one hundred times with different random geometries. We recorded standard deviations to validate how repeatable and systematic the results are. Three different cases with same data were benchmarked:

1. The implemented algorithm. Before benchmarking, all data is stored in the internal data storage of the toolkit. The algorithm is responsible for reading the geometries from the data storage based on the identifiers of the features.
2. Corresponding GEOS algorithm. The same geometries as in the first case are converted into the geometry model and structure of GEOS before benchmarking. GEOS is accessed through its C-interface.
3. GEOS algorithm wrapped inside the toolkit model compliant interface. The internal geometry model is converted into the GEOS geometry model and structure during benchmarking. Similarly, if the operation creates new geometries, these are added to the internal data storage during benchmarking.

10.3.3 Adjusting the Size of the Internal Data Storage

After the first benchmark iterations on the simulator it became evident that something was wrong with the algorithms that created new geometries. The problem appeared as an unordinary high standard deviation and sluggish performance, which declined after each run. *Figure 10.4* shows some of the results of the benchmarking at that moment. The algorithms affected in the figure are polygon centroid calculations and convex hull calculations, which both create new geometries as a result.

The reason turned out to be the large number of times the memory-mapped files (on which the toolkit internal data storage is based) were resized during the tests. This had a severe impact on the on performance, and was caused by the inser-

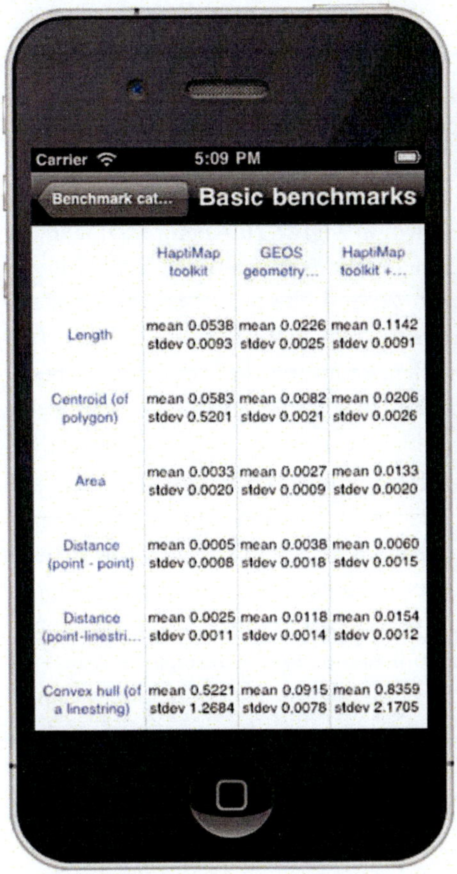

Fig. 10.4. Benchmarking with a small internal memory reallocation size on the iPhone simulator.

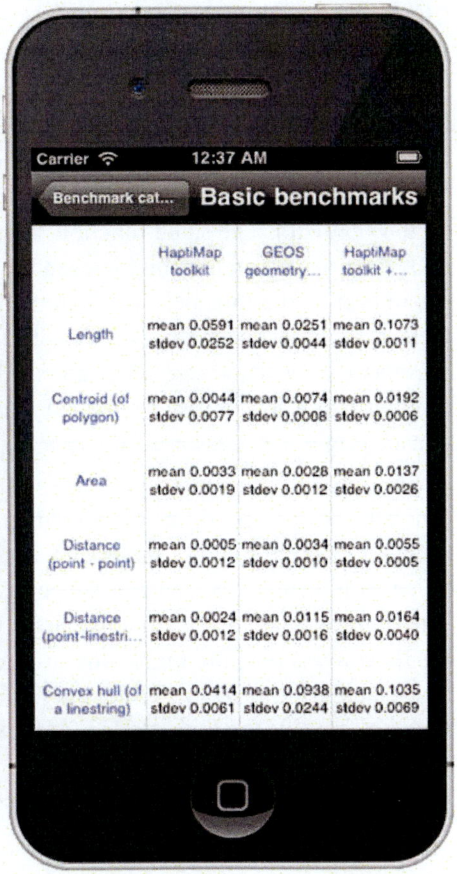

Fig. 10.5. Benchmarking results on the iPhone simulator on a desktop computer after the memory allocation fix.

tion of a large number of new geometry elements. The increments by which the mapped file sizes were increased were not designed for the benchmarking, instead, for normally sized input data (a couple of hundred features at a time). Thus, to get more realistic results, we increased the reallocation limit. The observation suggests that the size increment might need to be a variable that can be adapted for different types of applications. For instance, if the application relies heavily on the internal data storage and an intense amount of features are added to the data storage, then the size increment should be large. In general, there is much scope for optimizing the manner in which geometry elements are inserted and deleted from the memory-mapped files in order to improve performance.

10.4 Results

The results clearly show that implementing the algorithms from the beginning using the same formulas led to, in general, better performances in comparison to GEOS. Nevertheless, in cases with a high degree of variation, like the length calculation, the performance was poorer. *Figure 10.5* presents the results of one benchmarking on the simulator. The same functions are shown in *Figure 10.4* before the memory allocation fix. A comparison between the benchmark numbers of the figures shows the impact of the memory size increment. In general, the results on the simulator were five times better than the results from the real mobile device. *Figure 10.6* presents the results on the iPad tablet. The benchmark results were slightly better in comparison to the result obtained from the real iPhone 4, which is reasonable considering the processor differences.

Our most significant finding was the difference between toolkit functions and functions wrapping GEOS functionality. The result was independent of the environment (mobile phone / tablet). In each benchmark, the difference was at least twice as high for the wrapped functions, but it could be up to ten times higher. The simpler the algorithm and the larger the percentage of time taken for geometry model conversions, the higher the difference became. The difference between the time taken to perform the GEOS algorithms versus the time for performing wrapped functions reveals the time needed for the conversions. We obtained a third result by comparing the time needed to implement the algorithms. It took much more time to implement complex algorithms because of the required data structures and the

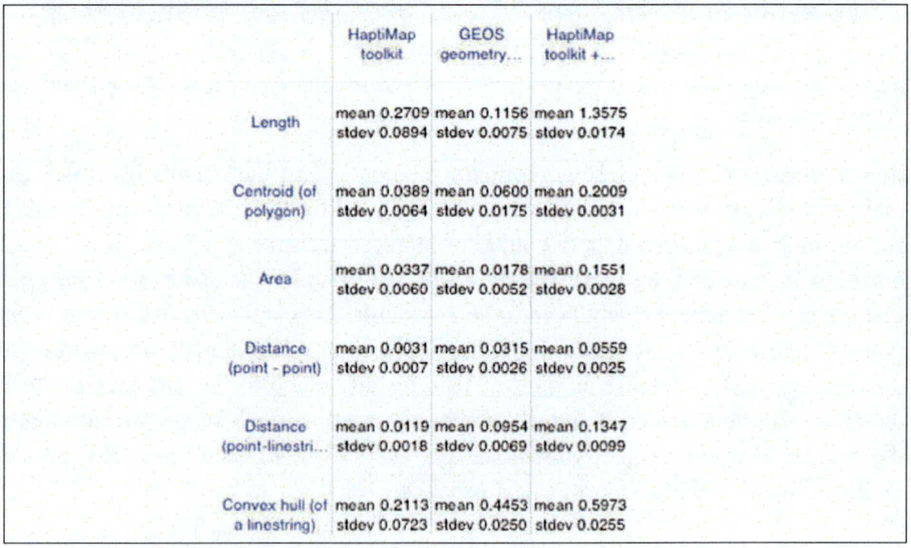

	HaptiMap toolkit	GEOS geometry...	HaptiMap toolkit +...
Length	mean 0.2709 stdev 0.0894	mean 0.1156 stdev 0.0075	mean 1.3575 stdev 0.0174
Centroid (of polygon)	mean 0.0389 stdev 0.0064	mean 0.0600 stdev 0.0175	mean 0.2009 stdev 0.0031
Area	mean 0.0337 stdev 0.0060	mean 0.0178 stdev 0.0052	mean 0.1551 stdev 0.0028
Distance (point - point)	mean 0.0031 stdev 0.0007	mean 0.0315 stdev 0.0026	mean 0.0559 stdev 0.0025
Distance (point-linestri...	mean 0.0119 stdev 0.0018	mean 0.0954 stdev 0.0069	mean 0.1347 stdev 0.0099
Convex hull (of a linestring)	mean 0.2113 stdev 0.0723	mean 0.4453 stdev 0.0250	mean 0.5973 stdev 0.0255

Fig. 10.6. Benchmarking results on the iPad 2 tablet.

number of boundary conditions that needed to be taken into account. However, the time difference between running these algorithms diminished.

The benchmarking also proves that the approach of using memory-mapped files is feasible. Similarly, the use of external data sources, in addition to the vector data used by the toolkit, was empirically demonstrated through application development *(Figures 10.1* and *10.2)*. However, even if the API of the toolkit is usable, we could not yet affirm that the use of the internal API is simpler or more user-friendly in comparison to other interfaces. One of the topics for future research might well involve usability; however, such a study must wait for more developers to become interested in the toolkit so that we could obtain statistically significant results.

10.5 Conclusion

In this paper we have presented an overview of geospatial data handling for the open-source, multi-platform HaptiMap toolkit. The toolkit enables developers to advance accessibility features in mobile map-based applications. The toolkit does not constrain the internal data model of an application which uses it, and the toolkit allows applications to visualize all of the same data that they would before employing the toolkit. The toolkit is composed of three logically separate layers and a set of plug-ins for external data sources. One of the layers contains a module to solve computational geometry problems. The module is mainly provided for human-computer interaction modules that require support on solving computational geometry tasks locally.

In addition, we have performed benchmarking to gain confidence in our solution concerning having access to the internal data storage and performing computational geometry tasks with our simplified data model. In this paper we have presented the results of the benchmarking, which answers our research question concerning performance. The benchmarking proves our approach to be significantly faster than the alternative of using wrapped functions. Moreover, through the approach it is possible to publish the toolkit without mandatory external dependencies, and to take into account special requirements for geometry computations coming from HCI developers.

However, providing our own algorithms takes a greater toll in the form of additional work as the algorithms become more sophisticated, and, at the same time, the performance addition decreases. Hence, a mature computational geometry library should be wrapped and made available as an optional extension to solve complex tasks.

The research work is continuing, with further developments, and the toolkit is advancing towards publishing at the latest stage of the project.

Acknowledgments

This survey was performed as part of the HaptiMap-project (www.haptimap.org) that is coordinated by the Rehabilitation Engineering Research Group of Lund University's Department of Design Sciences. The authors are grateful to the European Commission, which co-funds the HaptiMap-project (FP7-ICT-224675).

References

Anastassova M, Roselier S (2010) A Preliminary User Evaluation of Vibrational Patterns Transmitting Spatial Information. In Proceedings of the MobileHCI 2010, September 7–10, Lisboa, Portugal

Blattner MM, Sumikawa DA, Greenberg RM (1989) Earcons and icons: Their structure and common design principles, Human-Computer Interaction 4:11–44

Bosman S, Groenendaal B, Findlater JW, Visser T, de Graaf M, Markopoulos P (2003) GentleGuide: An Exploration of Haptic Output for Indoors Pedestrian Guidance. In Chittaro L (Ed.), Human-Computer Interaction with Mobile Devices and Services. 5th International Symposium, Mobile HCI 2003, Udine, Italy, September 8–11, 2003, Proceedings, pp. 358–362

Chittaro L, Burigat S (2005) Augmenting Audio Messages with Visual Directions in Mobile Guides: an Evaluation of Three Approaches. In MobileHCI '05: Proceedings of the 7th international conference on Human computer interaction with mobile devices & services. New York, NY, USA, pp. 107–114

eSpeak (2011) eSpeak: Speech Synthesizer. http://espeak.sourceforge.net/

Gaver WW (1986) Auditory icons: Using sound in computer interfaces. Human-Computer Interaction 2:167–177

Graham R (1972) An Efficient Algorithm for Determining the Convex Hull of a Finite Point Set. Info. Proc. Letters 1:132–133

Henning M (2007) API Design Matters. ACM Queue 5 (4), Mai/June, 2007

Holland S, Morse DR, Gedenryd H (2002) AudioGPS: Spatial audio navigation with a minimal attention interface. Personal and Ubiquitous Computing 6(4):253–259.

ISO – International Organization for Standardisation (2004) ISO 19125:2004. Geographic information – Simple feature access – Part 1: Common architecture

Kelly PD, Ferguson S, Rafferty K, Peng J-X (2011) Use of Geographical Information for Non-Visual Perceptualisation. In Proceedings of Geographical Information Systems Research UK (GISRUK 2011), Portsmouth, UK, 2011.

Klante P, Krösche J and Boll S (2004) AccesSights – A Multimodal Location-Aware Mobile Tourist Information System. In Computers Helping People with Special Needs. Proceedings of the 9th International Conference, ICCHP 2004. Lecture Notes in Computer Science, vol. 3118/2004, July 2004, pp. 287–294

Loomis JM, Golledge RG, Klatzky RL, Speigle JM, Tietz J (1994) Personal Guidance System for the Visually Impaired. In Assets '94: Proceedings of the first annual ACM conference on Assistive technologies. New York, NY, USA, pp. 85–91

Magnusson C, Beidegard B, Rassmus-Gröhn K (2009) Soundcrumbs – Hansel and Gretel in the 21st century. In Proceedings of the 4th international workshop on Haptic and Audio Interaction Design (HAID'09)

Magnusson C, Brewster S, Sarjakoski T, Roselier S, Sarjakoski LT, Tollmar K (2009) Exploring Future Challenges for Haptic, Audio and Visual Interfaces for Mobile Maps and Location Based Services. Proceedings of the CHI 2009 –LocWeb Workshop, April 4–9, Boston, MA, USA

McGookin D, Brewster S, Priego P (2009) Audio Bubbles: Employing Non-speech Audio to Support Tourist Wayfinding. In Proceedings of the 4th international workshop on Haptic and Audio Interaction Design (HAID'09)

Pielot M, Boll S (2010) Tactile Wayfinder: Comparison of Tactile Waypoint Navigation with Commercial Pedestrian Navigation Systems. In proceedings of the 8th International Conference on Pervasive Computing, May 8, Helsinki, Finland

Pielot M, Poppinga B, Heuten W, Boll S (2011) 6th Senses for Everyone! The Value of Multimodal Feedback in Handheld Navigation Aids. In Proceedings of the ICMI '11: 13th International Conference on Multimodal Interaction.

Schafer RM (1977) The Sounscape: Our Sonic Environment and the Tuning of the World. Rochester:Destiny Books, 1994 (1st. ed. 1977)

Scheibel T (1997) Sharing memory with memory-mapped files. Performance Computing/ Unix Review 15(11):47–50, October 1997

SpatiaLite (2011) SpatiaLite 2.3.1 http://www.gaia-gis.it/spatialite/

Strachan S, Eslambolchilar P, Murray-Smith R (2005) gpsTunes – controlling navigation via audio feedback. In MobileHCI 2005 1:275–278, Salzburg, Austria, ACM

Van Erp JBF, Van Veen HAHC, Jansen C, Dobbins T (2005) Waypoint Navigation with a Vibrotactile Waist Belt, ACM Transactions on Applied Perception 2(2):106–117

Walker NB, Nance A, Lindsay J (2006) Spearcons: speech-based earcons improve navigation performance in auditory menus. In Proceedings of the International Conference on Auditory Display, pp. 95–98

Chapter 11

Using OSM for LBS – An Analysis of Changes to Attributes of Spatial Objects

Peter Mooney[1] and Padraig Corcoran[2]

[1] Department of Computer Science, National University of Ireland Maynooth, Co. Kildare, Maynooth, Ireland
peter.mooney@nuim.ie
[2] School of Computer Science & Informatics, University College Dublin, Ireland
padraig.corcoran@ucd.ie

Abstract

The quality of volunteered and crowd-sourced spatial data is not in most cases audited prior to being made accessible to end-users. Studies have shown that this spatial data varies significantly in terms of its geometric quality, its semantic consistency, in terms of its comprehensiveness of coverage and in terms of its currency. Subsequently it often compares poorly with the authoritative data capture and mapping undertaken by national mapping agencies and commercial companies. In this paper we highlight a specific type of problem encountered with volunteered geographic information (VGI) – the naming of real-world features. Many Location-based Services (LBS) applications are using VGI as a spatial data source. Examples include: www.mapswithme.com the offline travel guide and routing application OpenRouteService (www.openrouteservice.org). The volatility in VGI, as shown by the results in this paper, will require LBS developers to carefully consider how they manage and use the spatial data sources generated by VGI and crowd-source paradigms.

Keywords: VGI, OpenStreetMap, Crowd-sourcing, Spatial Data

G. Gartner, F. Ortag (Eds.), *Advances in Location-Based Services*, Lecture Notes in Geoinformation and Cartography, DOI 10.1007/ 978-3-642-24198-7_11,

11.1 Introduction

The explosive growth of location-aware devices, wireless communications, and mobile databases has resulted in the realization of location-based services as commercial products and research prototypes. (Mokbel and Levandoski, 2009). These technologies now allow citizens to capture information about their position in space. This includes the capture of both the geometry and attribution of the landscape or environment that they are: currently living in, passing through, or otherwise interested in. OpenStreetMap (OSM) is an example of a collaborative initiative where such spatial data and information can be collated to build-up comprehensive databases operating at all levels from the local and specialist to the global and general. OSM is probably the most famous example of a collaborative, crowd-driven, social-network project for spatial data and information (Over et al 2010, Mooney et al 2010b). However the quality of volunteered geographic information (VGI) and crowd-sourced data is not in most cases audited prior to being made accessible to the public. The data varies significantly in terms of its geometric quality, its semantic consistency, in terms of its comprehensiveness of coverage and in terms of its currency. Subsequently it compares poorly with the authoritative spatial data capture and mapping regimes undertaken by national mapping agencies (NMA). However, despite the lack of resources compared to NMA, crowd-sourced and volunteered data is in many cases more up-to-date than the authoritative source of information (as in the case of the UK as described in Haklay, (2010)), can incorporate features of interest not covered by mapping agencies and is often broader and richer in the meta-data captured. In this paper we analyse the naming of spatial objects in OSM in this collaborative environment. The data in OSM is most commonly gathered by OSM contributors surveying areas with GPS then uploaded to OSM using one of the many editors (Potlatch, JOSM, Merkator, etc) or by using an editor to trace the outline of geographic features from aerial imagery. In both cases tags or attributes (key-value pairs) can be optionally associated with the geographic object. The OSM community maintain a community endorsed ontology of key-value pairs on the "Map Features" page (OSM, 2011)

11.1.1 Tagging Objects in OSM

The name attribute of objects in spatial databases is arguably one of the most important for Location-based Services (LBS) applications. Accessing information by its geographic reference is natural and useful in several contexts: for example when looking for information resources on a city, town, or area for tourism activities from mobile devices it is common practice to attempt to access information by specifying the geographic place of interest. Edwardes (2009) states that "one of the quintessential ways in which people make sense of space is by naming it" and that identification of places is "perhaps the most important perspective in LBS". More generally, it has been estimated that about 15% of the queries submitted to general purpose search engines contain geographic names (Bordogna et al, 2011). Searches for geographic information on the Internet often fail due to people referring to locations, such as local neighbourhoods, with commonplace or vernacular names that are not recognised by conventional administrative gazetteers (Twaroch et al, 2009). de Longueville et al (2010) comment that in many real-world applications, vernacular names, which do not correspond to those official place names in gazetteers, are often used to name places by people and those who contribute to VGI projects. In fact the authors use the term "open gazetteer" to designate the concept of gazetteer enriched with vernacular place names

Joshi et al (2010) remarks that one of the most "potent forms of noise in collaborative environments is the incorrect or alternate spellings associated with place-names". In the English language this form of noise is compounded by the inherent polysemy and synonymy of words in the English language.

According to the TagInfo service in May 2011 (TagInfo, 2011) there are just over 26 million objects in the OSM global database with a name tag assigned. In OSM the name tag is a free text field. When a contributor adds or edits an object in OSM they can optionally create, edit, or delete a name tag. The highway tag is used to indicate: roads, streets, paths, lanes, etc. This tag is specified formally in the ontology provided on the OSM Wiki pages called "Map Features". This specifies values for the highway tag from (motorways, Interstates, *Autobahnen*) to (laneways, paths, tracks, trails). Generally the highway-tag represents the importance of the highway in the street/road network grid. Consequently contributors should understand where in the street/road network hierarchy the feature they are working on fits. All tagging in OSM should be performed in accordance with the community agreed ontology on the "Map Features" page. The OSM editor software packages all provide functions in the user interfaces to make selection of keys, and where appropriate corresponding values, easy and partially automated. However contributors have the ability to ignore these suggestions and provide their own tags.

In studies on Wikipedia it was found that most Wikipedians contribute to a relatively small set of articles each. Their contribution was biased towards one or very few article(s) (Zhang et al, 2010). At the same time, each article's contributions are often championed by very few active contributors including the article's creator. Stein and Hess (2007) found in an analysis of the German Wikipedia that the number of contributors and who those contributors were was very strongly correlated with article quality.

11.1.2 Research Contributions of this Work

The research focus of this paper is as follows. Current research on OSM data quality and user behaviour is predominantly focussed on the "current" snapshot of the OSM database for a given region or country. Fresh updates to these databases can be downloaded as frequently as every 3 hours. However this approaches makes the assumption that the spatial data (geometries and attributes) within the OSM database have been collected and managed in a process similar to NMA of: survey/collect data, assimilate, combine, correct, check, and release. Mooney and Corcoran (2011c) argues that is not the case and that edits and updates are made in often unpredictable ways. In LBS some of the most popular applications involve queries of the form "How do I get to location X?" or "Plan me the shortest route to street/place X". In this paper we provide an analysis of the entire version editing history of over 24,000 objects in 4 OSM databases in Europe. We show that significant changes can occur over time as a contributor or contributors edit/update object's geometry or spatial attributes. Specifically we analyse changes to objects with a "name" attribute and objects with a "highway" attribute. We highlight, and attempt to quantify, the uncertainty over the names assigned to objects and/or the assignment of designation status to the highway attribute. We are not aware of any other work of this type on VGI/OSM which specifically addresses these problems through analysis of the full version history. The key outcomes are as follows: (1) uncertainty in name or highway attribute value is not strongly correlated to increasing number of contributors editing a given object, (2) approximately 10% of objects in the OSM databases analysed exhibited changes in the name or highway attribute, and (3) this provides a warning to LBS developers that OSM may need to be supported, in parallel, by gazetteers etc to ensure accuracy of spatial queries.

The remainder of the paper is organised as follows. In *Section 11.2* we give a brief overview of the current literature on this topic. *Section 11.3* outlines the experimental set-up. The results of the analysis performed on the four OSM databases are provided in *Section 11.4.* The paper closes with *Section 11.5* which provides a summary of the main conclusions from this work and offers some possibilities for future work.

11.2 Overview of Related Work

OpenStreetMap (OSM) provides a highly dynamic source of spatial data for use in Location-based Services (Jacob et al, 2010). The purpose of this paper is to provide information on how geometries and attributes change in the OSM database and to comment on the effect this might have on LBS using OSM as the primary source of spatial data. Changes occur in OSM due to the upload of new spatial data and/

or the updating of existing contributions (geometry and spatial attributes) by the original creator or another contributor. Some of the reasons for these changes could be: contributor disagreement, changes to geometry (shortening of roads, resizing, etc), actual real world changes reflected in the data, combination of polylines/polygons into multipolygon relations, mistakes being made by contributors through not understanding spatial data handling or incorrect use of the editor software.

Some studies have highlighted problems with OSM in terms of spatial data quality. In Mooney et al (2010a, 2010b) the authors provide analysis of how the representation (points used to represent features, tagging, etc) of features in OSM can vary greatly from country to country and within object classes. This work also highlighted problems with the quality of the shape representation of polygon features representing natural features in OSM when compared with NMA data for the same features. However other studies such as Over et al (2010), Haklay (2010), Girres and Touya (2010) compares OSM very favourably, in terms of geometric accuracy, against road and street network databases from National Mapping Agencies and commercial sources. Fritz et al (2009) even suggest at using OSM and other crowd-sourced VGI as an alternative approach for "validating and calibrating global land cover". Gazetteers are another area where the crowdsourcing of information and knowledge could be beneficial. Keßler et al (2009) describe a bottom-up approach for gazetteer building based on geotagged photos harvested from the web which could be used to compliment authoritative

gazetteers. The authors argue that most existing gazetteers are managed strictly top-down where entries can only be added or changed by the responsible toponymic authority. In the VGI paradigm this process is completely reversed with contributors capable of changing entries over time. Of course this has the advantage of problems with mistakes and vandalism. However clustering of VGI data (in the case of Keßler et al (2009) geotagged photographs) a more robust gazetteer can be generated. Rice et al, (2011) build a pedestrian information system for visually impaired users by fusing together user-contributed geospatial data and a localized gazetteer system. The authors feel that the VGI nature of the geospatial data used "leverages local geographic expertise and offers significant advantages in dealing with hazard information (for visually impaired users) in real-time".

OSM and Wikipedia are similar in that they are both crowd-source Wikis allowing collaboration on knowledge and information generation. It is therefore obvious to look to research work on Wikipedia to see if there are similar issues occurring in terms of: contributor behaviour, editing, etc. Wikipedia has grown to be the world largest and busiest free encyclopedia, in which articles are collaboratively written and maintained by volunteers online (Hu et al, 2007). They suggest for Wikipedia that given it's collaborative nature the "user interaction data" must be analysed with article quality metrics to obtain a good overall understanding of the quality of an article.

Neis (2011) reports during the first two weeks of May 2011 17000 new "Unconnected Roads" errors were introduced into the OSM Europe database. This is of concern for LBS applications using OSM databases as the street/road network data source. However this problem of unconnected roads is complimented by Neis stating that in OSM Europe for the months (February, March, April 2011) there has been an increase of "about 2850000 new OSM way segments" (polylines) for routing. So while there are errors in the OSM Europe database (and others) this is somewhat offset by continued contributions. Overall literature is beginning to appear on analysis of the quality and usability of OSM data. However there is little literature providing an analysis of OSM through investigation of the history of edits and contributions to the OSM database.

11.3 Experimental Setup

Four OSM databases were accessed for this research: Ireland, United Kingdom, Austria, and Germany. The raw data was downloaded from the Geofabrik web service during April 2011. The raw data is made available in OSM-XML format. The raw data contains the most up-to-date version of all features in the OSM global database for that country. All tags (annotations) are also included in the XML. For this paper we focused on "high edit" features: that is features (polylines and polygons) that have been edited by OSM contributors 15 times or more. These "high edit" features were chosen in an attempt to select a subset of features from OSM which have undergone collaborative editing from multiple contributors or editors. From the four databases selected this provided our study with over 24,000 spatial objects and these objects are summarised in *Table 11.1*. Objects with feature types highway, land-use, natural, amenity, and waterway were chosen. According to the Tag-Info web-service (taginfo.openstreetmap.de) these are the 5 most popular feature types in OSM.

Processing the OSM-XML is made difficult by the size of the downloaded files (Mooney and Corcoran, 2011). For example the uncompressed OSM-XML file for Germany is approximately 14Gb in size. Conventional desktop tools for XML data handling cannot process this volume of data. Using Linux command-line scripting (grep and sed) combined with PHP scripts the following steps are performed. Firstly all "ways" (polygons and polylines) in the OSM-XML are extracted into smaller, more manageable sized, files (~ 100Mb per file). These files are then processed by the PHP scripts to extract the "high edit" objects. The most time-resource consuming part of the data gathering process is the download of the entire version history, in OSM-XML format, for each of the 24,000 spatial objects. The OSM API (Application Programming Interface) is used for this purpose. When the version history for an object is downloaded the OSM-XML is processed by a PHP script and: the tags, contributor history, geometry history, and other information for the object is stored in the PostGIS database for further analysis.

11.4 Experimental Analysis

In this section we provide results of the experimental analysis of the four OSM databases described above. For the purposes of analysis we grouped the UK and Ireland databases together. The four databases were chosen for the following reasons: (1) the authors are from Ireland and would have local knowledge understanding of the dataset, (2) the UK, Germany, and Austria are the three most active regions of OSM in Europe, and (3) Austria has used government donated data in OSM.

Table 11.1. Summary of three OSM databases used in this case-study

Attribute	Germany	Austria	UK + Ireland
Total Objects	10,603	3,367	10,693
Total Highways	902	2,359	8,999
Objects with "name" tag	5,062	2,571	7,642
"name" changes	1110	767	1815
Highway change	408	1248	3703

In *Table 11.1* the summary of the contents of the three OSM databases is provided. Each database contains objects representing: highways, land-use, natural features, amenities, and waterways. The UK-Ireland database is dominated by highway features. The Germany and Austria database provides a better distribution of object type. The fourth row of the table provides a count of the number of objects where a "name" tag is assigned to the object. The "name" changes row shows the total number of objects where the name of the object changed one or more times over the lifetime of the object. The "highway-change" row indicates the number of highway objects where the designation (highway status ie motorway, secondary, path, lane, etc) was changed one or more times.

11.4.1 Contributor Behaviour

In the PostGIS database we store each version of all objects as an atom of a three-dimensional (3-D) unit geographic information 'location (geometry), time, attributes'. The term "place name" has several synonyms, including "toponym", "geographical place name", and "geographic place name". Confusion, uncertainty, and misunderstanding may occur when the name for an entity is spelled in different ways, when different names are used for the same place, when the same name is used for different places, or when a name is applied to a feature in an unexpected or different way from the general understanding of how it should apply. The term "allonym" is occasionally used to refer to two or more names for the same place; it essentially means place–name synonym (Beall, 2010). *Tables 11.2, 11.3,* and *11.4* show examples of collaborative editing to three different features in Austria, UK, and Germany. In *Table 11.2* a street in the city of Atmont, Austria, is assigned a slightly different name tag by 4 different users (see User_ID) column. The final row of the table shows the current value assigned to the name tag and the version number of the object in the OSM database (indicated in brackets). *Table 11.3* shows an example of a street in Birmingham, UK. Two contributors are involved in the tagging of this object. There are several variations on spelling of the streetname. The example in *Table 11.4* is from Hamburg, Germany. In this example five contributors are involved. The interesting aspect of this example is the number of changes in designation to the "highway" object. This introduces uncertainty related to the actual physical designation of this object.

Table 11.2. OSM_ID 24228123 – street in Atmont City, Austria

Version	Name Tag	Creation Time	User_ID
1	Ennstal Bundeststrasse, Ennsradweg	2008-05-05 15:10:06	16170
7	Ennstal Bundeststraße, Ennsradweg	2008-09-13 20:20:55	45347
8	Ennstal Bundeststraße	2008-10-25 09:05:33	12408
10 (15)	Hauptstraße	2009-07-19 21:24:51	6470

Table 11.3. OSM_ID 24276789 in North-east Birmingham UK

Version	Name Tag	Creation Time	User_ID
2	Oakthorp Drive	2008-05-08 19:39:45	35691
6	Over Green Drive	2008-05-09 08:50:30	35691
9	Oak Thorp Cr	2008-05-09 08:52:52	35691
10	Oak Thorp Dr	2008-05-0908:53:10	35691
15	Oak Thorpe Dr	2008-05-11 13:54:37	35691
18	Oak Thorp Drive	2010-02-07 14:38:14	9065
19 (current)	Oak Thorpe Drive	2010-08-24 11:32:25	35691

Table 11.4. OSM_ID 9782645 in Hamburg, Germany

Version	Name Tag	Creation Time	User_ID
2	Unclassified	2007-10-18 11:10:53	4902
3	Secondary	2008-01-11 15:15:07	21021
4	Unclassified	2008-01-11 15:25:52	21021
13	Construction	2009-10-22 12:47:15	124032
16	Secondary	2010-02-17 11:36:30	211280
17	Unclassified	2010-02-18 09:48:43	211280
18	Pedestrian	2010-02-22 15:21:24	211280
19 (current 23)	Tertiary	2010-02-25 16:09:54	44838

11.4.2 Changes to "Name" Tags and Highway Designation

In an attempt to quantify the types of changes that the value of the name tag for objects in the OSM databases undergo we applied two well known and robust string matching metrics from the domain of text similarity matching. The Jaro–Winkler distance (Jaro, 1989) is a measure of similarity between two strings. The higher the Jaro–Winkler distance for two strings is, the more similar the strings are. The literature indicates that the Jaro–Winkler distance metric is most suitable for short strings such as person names or placenames. The Jaro-Winkler score is normalized such that 0 equates to no similarity and 1 is an exact match (Top et al, 2007). The Levenshtein distance between two strings is defined as the minimum number of character edits required to transform one string into the other (Yujian and Liu, 2007). The allowable edit operations are: deletion, insertion, or substitution of a single character at one time. The Levenshtein distance has a lower bound of zero if and only if the strings are identical whereas the upper bound is either at maximum the length of the longer string or at minimum the difference of the lengths of the two strings (Ackroyd, 1980). Both metrics are used for similarity searching in databases. For example for the strings "A171" and "Scalby Road" the Jaro-Winkler distance is 0 (no similarity) while the Levenshtein distance is 11 indicating the number of operations required. Another example is "Western Road" and "Western Avenue". The Jaro-Winkler distance shows good similarity at 0.74 while the Levenshtein distance is 6. These two metrics are suitable to take into account spelling errors introduced by contributors when naming objects (or renaming). For example in *Table 11.3* the values assigned to the street over the editing period are very similar.

For each object in the OSM databases we extracted those objects where the name tag value was changed at least once as described in *Table 11.1.* We calculated the Levenshtein and Jaro-Winkler distance metrics for each name transformation pair, in order – for example "Western Road" changing to "Western Avenue". For each object when calculate the mean Levenshtein and Jaro-Winkler distance for each object. *Figures 11.1* and *11.2* below show mean Levenshtein and Jaro-Winkler distances for each object plotted in scatter plot format for the UK and Germany databases. The pattern of distribution is very similar. There are instances where there are clusters of very similar name transformations – those with mean Jaro-Winkler close to 1 having very small (<5) mean Levenshtein distances. On the other hand there are examples of some completely dissimilar name transformations – those with mean Jaro-Winkler distance at 0 and large values for mean Levenshtein distances.

As shown in *Tables 11.2, 11.3* and *11.4* the changes to names of objects or the designation values for highways can happen in the presence of a single contributor or multiple contributors. We calculated the a number of statistics to investigate if the number of changes to names or objects or highway designation increased as the number of contributors increased. We calculated the correlation and also the

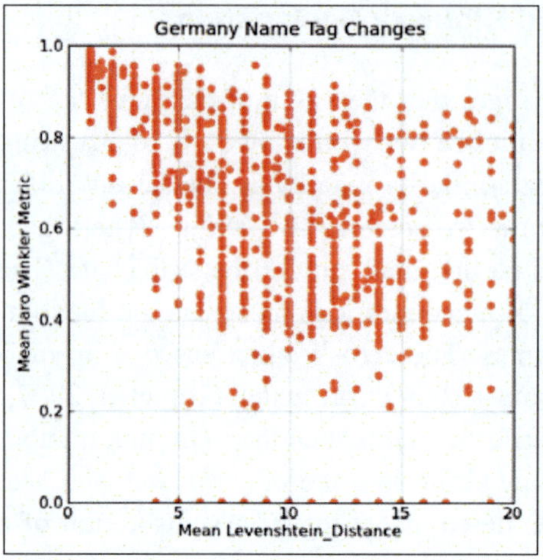

Fig. 11.1. Mean Levenshtein Distance plotted against the Mean Jaro-Winkler distance for the changes to "name" tags for objects in the Germany OSM Database

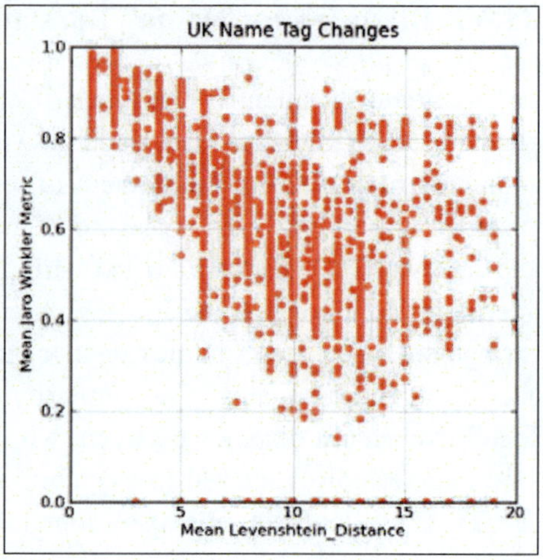

Fig. 11.2. Mean Levenshtein Distance plotted against the Mean Jaro-Winkler distance for the changes to "name" tags for objects in the UK Ireland Database

Spearman correlation (rho ρ and p-value). The results are summarised in *Table 11.5*. Unfortunately there does not appear to be any significant statistical relationship between the number of contributors and the number of changes to name attributes. We calculated the two-sided p-value for a hypothesis test where the null hypothesis is that two sets of data (number of contributors, number of changes performed by contributors) are uncorrelated. In all cases the p-value exceeds 0.05 so we must accept the null hypothesis. While this is somewhat disappointing it is not unexpected. The examples in *Tables 11.2* and *11.3* show the effects of multiple contributors and then a single contributors.

Table 11.5. Analysis of the effects of the number of contributors on the changes to name attributes on objects in the case study databases

Database	N	corr	rho ρ	p-value
Germany	1110	0.022	0.045	0.105
UK Irl	1815	-0.171	-0.013	0.561
Austria	767	0.088	0.078	0.051

In *Figure 11.3* we show a simple plot of the number of unique contributors for the N = 1815 objects used for the analysis in *Table 11.5* plotted against the number of unique "name" tag values for those objects. It is immediately obvious why the correlation tests return such inconclusive results. Small numbers of contributors per object (single, two or three) appear just as likely to introduce uncertainty into the accuracy of the value of a "name" tag value as the arrival or introduction of additional contributors to the edit history of an object.

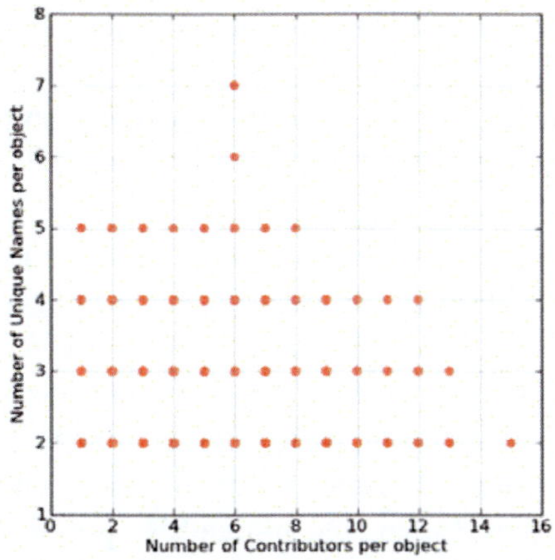

Fig. 11.3. The number of unique contributors plotted against the number of unique names for objects in the UK Ireland database

Table 11.6 provides the same analysis for objects with the highway tag and the effects of the number of contributors on the number of changes to highway designation. Again we must accept the null hypothesis.

Table 11.6. Analysis of the effects of the number of contributors on the changes to highway designation on objects in the case study databases

Database	N	corr	rho ρ	p-value
Germany	408	0.04	0.025	0.604
UK Irl	3703	0.03	0.033	0.019
Austria	1248	0.041	0.045	0.105

Some specific examples will help to highlight the highway designation problem. In the Austria database there are 93 highway objects who have their designation changed from "primary" to "secondary" (or vice-versa), in the German database this is 46, while in the UK and Ireland this is 350. We feel that this aspect of the collaborative contributions to OSM requires further investigation to explain the rationale behind these decisions.

11.5 Conclusions

In this paper we have analysed "high edit" geographical features from four OSM databases from Europe. "High-edit" features, those with at least 15 versions in their history, were chosen to provide us with features which exhibit the edit patterns of a collaborative crowd-sourced environment.

The results of the experimental analysis on the number of contributors examined against the corresponding number of changes to objects do not reveal any significant patterns. In the data for this work we found that there were many different "patterns" of contribution: single contributors making multiple changes, multiple contributors making multiple changes, and a mixture of these. The analysis here shows significant changes to the names of objects. There is also significant changes to the designation of highways. Both of these issues should be considered very carefully before OSM is used as the primary source of data for LBS applications. Codescu et al (2011) provides a recent and valuable reference for our work. In their paper they develop a web service focusing on finding locations not only by their address, but by systematically relating the places to activities that a person could perform there. This is helpful if a person wants to explore a new city, or plans leisure activities. The authors remark that "OpenStreetMap provides a rich set of tags that can be used for activity-oriented search". However, they acknowledge the evolving nature of tags in social media and collaborative web projects by integrating several ontologies that are related to each other and develop "matching tools to cope with the evolving nature of the tags".

A potential drawback of the OpenStreetMap crowdsourced model is the lack of a specific set of "administrators" or "moderators" for the spatial content uploaded and edited to the OSM database. It could be easily argued, based on the statistics Neis (2011) provides for short periods of time (3 months) that a very large number of administrators or moderators would be required to check and verify uploaded data. This might not be feasible or may as Girres and Touya (2010) warn "kill the joy" of contribution to VGI projects such as OpenStreetMap. While Allen (2010) outlines the key advantages Web 2.0 and crowdsourcing has brought to knowledge sharing he states the open question if lack of coordination and quality issues actually provide barriers to knowledge sharing and usage. In terms of somehow automating quality control OSM editor software could be extended to check for geometric inaccuracies such as self intersecting polygons or topological inconsistencies. In order to bring consistency to tagging a more fundamental review of the contribution structure to OSM is probably needed. A potential solution to the problems outlined in this paper could be along the lines of the moderator structure in Wikipedia (Zhang et al, 2010).

From a Location-based Services (LBS) viewpoint the results of this research should not discourage the use of OSM as a data source for LBS applications. As Haklay (2010) and Girres and Touya (2010) show the geometric accuracy of OSM is very good, if not better, when compared to national mapping agency data. However this may be isolated to large urban areas. Of particular concern for LBS developers is the sometimes unpredictable changes in "names" and "highway" tags – for example a highway object having it's designation changed from primary (1st class road) to tertiary (3rd class road). This type of change would have unexpected effects for applications such as route finding in LBS.

Another interesting aspect of future work is to investigate if there is a trend in the general, world-wide, OSM data in regards to stability of place-names. This would involve investigation of the entire OSM global database. As the community in OSM grows larger, and potentially more diverse, do tag values for attributes such as "name" or "highway" become more standardized over time? This would assist in development of stability metrics for OSM data which could indicate the rate of change of edits to OSM globally or on a national or local scale.

Acknowledgements

Dr. Peter Mooney is a research fellow at the Department of Computer Science NUI Maynooth and he is funded by the Irish Environmental Protection Agency STRIVE programme (grant 2008-FS-DM-14-S4). Dr. Padraig Corcoran is a post-doctoral researcher at University College Dublin and is funded by the IRCSET EMPOWER Postdoctoral Fellowship Scheme. The authors gratefully acknowledge these supports.

References

Ackroyd, M.; "Isolated word recognition using the weighted Levenshtein distance," Acoustics, Speech and Signal Processing, IEEE Transactions on, vol.28, no.2, pp. 243-244, Apr 1980

Allen, J.P.; (2010), "Knowledge-Sharing Successes in Web 2.0 Communities," Technology and Society Magazine, IEEE, vol.29, no.1, pp.58–64, Spring 2010

Beall, J (2010) Geographical research and the problem of variant place names in digitized books and other full-text resources, Library Collections, Acquisitions, and Technical Services, Volume 34, Issues 2–3, Pages 74–82

Bordogna, G., Ghisalbertib G., Psaila, G., Geographic information retrieval: Modeling uncertainty of user's context, Fuzzy Sets and Systems (2011), doi:10.1016/j.fss.2011.04.005

Codescu, M., Horsinka, G., Kutz, O., Mossakowski, T. and Rau, R. 2010. DO-ROAM: Activity-Oriented Search and Navigation with OpenStreetMap, Fourth International Conference on Geospatial Semantics. Brest, France.

De Longueville, B., Ostländer, N. and Keskitalo, C. (2010) Addressing vagueness in Volunteered Geographic Information (VGI) – A case study. International Journal of Spatial Data Infrastructures Research, Vol 5 ISSN 1725-0463

Edwardes, A.J. 2009. Geographical perspectives on location for location based services. In Proceedings of the 2nd International Workshop on Location and the Web (LOCWEB '09). ACM, New York, NY, USA,, Article 5, 4 pages.

Fritz S., McCallum I., Schill C., Perger C., Grillmayer R., Achard F., Kraxner F., Obersteiner M. Geo-Wiki.Org: The Use of Crowdsourcing to Improve Global Land Cover. Remote Sensing. 2009; 1(3):345–354.

Girres, J.-F., and Touya, G. Quality assessment of the french openstreetmap dataset. Transactions in GIS 14, 4 (2010), 435459.

Haklay, M. How good is volunteered geographical information? a comparative study ofopen-streetmap and ordnance survey datasets. Environment and Planning B: Planning and Design 37, 4 (2010), 682–703.

Meiqun Hu, Ee-Peng Lim, Aixin Sun, Hady Wirawan Lauw, and Ba-Quy Vuong. (2007). Measuring article quality in wikipedia: models and evaluation. In Proceedings of the sixteenth ACM conference on Conference on information and knowledge management (CIKM '07). ACM, New York, NY, USA, 243–252.

Jacob, R., Mooney, P., Corcoran, P., and Winstanley, A. C. (2010). Haptic-GIS: exploring the possibilities. SIGSPATIAL Special 2, 3 (November 2010), 13–18.

Jaro, M. A. (1989). "Advances in record linkage methodology as applied to the 1985 census of Tampa Florida". Journal of the American Statistical Society 84 (406): 414–20.

Joshi, D., and Luo, J., and Yu, J., and Lei, P., and Gallagher, A. (2010). Rich location-driven tag cloud suggestions based on public, community, and personal sources. In Proceedings of the 1st ACM international workshop on Connected multimedia (CMM '10). ACM, New York, NY, USA, 21–26.

Carsten Keßler, Patrick Maué, Jan Torben Heuer and Thomas Bartoschek, Bottom-Up Gazetteers: Learning from the Implicit Semantics of Geotags GEOSPATIAL SEMANTICS, Lecture Notes in Computer Science, 2009, Volume 5892/2009, pages 83–102

Mokbel, M and Levandoski, J. (2009). Toward context and preference-aware location-based services. In Proceedings of the Eighth ACM International Workshop on Data Engineering for Wireless and Mobile Access (MobiDE '09). ACM, New York, NY, USA, 25–32.

Mooney, P., Corcoran, P., and Winstanley, A. C. A study of data representation of natural features in openstreetmap. In Proceedings of the 6th GIScience International Conference on Geographic Information Science (2010a), GIScience 2010, University of Zurich, p. p150.

Mooney, P., Corcoran, P., and Winstanley, A. C. Towards quality metrics for openstreetmap. In Proceedings of the 18th SIGSPATIAL International Conference on Advances in Geographic Information Systems (New York, NY, USA, 2010b), GIS '10, ACM, pp. 514–517.

Mooney, P and Corcoran, P., Accessing the history of objects in OpenStreetMap. Proceedings of the 14th AGILE International Conference on Geographic Information Science, Utrecht, The Netherlands, Eds: Stan Geertman, Wolfgang Reinhardt and Fred Toppen p 141 April 2011.

Neis, P. (2011) OpenStreetMap Routing View Europe (May 2011) Online Blog http://neis-one.org/2011/05/routing-view-2011-05/ (Last checked May 2011)

Over, A. Schilling, S. Neubauer, A. Zipf, Generating web-based 3D City Models from OpenStreetMap: The current situation in Germany, Computers, Environment and Urban Systems, Volume 34, Issue 6, GeoVisualization and the Digital City – Special issue of the International Cartographic Association Commission on GeoVisualization, November 2010, Pages 496–507

OSM (2010) The Map Features Page on OpenStreetMap Wiki http://wiki.openstreetmap.org/wiki/Map_Features (Last Checked May 2011)

Stein, K and Hess, K. 2007. Does it matter who contributes: a study on featured articles in the german wikipedia. In Proceedings of the eighteenth conference on Hypertext and hypermedia (HT '07). ACM, New York, NY, USA, 171–174.

Rice, Matthew T. and Hammill, William C. and Aburizaiza, Ahmad O. and Schwarz, Sara and Jacobson, R. Daniel and Ruas, Anne. Integrating User-contributed Geospatial Data with assistive Geotechnology Using a localized Gazetteer. Advances in Cartography and GIScience. Volume 1 Lecture Notes in Geoinformation and Cartography 2011, Springer Berlin, pages 279–291

TagInfo The TagInfo Application (2011) http://taginfo.openstreetmap.de (Last Checked May 2011)

Top, P.; Dowla, F.; Gansemer, J.;, "A Dynamic Programming Algorithm for Name Matching," Computational Intelligence and Data Mining, 2007. CIDM 2007. IEEE Symposium on, vol., no., pp.547–551, March 1 2007-April 5 2007

Twaroch, F.A and Purves, R.S. and Jones, C.B (2009) Stability of Qualitative Spatial Relations between Vernacular Regions Mined from Web Data, International Workshop on Geographic Information on the Internet, Toulouse, France

Twaroch F.A., Jones, C.B. and Abdelmoty, A.I.: (2008) Acquisition of a Vernacular Gazetteer from Web Sources. In: Workshop on Location on the Web 2008, Locweb 2008, Beijing, China, pp. 61--64.

Yujian, Li; Bo, Liu;, "A Normalized Levenshtein Distance Metric," Pattern Analysis and Machine Intelligence, IEEE Transactions on, vol.29, no.6, pp.1091–1095, June 2007

Yi Zhang, Aixin Sun, Anwitaman Datta, Kuiyu Chang, and Ee-Peng Lim. (2010). Do Wikipedians follow domain experts?: a domain-specific study on Wikipedia knowledge building. In Proceedings of the 10th annual joint conference on Digital libraries (JCDL '10). ACM, New York, NY, USA, 119–128.

Chapter 12

Combining VGI with Viewsheds for Photo Tag Suggestion

Barend Köbben, Otto Huisman and HsiangHsu Lin

Faculty of Geo-Information Science and Earth Observation,
University of Twente, Enschede, The Netherlands
{kobben, huisman}@itc.nl, hsiang23767@alumni.itc.nl

Abstract

In this research we attempt to develop an improved method for tagging of digital photos, making the tagging process simpler and more accurate for users, and thus ultimately increasing the quality of the VGI data as a whole. Our method combines Volunteered Geographic Information (VGI) in the form of tags on photo sharing websites (such as Flickr and Panoramio) with a visibility analysis of the new photo to be tagged. This analysis is performed by combining the photo metadata with a Digital Surface Model and building footprints. This approach is used to derive a ranked set of suggestions for a given photograph. To deploy the system, a webservice was implemented.

Keywords: VGI, photo-tagging, visibility analysis, DSM, EXIF, webservice

12.1 Introduction

Recent developments in Internet and communications technologies, combined with the uptake of social media, are encouraging more and more people to share information through the World Wide Web. A substantial portion of this information has a spatial component. Goodchild coined the term Volunteered Geographic Information (VGI) to define this new phenomenon, describing it as "the widespread engagement of large numbers of private citizens, often with little in the way of formal qualifications, in the creation of geographic information" and noted that "collectively,

G. Gartner, F. Ortag (Eds.), *Advances in Location-Based Services*, Lecture Notes in Geoinformation and Cartography, DOI 10.1007/ 978-3-642-24198-7_12,

they represent a dramatic innovation that will certainly have profound impacts on geographic information system and more generally on the discipline of geography and its relationship to general public." (Goodchild 2007, p.212)

Two key effects emerge from this phenomenon. The first of these is that new virtual geographies are being defined by the collective actions and choices of individuals. Websites such as Flickr and Panoramio offer free space for users to upload their photos and enrich these with comments, blogs and tags; these tags can include geo-referencing information, as informal, plain-text toponymes or from more formal systems such as GPS coordinates. The result is a large body of fairly unstructured but potentially rich geographic information. The second effect is that the volunteered data can be, and is, mined: Both scientists and commercial companies help users by creating useful applications to act on internet and contribute data through internet, in the process helping themselves to collect data. For example, ranking VGI data by canonical view (Yang et al. 2010) has been used to extract popularity of tourist destinations from tagged photos.

In this paper, we consider the case of individual tourists using the existing body of VGI photos to tag their own photographs. In general, users of photo-sharing sites upload these only after having returned from their trip. Being provided with information about objects likely to be in their photos, and the tags put by other users on these objects should make the task of tagging easier and more accurate and should help to ultimately increase the quality of the VGI data as a whole. The main problem in developing such a procedure is how to identify objects that might exist in a given spatially referenced photograph.

There are several methods to solve this problem, some of which we discuss in *Section 12.2.* For this research we designed a methodology that is presented in *Section 12.3:* We focus on using a visibility analysis that takes into account the photo's metadata, combined with building outlines as well as a Digital Surface Model, in order to determine the objects that are likely to appear in the photo. We then use a cluster analysis of the VGI photos found within the locations of those objects to suggest tags. We have tested our methodology in an experimental prototype *(Section 12.4)* and implemented a tagging suggestion webservice *(Section 12.4.1)*. Finally, the results are discussed in *Section 12.5*.

12.2 Related Work

Any system to help users determine the location and/or the objective in a photograph needs some sort of annotation to recognize or read the characteristics of the photo. The prominent standard in this area is the EXchangeable Image File format (EXIF), used in virtually all digital cameras. It was originally developed by

Japan Electronic Industries Development Association (EXIF 2002). Essentially, it is a metadata standard for digital photographs. EXIF is used within many different image formats, including JPEG and TIFF. It can store a multitude of information: general photographic attributes such as date and time, lens information, focal length, CCD information and image resolution, as well as geo-location information such as GPS location and compass direction.

Viana et al. (2008) argue that there are two main categories of photo annotation: context-based and content-based. The characteristic of content-based algorithms is analysis of the image itself. When the system can determine what the 'objective' (the main object(s) in the photo) is, and its location, it can predict the location of the camera. Although content-based algorithms to automatically generate photo annotations were developed several years ago and were used in real applications, there is still a range of factors that badly affect the results (Naaman et al. 2004). Also, the accuracy of content-based annotation is considered an important issue. To overcome several of these issues, context-based algorithms can be used.

The main challenge of context-based algorithms is trying to determine precise (or formal) geo-locations from the context of photos that include only informal or relative annotation. Thus, a tag such as "Louvre museum" could be transformed into the more precise address "Musée du Louvre, Place des Pyramides, 75001 Paris, France" and ultimately to a formal latitude and longitude. Researchers have used simple geographic mining methods to extract the popular objects within a certain area from the tag clouds of large bodies of photos. To improve these methods, Moxley et al. (2008) used a new approach in their SpiritTagger. Based on the ideas of Social Network Systems, they considered other users' tag distributions, and if there are multiple similar annotations within the defined search radius, only the prominent ones are collected.

Iwasaki et al. (2005) concentrated on the context of the actual digital photograph and used the photo direction and the photometric subject distance to predict the major objective in photos. The general procedure adopted here is using these to determine the location of the photographer and then translating the subject distance into map units to search within that radius for possible objects in a reference database.

There is a long history of using viewsheds for visibility analysis. Recently, Bartie et al. (2010) proposed extensions to existing visibility models, geared specifically to Location Based Services and the visibility of landmarks in urban environments.

12.3 Methodology

Our approach seeks to combine the geographic mining of existing VGI tags with the photo context method of Iwasaki et al. (2005). Theirs and other similar approaches use a search radius around the photographers' location to find possible candidate objects, but this is rather coarse. We propose a more elaborate visibility analysis in this research that tries to estimate the buildings or objects truly visible in the actual view field of the camera, taking into account the camera viewing angle as well as a Digital Surface Model and building outlines. Then we mine the tags connected to these objects, by performing a cluster analysis on the existing body of VGI picture data. This results in a suggestion of possible photo tags to the user. An overview of the workflow we propose is shown in *Figure 12.1*. The data we need to prepare for several key components and the processing steps taken are listed below:

1. **Determine theoretical field of view:** We have to first calculate the view angle for each individual picture, because in a camera it is not fixed, as it is in the view through human eyes. The formula for the calculation is shown in *Equation 12.1,* where θ is the view angle, l_d is the CCD size (or image dimension) and l_f is the focal length:

$$\theta = 2 \arctan \left(\frac{l_d}{2\, l_f} \right) \tag{12.1}$$

   l_d and l_f are determined using the EXIF metadata attributes *CCDSize* and *FocalLength*, respectively. The view angle then is combined with the camera position (*GPSLatitude* and *GPSLongitude* from the EXIF) and direction (*GPSImgDirection*) to calculate an initial or theoretical field of view FOV_t in *Figure 12.1).* This FOV_t would only be realistic if the camera were placed in an empty and level field. In the real world, the terrain relief as well as buildings, trees and other objects surrounding the camera would block parts of the view.

2. **Use DSM to calculate realistic field of view:** To go from the initial FOV_t to a more realistic view, a Digital Surface Model (DSM) is an important input. It differs from a DEM, a Digital Elevation Model, in that it does not model the *z*-value of ground elevation only, but includes the height of objects on the earth's surface such as buildings, trees, et cetera. A DSM can be obtained from (digital) photogrammetry, remote sensing imagery, laser-scanning data, or by combining a DEM with 3D building and other object data. We use a so-called "viewshed analysis", a process implemented in most GISs, to calculate the realistic field of view. The input needed for this is the same as for determination of FOV_t with the addition of the DSM and the observer height. The result is a realistic field of view FOV_r that is actually a subset of FOV_t. Although we expect this to be a fairly accurate depiction of the part of the world that is visible in the photograph, it is too restricted for our purpose: We will find in the field of view parts of objects, mostly buildings, that as a whole are likely candidates for useful photo tags. If one would see in the picture only a small corner of the Louvre Museum, all tags placed within the whole of that buildings footprint would be relevant for our picture, as they would describe the object of which we see only a small part.

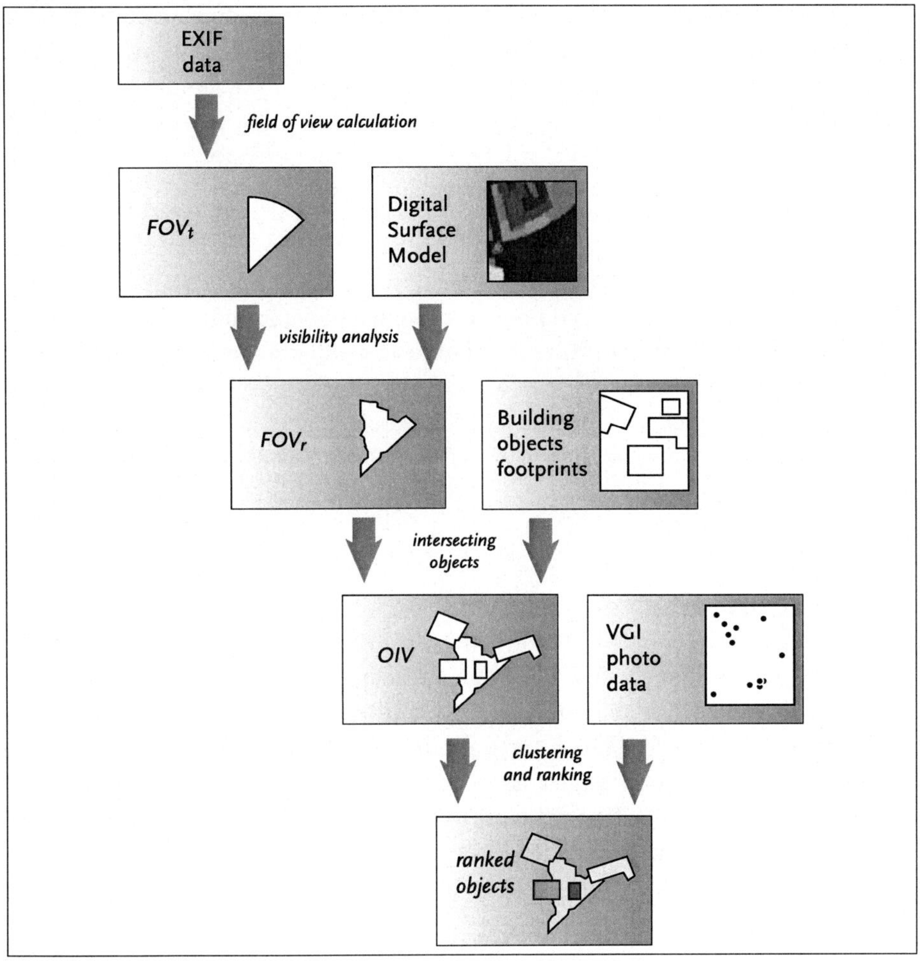

Fig. 12.1. The workflow of the visibility and clustering analysis.

3. **Adding intersecting object footprints:** To overcome the limitation mentioned above, we use further GIS processing to combine the FOV_r with the buildings and other objects that are intersecting it. The result is a set of objects visible in the view, *OIV* in *Figure 12.1.*
4. **Clustering VGI data and ranking:** The next step is collecting the tags relevant for the field of view and the objects in the view. Obviously, tagged photos contributed by users on sites like Flickr and Panoramio are not randomly distributed over space. One can expect to find clusters for special landmarks, famous buildings, popular plazas, et cetera. Generally, a cluster means there are similar things occurring within a given area: "clustering of a spatially-referenced feature is broadly defined by the term *unusual aggregation of events*" (Lawson 2010, p. 232). Different data types and different analytical purposes have given rise

to many different clustering analysis algorithms, which range from simple to complicated, and most of which are available in typical GISs. We experimented with several clustering algorithms, and ended up using a simple frequency, a count of occurrences of VGI photo points within each viewshed and object polygons. We then translate the absolute frequencies into relative values by calculating for each polygon object the percentage *p* as:

$$p = \frac{n}{N} 100 \tag{12.2}$$

where *n* is the number of VGI photos within the polygon and *N* the total number of VGI photos within *OIV* (the set of objects visible in the view). Thus we end up with a ranking of the various objects, which can be symbolised with colour values so users can see which objects have been tagged most. By exploring the tags within these objects, they can find likely candidates for tagging their own photograph.

12.4 Experiment

We chose to test our methodology in a practical and pragmatic manner: We wanted to use only readily available, off-the-shelf software and we used as a study area our hometown, Enschede. It is located in the East of the Netherlands, close to the German border. It is not a major tourist destination, but it has its share of national and international visitors. The main reason for using it for our study area was the ready availability of the base data we needed, such as the DSM and building footprint layers. Photos were taken at chosen locations in the city centre, e.g. of the City Hall (see *Figure 12.2)*.

The steps of the methodology proposed in the previous section were implemented as follows: We tested with two types of camera, an Apple iPhone4 and a Fujifilm F200 EXR. The former is a smartphone equipped with a digital camera, GPS and digital compass and therefore all necessary EXIF data is registered by the device itself. The latter is a standard digital camera. The position and direction data was recorded separately with an eTrex GPS with built-in digital compass and afterwards integrated in the EXIF data. We did not separately determine a theoretical field of view FOV_t for each individual picture, but directly created the realistic field of view FOV_r by combining it with a surface model of the city of Enschede. The source of this was high-resolution airborne laser altimetry data acquired in 2007 with a point density of 20 points/m^2. The DSM has been derived at ITC, using the method described in Vosselman (2008). It can be seen depicted in grey scales in the lower half of *Figure 12.2* (right). For our experiment, we only used the portion of this DSM that covered the centre of Enschede.

Fig. 12.2. Example of one of the experiments. Original picture of Enschede City Hall (left) and resulting ranked polygons depicted on top of DSM and building footprint layers (right).

We then used the *Viewshed* toolbox in ESRIs ArcGIS. All parameters and input data needed for this process came from the EXIF data and the DSM, with the exception of the observer height. This we fixed at an arbitrary 2 meters above the height of the DSM at the observer location. The result of the *Viewshed* is a raster layer, which was converted using the *Raster to Polygon* toolbox.

To create *OIV*, the set of objects visible in the view, we needed to add the intersecting object footprints to FOV_r. For this we used the buildings footprint data from the Dutch National Mapping Agency, the "Topografische Dienst Kadaster". This data was created for a nominal scale of 1:10,000, and the version we used was published in 2002. It can bee seen as the yellow lines in *Figure 12.2* (right). We employed a combination of the ArcGIS toolboxes *Clip*, *Spatial Join* and *Union*.

The methods used in this step, as well as the previous viewshed calculation, were fairly simple. In literature we did encounter several other, more sophisticated, methods. For example, Bartie et al. (2010) propose calculating distance factors into perceived area calculations, calculating a clearness index for objects and calculating how much of an object is on the skyline. We chose not to use these and other techniques, firstly because the additional information is not vital to our methodology, and secondly because they use complicated algorithms not available in current off-the-shelf software.

For the next step of clustering and ranking the VGI data we collected photo data from the sites www.flickr.com and www.panoramio.com. We only included photos that were geo-tagged within our area of interest, i.e. that included georeference latitude and longitude that was within the boundaries of Enschede centre. The total number of photo points was 953. The frequency count and percentage calculation were performed using the ArcGIS toolboxes *Spatial Join* and *Summary Statistics*. Thus we ended up with the ranked object polygons, the coloured areas in *Figure 12.2* (right). The red polygon is the most likely objective in this particular photo, and is in fact Enschede City Hall. It is one of the more famous buildings in the Enschede area and consequently many visitors take photos of it.

12.4.1 Implementation as a Webservice

From our experiment, we concluded that the methodology of combining the visibility analysis with the VGI clustering and ranking works as expected. However, the implementation using ArcGIS tools is not useful for the intended users: individual tourists using the existing body of VGI photos to tag their own photographs more easily and accurate.

We therefore tested if the experiment could be implemented as a webservice, where photographers are presented with a webpage to upload the data for the photo they want to tag, and have the ranked object polygons returned to them.

Fig. 12.3. Screen dump of Google Earth with Panoramio layer turned on and the resulting KML file overlaid.

We used the popular *OpenLayers* Javascript API to develop a webpage that lets the user input a photo-point by adding the relevant EXIF data or alternatively by clicking on a location in a map. The data is then processed server-side by a Python script that uses ESRIs *arcpy* module to run the ArcGIS models that implement the actual analysis steps documented above.

Some additional post-processing is done to return the result as a KML file, which then could be shown in a webpage on top of popular base maps (such as *Google Maps*, *Bing Maps* or *OpenStreetMap*) and could also be downloaded to be added to *Google Earth* (as shown in *Figure 12.3)* or other stand-alone viewers that support KML. In that way, the user can freely explore the existing tagged photos that coincide with the highly-ranked viewshed polygons, in order to find possible candidates for tags of his or her own photograph.

It has to be noted that this implementation was used purely as a proof of concept. The system as it now stands is not suitable for real-world use, mainly because of the very limited spatial extent (the centre of Enschede) for which it can be used. This is due to the availability of DSM and building footprint data for only that area.

12.5 Discussion and Conclusion

In implementing the experiment and the webservice described above, we encountered some minor problems. The accuracy of the GPS in the iPhone and the digital compass in the eTrek were rather limited. Furthermore, the building layer was relatively dated and footprints did not always match the newer DSM and the actual situation. Also, this layer shows building outlines only, not individual shops and houses which make up the buildings themselves, and does not include other environmental objects such as statues or other street objects.

A more fundamental problem is the dependency on the DSM and building data already mentioned in *Section 12.4.1*. This data was readily available to us in our limited experiment area, but for a real-world implementation one needs datasets that cover a wider area. For the buildings the freely available OpenStreetMap data (www.openstreetmap.org) should be accurate and complete enough in many parts of the world, and certainly in Europe. But currently DSM data is not readily available for such large areas, not of a sufficient spatial resolution and not for free certainly.

However, this availability is fast changing with the advent of digital multi-ray photogrammetry and fully automated ortho-rectification software and automated 3D model generation algorithms. This technology is for example used in the GlobalOrtho project of Microsoft Bing Maps for which the Vexel company is using their UltraCamG cameras to create seamless 3D models for the whole US and Western Europe in the next two years or so (Wiechert & Gruber 2009).

There are certainly elements for improvement, which we would like to explore in the future. The most obvious of those is that currently the actual tag suggestion is rather crude: the user is simply presented with all tags found within the ranked polygons. A more sophisticated system could process this list using text mining and semantic methods to do things such as removing duplicates and ranking by occurrence, thus presenting a more structured list of tags.

But regardless of the problems discussed above, the method developed works well and seems to offer an effective way of using existing VGI photo data to make the task of tagging both easier and more accurate, thus ultimately increasing the quality of the VGI data as a whole.

References

Bartie P, Reitsma, F, Kingham, S and Mills S. (2010), Advancing visibility modelling algorithms for urban environments, Computers, Environment and Urban Systems 34 (6), 518–531.

EXIF (2002), Exchangeable image file format for digital still cameras: EXIF version 2.2, Technical report, Standard Association of Japan Electronics and Information Technology Industries.

Goodchild M F (2007) Citizens as sensors: The world of volunteered geography, Geojournal 69(4), 211–221.

Iwasaki K, Yamazawa K, Yokoya, N. (2005), An indexing system for photos based on shooting position and orientation with geographic database, in IEEE International Conference on Multimedia and Expo, IEEE Computer Society, Los Alamitos, pp. 390–393.

Lawson A (2010) Hotspot detection and clustering: ways and means, Environmental and Ecological Statistics 17 (2), 231–245.

Moxley E, Kleban J, Manjunath B S (2008) Spirittagger: a geo-aware tag suggestion tool mined from Flickr, in Proceeding of the 1st ACM international conference on Multimedia information retrieval, MIR '08, ACM, New York, pp. 24–30.

Naaman M, Harada S, Wang Q, Molina H, Paepcke A. (2004) Context data in geo–referenced digital photo collections, in MULTIMEDIA '04: Proceedings of the 12th annual ACM international conference on Multimedia, ACM, New York, pp. 196–203.

Viana W, Filho J B, Gensel J, Villanova-Oliver M, Martin H (2008) Photomap: from location and time to context-aware photo annotations, Journal of Location Based Services 2(3), 211–235.

Vosselman G (2008) Analysis of planimetric accuracy of airborne laser scanning surveys, in ISPRS 2008 – Proceedings of the XXI congress : Silk road for information from imagery, ISPRS Comm. III, WG III/3, Beijing, pp. 99–104.

Wiechert A, Gruber M (2009) Aerial perspective: Photogrammetry versus lidar, Professional Surveyor Magazine 29(8).

Yang L, Johnstone J, Zhang C (2010) Ranking canonical views for tourist attractions, Multimedia Tools and Applications 46(2), 573–589.

Chapter 13

A Preference SQL Approach to Improve Context-Adaptive Location-Based Services for Outdoor Activities

Florian Wenzel[1], Martin Soutschek[2] and Werner Kießling[1]

[1] Chair for Databases and Information Systems, University of Augsburg, Germany
{wenzel, kiessling}@informatik.uni-augsburg.de
[2] Alpstein Tourismus GmbH & Co. KG, Immenstadt, Germany
martin.soutschek@alpstein.de

Abstract

Personalized and context-aware information filtering poses a major challenge for the development of meaningful location-based recommendation services, and even more for complex mobile apps for planning outdoor activities. To this end search engine technology has to be extended by sophisticated preference handling capabilities for large amounts of application data, which is typically stored in relational geo-databases like PostGIS. In this paper we propose to employ the Preference SQL query language with its Best-Matches-Only query model as one key enabling technology. Moreover, we introduce a novel location-based preference constructor for querying such geo-databases. We further demonstrate how this extended Preference SQL system can be integrated with an existing commercial web platform for planning outdoor activities such as hiking or mountain bike tours. By performing personalized and context-aware query composition, employing a preference repository attached with social networks and using online information like weather conditions, best matching tour recommendations can be computed with one single query and delivered to mobile smart phones. We are highly confident that forthcoming practical use case evaluations will show the effectiveness and efficiency of our approach for a wide range of application scenarios.

Keywords: preference handling, Preference SQL, personalization, mobile Location-Based Services, context-aware search

G. Gartner, F. Ortag (Eds.), *Advances in Location-Based Services*, Lecture Notes in Geoinformation and Cartography, DOI 10.1007/ 978-3-642-24198-7_13,

13.1 Introduction

The success of the iPhone and the Android platform is fueling a new wave of developments and expectations in mobile *Location-Based Services* (LBS). In an even broader perspective, Google and other companies have announced a *mobile first* strategy for new products and services. A core challenge is the development of smart mobile services, which provide personalized and relevant contextual information, even if users do not actually search for that specific information. Smart services should deliver "perfect" results. But what can be considered as a perfect result?

13.1.1 Related Work

Van den Eijkel et al. (2001) describe personalization as the customization of services to individual user characteristics and information needs. They argue that personalization improves the usability of services, leading to more satisfying results for the user. User satisfaction is a function of expectations and the degree to which services either meet user expectations, fall short of meeting them, or exceed them, resulting in the highest levels of satisfaction (Hoffmann & Bateson, 2008). Although a high level of user satisfaction may not necessarily be equivalent to a perfect result for an activity, the more satisfying a result appears to be for the user, or the closer a result meets the goal-oriented activity of a user, the more perfect a service may be perceived. As such, user characteristics and personal preferences play a key role for smart LBS. Poslad et al. (2001) are supporting this view by considering personalization as a central requirement for creating meaningful LBS. However, Mokbel and Levandoski (2009) point out that commercial state-of-the-art LBS do not yet deliver smart results for the user and consequently fail to meet the high expectations put into smart mobile services.

Two core processes need to be considered for providing such smart mobile services (Williams et al. 2002), (1) *information filtering* and (2) the *presentation of information* to the user. While filtering processes cover the analysis of context parameters and aim to identify relevant information, presentation processes control how relevant information is presented to the user through adaptive user interfaces.

Information filtering performs selective dissemination of new information by selecting relevant information for long-term users with the help of user models and profiles (Hanani et al. 2001). While filtering methods may include various techniques such as cognitive or content-based filtering, social or collaborative filtering and assisted browsing, representation of user interests and modeling of user profiles is a key aspect in all systems (Tryfonopoulus et al. 2009). Well-proven approaches exist to determine the absolute value of content for a specific user, given formal methods of information retrieval such as TF-IDF or techniques of machine learning such as clustering, decision trees and artificial neural networks (Adomavicius, Tuzhilin, 2005). However, for a truly personalized service, two additional aspects have to be considered. (a) The relative value of information with regard to elaborate user profiles is oftentimes far more important than absolute rankings (Cohen et al. 1999). (b) The heterogeneity of data sources such as web documents, social network profiles or social media content leads to an extensive volume of semi-structured data that has to be normalized and stored in an aggregated fashion in order to be of use for the user. In the case of LBS this list of data sources can be extended to API content from service providers, tourism databases and GIS data.

13.1.2 A Preference-Based Approach for Information Filtering

In this paper we adopt a database approach for preference based information filtering as introduced by Kießling (2002) and described in Kießling et al. (2011) that is able to consider these additional aspects. For a broader perspective on preference research, Brafman and Domshlak (2009) as well as Stefanidis et al. (2011) provide valuable surveys. With Preference SQL we present a database middleware and query language that efficiently evaluates preference statements and allows for a best-matching evaluation of database content via strict partial orders. By using a database approach, the presented filtering system is able to handle atomic data attributes as well as complex geographic types and is extensible for further data integration. The presented theory is applied to the domain of LBS for outdoor activities, since this application poses additional challenges such as a complex context with multiple sensor data and further dependency on the type of activity such as hiking, mountain biking or skiing. A running example demonstrates the need for a highly personalized service.

Running Example: Paul is on a spontaneous weekend trip to the Alpine region and wants to find a hiking tour that is close to his current location. Besides being nearby, the tour should have additional attributes: the duration should be in the range of 4 to 6 hours, the level of difficulty should be *hard* (given possible levels *easy, medium, hard*) and the overall rating should be as high as possible. As a tiebreaker between equally suited tours, Paul favors tours that require an experience level of 3.

Today, commercially available LBS for outdoor activities don't allow the expression of user wishes in such a complex and distinct fashion. They typically provide interfaces for Parametric Search that enable users to search for tours by defining filter criteria such as distance, length, duration, and level of difficulty or scenic beauty. Unless a perfect result matching all user specifications is found, either no results are delivered by the service, or the user is flooded with any result that meets one of the specified attribute values. Both outcomes are not satisfying, since the service did not take into account that users may compromise one criterion for another, effectively negotiating personal preferences to achieve a next-to-perfect result. The novel approach presented in this paper proposes an architecture that improves information filtering by offering filter technologies that may be built from the perspective of a user.

The remainder of this paper is structured as follows: In *Section 13.2* we present a model for preference handling in outdoor domains as well as Preference SQL as query language to express and efficiently evaluate preferences. Novel extensions are presented to model location-based preferences that are of interest considering the given outdoor scenario. *Section 13.3* describes the evaluation of user preferences from a database point of view and highlights the notable improvements that can be achieved for the user. Following, *Section 13.4* presents an ongoing project in order to show that Preference SQL can handle modeling demands concerning personalization and context-awareness in the outdoor domain. Finally, *Section 13.5* summarizes current achievements and points out emerging research questions.

13.2 Preference Handling for Outdoor Domains

Preferences are an efficient way to model user statements such as "I prefer *y* over *x*" and are formally defined as strict partial orders by the Preference Algebra of Kießling (2002). Chomicki proposed a similar approach (2003). A distinction can be made between base preference constructors formulated for single numerical or categorical attributes such as "rating" or "difficulty" and complex preference constructors, typically including multiple dimensions. The running example is such a complex preference statement. For LBS, base preferences on atomic attributes are joined by novel constructors that operate on geometry objects, taking advantage of the spatial functionality of the underlying database system for evaluation. The following subsections are covering the most important base and complex preference constructors for LBS in an informal fashion. For formal definitions the interested reader is referred to Kießling (2002).

13.2.1 Preference SQL

Just like Relational Algebra, which is the foundation of SQL, Preference Algebra gives rise to Preference SQL, which extends the SQL-92 Standard by a PREFERRING clause to explicitly express user dispositions. At the University of Augsburg, a Java-based middleware has been developed that parses, optimizes (Hafenrichter & Kießling 2005) and efficiently evaluates (Preisinger & Kießling 2007) this novel type of queries on top of database systems such as PostgreSQL, Oracle or MySQL. For a brief overview and downloads the interested reader is referred to the Preference SQL Reference (2011). In contrast to conventional relational database systems that only return perfect matches for a given SQL query, Preference SQL implements a *Best-Matches-Only* strategy and returns perfect matches if such tuples exist and else best matches but nothing worse.

Preference SQL supports a rich set of algebraic base preference constructors that model user wishes that apply to single attributes of a relation with numerical as well as categorical domains. For evaluation of base preferences, Preference SQL compares the attribute values of relevant database tuples with the preferred user values specified in the preference constructor and assigns a level value to each comparison. Based on these levels it can be determined which tuples of the database relation are best matches concerning the base preference. A level of zero defines a perfect match while a higher level is always less preferable. To combine preferences on single attributes to multidimensional preference statements, Preference SQL further provides complex preference constructors. The full Preference SQL syntax can be found at (Kießling et al. 2011).

13.2.1.1 Base Preference Constructors

All base preference constructors on numerical or categorical domains covered in this subsection are sub-constructors of the $SCORE_d$ constructor, which defines levels according to a scoring function *f(x)* depending on the type of the preference. The NEARBY constructor is covered in *Section 13.2.2*.

As illustrated in the taxonomy of base preference constructors in *Figure 13.1,* an optional d-value can be assigned to numerical base preference constructors, which allows the user to express a certain range of tolerance with respect to the desired numerical value. Given a d-value greater than zero and a scoring function *f(x)*, the level of *x* is defined as $level(x) = \left|\frac{f(x)}{d}\right|$.

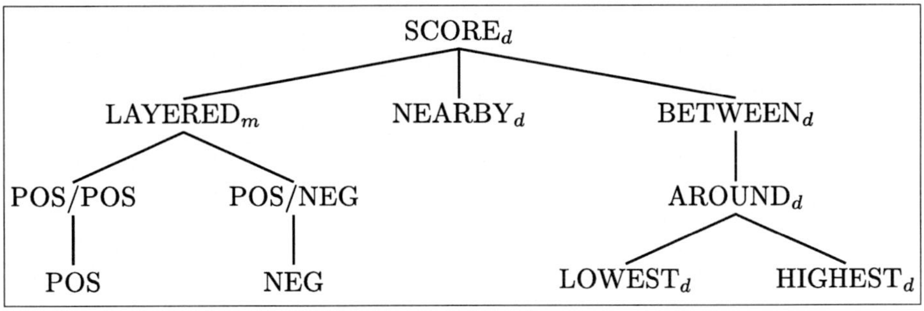

Fig. 13.1. Taxonomy of base preference constructors

BETWEEN$_d$(A, [low, up])

The BETWEEN constructor helps users to express the wish for numeric values of an attribute *A* that should fall into a specified interval *[low,up]*. If no such values exist then values with closest distance to the interval are preferred. Distance in this case acts as scoring function for level determination that can be further modified by the use of a d-parameter.

Example: Following the example of *Section 13.1,* Paul's wish for duration of 4 to 6 hours can be modeled as $\text{BETWEEN}_{d=1}$(duration, [4,6]).

AROUND$_d$(A, z)

The AROUND constructor defines the wish for a certain numeric value *z* of an attribute *A*. If no such value exists then the closest value to *z* is preferred. The level can thus be determined analogously to BETWEEN, setting low = up = *z*.

Example: Looking at Paul's complex preference statement, a wish for a specific experience level can be expressed by $\text{AROUND}_{d=1}$(experience, 3).

HIGHEST$_d$ (A), LOWEST$_d$(A)

A common preference on numerical values is to look for the highest or lowest value that is present in a given attribute domain dom(*A*). The HIGHEST and LOWEST constructors model such a behavior.

Example: Given the running example, the wish for a tour with highest overall rating would be stated as $\text{HIGHEST}_{d=1}$(rating).

LAYERED$_m$ (A, L)

It is common for users to specify certain values of a domain dom(*A*) of a categorical attribute *A* to be of special interest or to be especially disliked. For this case, the LAYERED_m-constructor partitions dom(*A*) into *m*+1 sets $L= L_1, ..., L_{m+1}$. The values contained in L_1 are preferred to all other values, values of L_2 are more preferable than any values from sets with index greater than 2. In general, a level is assigned to each element *x* of a set L_i as follows*: level(x)* = *i-1*. For convenience, one of the sets L_i may be called "others" containing all values of dom(*A*) that are not explicitly stated elsewhere. Special cases of LAYERED_m are represented in *Figure 13.1* with their own sub-constructors. POS(*A*, POS-set) allows a user to define a set of preferred values, NEG(*A*, NEG-set) lets him state a set of disliked values. Furthermore, the combinations POS/POS and POS/NEG are also available.

Example: For Paul, it is convenient to express his favor for a *hard* tour with the help of POS(difficulty, {'hard'}). If he further dislikes *easy* tours, he is able to state that fact via POS/NEG(difficulty, {'hard'}, {'easy'}).

13.2.1.2 Complex Preference Constructors

A complex preference constructor takes two or more preference constructors as arguments. These can either be base preference constructors or again complex preference constructors. This inductive definition allows the construction of elaborate preference terms, which typically incorporate multidimensional user preferences. Complex constructors determine the relative importance of involved preferences. Users often state this relative importance in two different fashions:

- Preference A is *as important as* preference B
- Preference A is *more important than* preference B

The following complex constructors are giving these two informal statements a formal foundation.

PARETO (A_1 x A_2, P_1, P_2)

For the complex PARETO constructor, involved preferences P_1 and P_2 on attributes A_1, A_2 are of *equal importance*. The concept of Pareto domination states that a tuple is better than another tuple if it is better in at least one component and at least equal in all other components. Considering two participating preferences, a database tuple t_1 is thus better than a tuple t_2 if it is better with respect to the first preference P_1 and equal or better concerning the second preference P_2. Alternatively, t_1 is also dominating t_2 if it is better concerning P_2 and equal or better with respect to P_1.

Example: To illustrate this formal property, consider Paul's preference for hard tours and highest customer rating. A tour is better than another tour if one of the following three cases occurs: (1) It is a better match concerning difficulty and has a higher user rating, (2) it is a better match concerning difficulty and has an equal user rating or (3) it has a higher user rating and is equally preferable concerning difficulty. In contrast, if one tour is a better match concerning difficulty but has a worse rating than another tour, both tours are presented to the user. This gives the user the choice to weigh alternatives, whether he wants to cut back on difficulty or to accept a lower customer rating.

PRIORITIZATION (A_1 x A_2, P_1, P_2)

For the complex PRIORITIZATION constructor, involved preferences P_1 and P_2 on attributes A_1, A_2 are of *ordered importance*. Only for tuples that are equally preferable regarding the first preference P_1, the other participating preferences are evaluated.

Example: As Paul already stated he has a very distinct opinion towards location, duration, difficulty and customer rating of outdoor tours in question. The experience level in contrast should only act as a tiebreaker und can thus be combined with a preference term containing the aforementioned preferences via a PRIORITIZATION constructor. This way, only for preference terms that are equally satisfying with reference to all other stated base preferences, the experience level determines which tour is better.

Given a relation "tours" with sample datasets of outdoor tours, Paul's preference statement can be transformed into a Preference SQL query with the help of the presented base and complex constructors.

```
SELECT * FROM tours WHERE region = 'Alpine'
PREFERRING (duration BETWEEN 4, 6, 1
                AND difficulty IN ('hard') NOT IN ('easy')
                AND rating HIGHEST 5, 1)
                PRIOR TO experience AROUND 3, 1
```

The POS/NEG preference is expressed by the keywords "IN" and "NOT IN". Other base preferences are indicated by their constructor name. HIGHEST requires the supremum of the attribute domain in question as additional parameter for evaluation. PARETO is indicated by the keyword "AND", the PRIORITIZATION constructor by the keyword "PRIOR TO" and d-values are added to base preferences separated by comma.

The presented constructors allow expressing preferences on atomic attributes of the database. However, further extensions are needed in order to include geospatial references into Preference SQL queries, such as Paul's wish to find tours that are in vicinity.

13.2.2 Location-Based Preference Constructors

The base preference constructors presented so far are already very useful in a traditional search engine environment. However, when it comes to LBS, novel constructors are needed to express preference statements on geographic objects depending on a certain user location. This gives rise to the concept of location-based preferences. Next, we present a new NEARBY constructor operating on attributes of data type *geometry*.

NEARBY$_d$(geometry, lon, lat, distance function)

Paul's wish to find a tour that is close to his current location which is denoted by the coordinates (lon,lat) can be expressed via the NEARBY constructor. In this case, geometry objects, like POIs, which are stored in a PostGIS database, with lowest distance to the current location, are preferred. The distance function to use for comparison is provided to the constructor as an additional parameter. This could be a net-cost function provided through an integrated routing service or a built-in function of the underlying database. In most cases, distance functions of PostGIS are applicable. The possibility to specify the distance function is crucial since it depends on the type of application.

Example: If it is raining, Paul might be looking for museums instead of outdoor activities. In this case, the Euclidean distance might be an acceptable distance function (given a dense urban network). PostGIS provides adequate functions to determine the maximum (*ST_maxDistance*) as well as the minimum (*ST_Distance*) distance between two geometries. If Paul is an optimistic user he might choose the minimum distance for guidance, guessing that the entrance is at the side of the building facing towards him. On the other hand, if he were pessimistic he would favor the maximum distance, since the entrance could as well be at the opposite side. In contrast, if he decides to go on a hike nevertheless, the net-costs might be a good choice since tours might be close with respect to Euclidean distance, but might require the passing of a deep valley first.

The use of the d-value is highly intuitive for preferences on geographic attributes since it models the fact that a deviation of d-units is tolerable. Given a d-value of 2000 would mean in Paul's case that all tours within a 2000-meter radius around his current location are assigned a level of 1, all between 2000 and 4000 meters a level of 2 and so on. If a tour exists right at Paul's current location, it gets a level of zero assigned.

Regarding only this base preference, the three closest tours with numbers 2, 3 and 5 in *Figure 13.2* would thus be considered as equally suited. They have a level of 2 while the tours with numbers 4 and 1 to the left would be less preferable, having a level of 3 and 4 respectively. In this case only the starting points of the tour are used for distance determination.

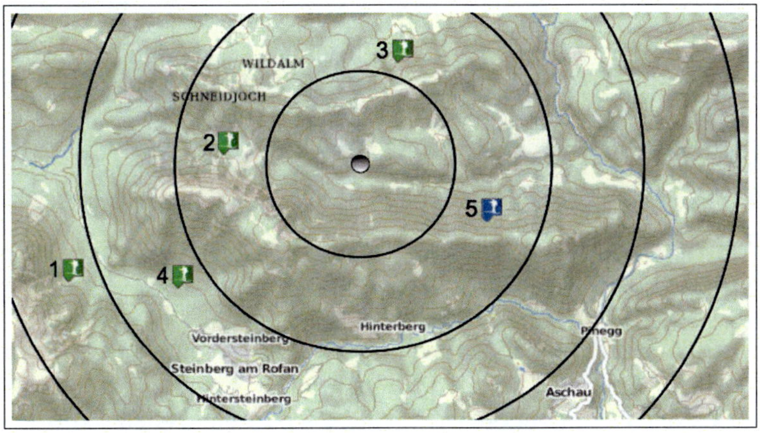

Fig. 13.2. Level of outdoor tours concerning NEARBY preference (adapted from outdooractive.com)

The Preference SQL query of the previous section can now be extended to include Paul's preference for nearby tours with a pessimistic distance function *ST_maxDistance* that is provided by PostGIS (PostGIS Spatial Database Extension 2011).

```
SELECT * FROM tours WHERE region = 'Alpine'
PREFERRING (route NEARBY 47.5494, 11.827211, 2000, ST_maxDistance
                AND duration BETWEEN 4, 6, 1
                AND difficulty IN ('hard') NOT IN ('easy')
                AND rating HIGHEST 5, 1)
                PRIOR TO experience AROUND 3, 1
```

The development of the NEARBY constructor is driven by the fact that Standard SQL doesn't provide such flexibility. It only allows returning objects within a given radius. As alternative, all objects can be retrieved, sorted by their distance to the current location. Hence, if an object is only slightly outside of a predefined radius, it is omitted. On the other hand, if all objects are returned, flooding occurs. With NEARBY, objects still have a chance to be presented to the user even if they are not in immediate vicinity. If they are more suitable concerning other attributes of the preference statement, they might still be returned.

13.3 Preference SQL Query Evaluation

With the given base and complex constructors Paul's preference for certain outdoor tours can be expressed as a complex preference term leading to the truly personalized search query presented in *Section 13.2.2.* As result of a preference query, Preference SQL returns a result set according to the *Best-Matches-Only* (BMO) query model.

13.3.1 The Best-Matches-Only Query Model

The preference selection returns all tuples of the database that are perfect matches given the specified complex preference term. If no such tuples exist, the best alternatives are returned but nothing worse. For more information on the BMO model we refer to (Kießling 2002, Chomicki 2003).

Previous subsections explained how best matches are defined for base as well as complex preferences. This BMO model preserves perfect matches. Even the d-parameter doesn't change that behavior since it doesn't affect the extent of level zero of any base preference. Hence, the user gets the best of both worlds. If perfect matches exist he doesn't get bothered with less than optimal alternatives. On the other hand, if no perfect matches exist, the user gets a BMO set and still has the choice to accept some tradeoffs.

To get an impression of the results of the aforementioned preference query, the running example is evaluated using the sample outdoor tour database described in *Table 13.1*. The "route" column in this case contains geometry objects representing the sequence of coordinates of the corresponding tour.

Table 13.1. Sample outdoor tour dataset

tours					
id	route	duration	difficulty	rating	experience
1	LINESTRING	9	hard	4	3
2	LINESTRING	3	easy	5	3
3	LINESTRING	5	hard	3	2
4	LINESTRING	7	hard	5	3

From *Section 13.2* we know how to evaluate base preferences based on level values. Levels concerning NEARBY can be determined looking at *Figure 13.2*. The calculated level values concerning all base preferences are presented in *Table 13.2* based on the given tour data.

Table 13.2. Level values concerning each base preference

id	route	duration	difficulty	rating	experience
1	4	3	0	1	0
2	2	1	2	0	0
3	2	0	0	2	1
4	3	1	0	0	0

First, we are evaluating the complex preference term that acts as the first argument of the PRIORITIZATION constructor and contains all preferences up to the keyword PRIOR TO.

```
SELECT * FROM tours WHERE region = 'Alpine'
PREFERRING  route NEARBY 47.5494, 11.827211, 2000, ST_maxDistance
            AND duration BETWEEN 4, 6, 1
            AND difficulty IN ('hard') NOT IN ('easy')
            AND rating HIGHEST 5, 1
```

This preference selection alone would return tours with id 2, 3, and 4. The tour with id 1 in contrast is dominated e.g. by the tour with id 4 since it is worse in at least one component and at best equal in all others. This result is already more intuitive than a standard SQL query with fixed search radius since it also returns tour 4 which besides being a little bit further away fits the other user preferences almost perfectly.

Now the full query is evaluated in order to retrieve the final results that are presented to Paul. For the remaining tours 2, 3, and 4 the experience level is evaluated, leaving tours 2 and 4 as final candidates. Since purely best matches are returned, Preference SQL is adapting to the quality of data. This phenomenon is illustrated by adding an additional tour 5 to the data set, resulting in the level values of *Table 13.3*.

Table 13.3. Table 3 Level values for newly added tour dataset

id	route	duration	difficulty	rating	experience
5	2	0	0	0	0

Now tour 5 is the only result of the user query, since it is as close as any other tour and perfectly fulfills all other base preferences.

The presented example only explained the evaluation of soft constraints that are stated in a Preference SQL PREFERRING clause. However, since the full SQL-92 Standard is supported, additional hard constraints can also be formulated in the WHERE clause. Preferences are then evaluated based on the result of this hard selection.

13.3.2 Performance Benchmark

With the functionality of NEARBY explained, the implementation details of location-based preferences as well as basic performance evaluations are presented. Geometry data is stored along with atomic attributes in a PostgreSQL database. The implementation takes advantage of the PostGIS extension that provides distance functions such as *ST_Distance* or *ST_maxDistance*. Since these functions are operating on all types of geometry objects, NEARBY can handle points as well as linestrings and polygons. For preference evaluation, such a function is called with the coordinates provided by the preference constructor as argument as well as the geometry column. This way, Preference SQL can use returned distance values as scoring function for preference evaluation.

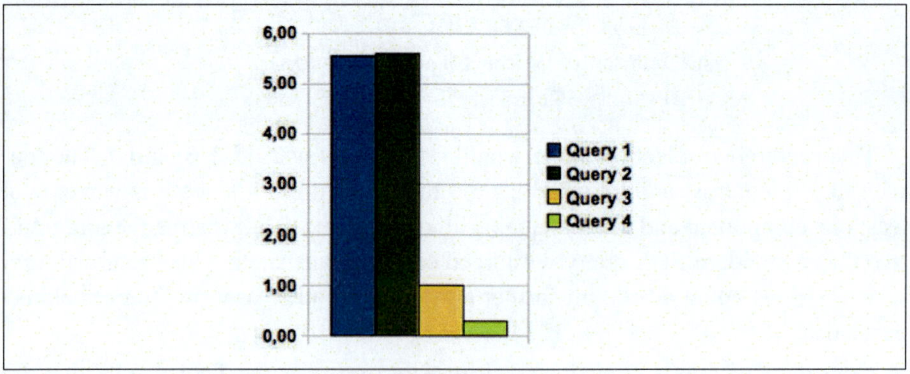

Fig. 13.3. Average query runtimes in seconds

Figure 13.3 shows average query runtimes for four Preference SQL statements. 50 test case data sets where generated. The test application as well as the Preference SQL Server and database system are located on a single server with 148 GB of RAM and two Quad-Core CPUs running at 2.53 GHz.

- Query 1 and 2 are evaluated on a relation of 23881 tours with complex geometries. Query 1 is issuing a NEARBY preference while Query 2 states the complex preference defined earlier in this section.
- Query 3 is equivalent to Query 1 as is Query 4 to Query 2. The only difference lies in the relation under consideration, which holds 1000 tours. Given the fact that the region with the most tours in the database holds 1300 entries, Query 3 and Query 4 are a more adequate representation of our running example.

Performance tests and first mobile prototypes show that, while single query runtime moderately increases compared to standard SQL queries, the overall user session time can be improved significantly. Instead of a repeated cycle of parameter changes and transmission of results via the mobile network inherent to Faceted Search, also known as Parametric Search (Sacco 2009), suitable results are often found with one single preference query. Users are thus unburdened of a time-consuming trial-and-error loop with suitable results being presented at first try via our Preference SQL approach.

13.4 A Location-Based Recommender Architecture

According to Nivola et al. (2008) planning a hike is one of nine thematic categories for user requirements when it comes to outdoor related LBS. Hence, to illustrate the integration of Preference SQL in a LBS we present PeToMoTo, an ongoing project of the University of Augsburg in cooperation with Alpstein Tourismus GmbH & Co. KG that supports hikers during the planning phase. The conceptual architecture is described in *Figure 13.4*. Recommendations are generated through explicit user statements, following a knowledge-based approach instead of collaborative filtering. Further details concerning this recommendation process are out of the scope of this paper.

Outdooractive.com (Outdooractive 2011) is an online platform featuring maps and tour information for all kinds of outdoor activities. The platform hosts an outdoor community where users can create personal profiles and link them with existing Facebook profiles (Facebook 2011). Users can also create and share own tours as well as rate and comment tours from other users. Based on the outdooractive.com-API, mobile Apps for iPhone and Android clients provide location-based search and navigation features and enable users to access maps and tours during their trips.

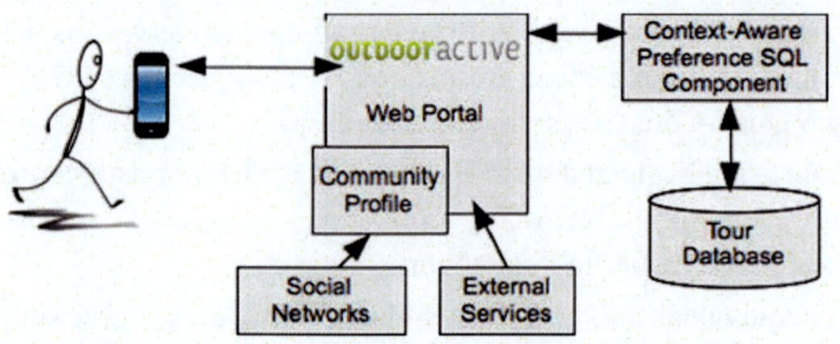

Fig. 13.4. Architecture of a LBS with Preference SQL

The context-aware Preference SQL component integrates into the existing platform as a middleware in-between portal and PostGIS database. Besides the core system described in *Section 13.2.1* it also hosts the *Preference Repository* (Holland & Kießling 2004) that stores user preferences and performs query composition based on the current user context. The stored preferences are either stated by the user directly at search time or come from an aggregation of the *outdooractive.com* profile together with other social networks if the user is logged in. The prevailing user context is determined by the current location of the user as well as the current time and date. With this information, external services can be queried in order to obtain additional context parameters such as weather or remaining hours of daylight.

Adaption to the given context can be achieved by three different means:

- Query composition
- Addition of base preferences
- Adaption of numerical base preferences

Through context dependent query composition, the preference for tour difficulty could be prioritized to all other user preferences in case of severe weather conditions. By addition of security relevant soft or hard constraints, such as a maximum tour difficulty, special demands of the outdoor domain are met. Finally, In the case of bad weather conditions, preferences for tour duration could be altered so that the preferred duration is multiplied by a factor, say 0.8. The reason for such an adaptation is the expert knowledge that physical performance decreases due to coldness and rain. The result of this procedure is a personalized Preference SQL query that is evaluated based on the given PostGIS tour database. Afterwards, best matches are returned to the *outdooractive.com* portal for presentation. The user in turn instantly obtains an adequate amount of personalized results.

13.5 Conclusion and Future Work

Providing excellent context-aware personalized information filtering is one of the major challenges for location-based mobile apps. In complex application domains like the planning of hiking or mountain bike tours, this problem becomes even more difficult. In this paper we presented a database approach using the Preference SQL query language as a promising solution to achieve this ambitious goal. Preference SQL provides a Best-Matches-Only evaluation of user preferences. We have demonstrated how to extend Preference SQL with novel location-based preference constructors that can be evaluated with standard geo-databases such as PostGIS as backend systems. Furthermore, we described the ongoing PeToMoTo project that integrates this technology into an existing commercial online platform to provide user-centric and context-aware recommendations.

The presented location-based preference is a first step to integrate spatial preferences into recommendation processes. Statements of domain experts indicate that a plurality of novel spatial preference constructors is needed to account for other application scenarios such as e.g. location planning. By using spatial functions of the underlying database, these new constructors can be implemented rapidly and for all types of geometries. In a similar fashion other extensions such as pgRouting or even external APIs can be used to incorporate network costs as distance function.

The next step in Preference SQL development will be extending the already implemented SQL 92 Standard by parts of the Standard SQL/MM syntax, an SQL extension to handle geographic data in relational databases. In case the starting point of a tour is of interest, a user could then apply a spatial function retrieving the first point of the tour linestring and state a preference on the result. Moreover, the combination of spatial preferences to spatial group queries, e.g. to find suitable restaurants for a group of people with each individual being at a different location (also known as Spatial Skylines), is another focus area.

Acknowledgement

The project PeToMoTo is funded by the German Federal Ministry of Economics and Technology (BMWi) according to a resolution of the German Bundestag, grant no. KF2751301. We also like to thank our anonymous reviewers whose comments helped to improve this paper.

References

Adomavicius G, Tuzhilin A (2005) Toward the Next Generation of Recommender Systems: A Survey of the State-of-the-Art and Possible Extensions. In IEEE Transactions on Knowledge and Data Engineering, vol. 17, no. 6, pp. 734–749.

Brafman R, Domshlak C (2009) Preference Handling: An Introductory Tutorial. In AI Magazine, vol. 30, no. 1, pp. 58–86, Menlo Park, CA, USA.

Chomicki J (2003) Preference Formulas in Relational Queries. In ACM Transactions on Database Systems (TODS), vol. 28, no. 4, pp. 427–466, New York, NY, USA.

Cohen W, Schapire R, Singer Y (1999) Learning to Order Things. In Journal of Artificial Intelligence Research, vol. 10, pp. 243–270.

Facebook (2011) [Online] Available: http://www.facebook.com [15 May 2011]

Hafenrichter B, Kießling W (2005) Optimization of Relational Preference Queries. In Proceedings of the 16th Australasian Database Conference (ADC), pp. 175–184, Darlinghurst, Australia.

Hanani U, Shapira B, Shoval P (2001) Information Filtering: Overview of Issues, Research and Systems. In User Modeling and User-Adapted Interaction, vol. 11, no. 3, pp. 203–259.

Hoffmann K, Bateson J (2008): Services Marketing: Concepts, Strategies & Cases, South-Western Cengage Learning, Mason, OH, USA.

Holland S, Kießling W (2004) Situated Preferences and Preference Repositories for Personalized Database Applications In Proceedings of the 23rd International Conference on Conceptual Modeling (ER), pp. 511–523, Shanghai, China.

Kießling W (2002) Foundations of Preferences in Database Systems. In Proceedings of the 28th International Conference on Very Large Data Bases (VLDB), pp. 311–322, Hong Kong, China.

Kießling W, Endres M, Wenzel F (2011) The Preference SQL System – An Overview. In Bulletin of the IEEE Computer Society Technical Committee on Data Engineering, vol. 34, no.2, pp. 11–18.

Mokbel M, Levandoski J (2009): Toward Context and Preference-Aware and Location-based Services. In Proceedings of the 8th International ACM Workshop on Data Engineering for Wireless and Mobile Access, Providence, RI, USA.

Nivala A, Sarjakoski T, Laakso K, Itäranta J, Kettunen P (2008) User Requirements for Location-Based Services to Support Outdoor Activities, in Gartner G, Rehrl K (Eds), Location Based Services and TeleCartography II, LNG&C Springer, pp. 167–184.

Outdooractive (2011) [Online] Available: http://www.outdooractive.com [15 May 2011]

Poslad S, Laamanen H, Malaka R, Nick A, Buckle P, Zipf A (2001): Crumpet: Creation of User Friendly Mobile services Personalized for Tourism. In Proceedings of the 2nd International Conference on 3G Mobile Communication Technologies, pp. 28–32, London, UK.

PostGIS Spatial Database Extension (2011) [Online] Available: http://postgis.refractions.net/ [15 May 2011]

Preference SQL Reference (2011) [Online] Available: http://trial.preferencesql.com [15 May 2011]

Preisinger T, Kießling W (2007) The Hexagon Algorithm for Evaluating Pareto Preference Queries. In Proceedings of the 3rd Multidisciplinary Workshop on Advances in Preference Handling (M-PREF in conjunction with VLDB ’07), Vienna, Austria.

Sacco G (2009). The Model. In Sacco G, Tzitzikas Y (Eds) Dynamic Taxonomies and Faceted Search. Springer, pp. 1–19.

Stefanidis K, Koutrika G, Pitoura E (2011) A Survey on Representation, Composition and Application of Preferences in Database Systems. In ACM Transactions on Database Systems (TODS), vol. 36, no. 3, New York, NY, USA.

Tryfonopoulus C, Koubarakis M, Drougas Y (2009) Information Filtering and Query Indexing for an Information Retrieval Model. In ACM Transactions on Information Systems, volume 27, issue 2.

van den Eijkel G, Lankhorst M, Bijlsma M, van Kranenburg H. (2001): Personalization of next-generation mobile services, Position paper Wireless World Research Forum, Munich, Germany.

Williams M, Halvatzaras D, McIntosh N, Mehrmann L, Pascottol R (2002): Role of a Personal Assistant in Delivering Personalized Context-Aware Services to Young People, In Proceedings of the IST Mobile Communications Summit 2002, pp. 513–517, Thessaloniki, Greece.

Section III

Mobile Sensors

Chapter 14

Discovering the Sensor Web through Mobile Applications

Theodor Foerster[1], Daniel Nüst[2], Arne Bröring[1,2,3] and Simon Jirka[2]

[1] Institute for Geoinformatics, University of Muenster, Germany
theodor.foerster@uni-muenster.de
[2] 52°North GmbH, Münster, Germany
{d.nuest, a.broering, s.jirka}@52north.org
[3] ITC, University of Twente, Enschede, Netherlands

Abstract

Sensor data is crucial for mobile applications to support the user in the field. Several mobile applications are available for accessing such sensor data. However, a comprehensive approach for discovering such sensor data in the Sensor Web according to the user's context (i.e. the location) has not been proposed yet. This article describes an approach for discovering data and services in the Sensor Web through mobile applications. The approach is demonstrated by an air quality scenario and is implemented based on Free and Open Source Software.

Keywords: Sensor web, mobile applications, sensor discovery, air quality

14.1 Introduction

The availability of mobile applications and mobile devices is increasing. Additionally, the computational power, the web-connectivity and the user experience of these devices improve constantly. At the same time, sensor data is becoming available through services based on Sensor Web technology. The Sensor Web enables data producers and users to publish and access sensor information via web- and standards based interfaces. Consequently, it has become one of the data hubs for pushing for instance environmental information to the user in an interoperable way.

G. Gartner, F. Ortag (Eds.), *Advances in Location-Based Services*, Lecture Notes in Geoinformation and Cartography, DOI 10.1007/ 978-3-642-24198-7_14, © Springer-Verlag Berlin Heidelberg 2012

The Sensor Web Enablement (SWE) initiative of the Open Geospatial Consortium (OGC) describes such a set of standards for accessing and publishing sensor data in a uniform way (Botts, Percivall, Reed, & Davidson, 2008). Examples of Sensor Web applications are hydrology (Jirka, Bröring, & Walkowski, 2010), environmental monitoring, risk monitoring (Jirka, Bröring, & Stasch, 2009; Klopfer & Simonis, 2009) or ocean observing systems (Raape et al., 2010). With the advancement of a revised set of standards towards a Sensor Web 2.0, sensor data has become discoverable and can be integrated on-the-fly in a plug & play manner (Bröring, Maué, Janowicz, Nüst, & Malewski, 2011).

Integrating the Sensor Web into mobile applications seamlessly, by making the services discoverable and easy to use is important for continuous information sharing especially in environmental applications. Users of mobile devices require information that fits their context and which they can use at their specific location. However, a comprehensive approach for discovering sensor data accordingly in the Sensor Web through mobile applications has not been proposed yet. In this article we refer to context as defined by Dey & Abowd (2000).

As a first step we target expert users who are familiar with the idea of the Sensor Web and concepts like observations and web services. We see a potential for this group in shifting research work to the actual locations of interest. Something that is only possible after the latest advancements in mobile technology.

This article presents a comprehensive approach to discovery on mobile devices that includes a mechanism to a) discover the relevant sensor data, b) retrieve the data using a light-weight sensor metadata protocol, and c) portray the data using a mapping application. Finally, the application is demonstrated in an air quality use case based on data from the European Environment Agency (EEA).

Section 14.2 describes the related work with a focus on Sensor Web technology and mobile applications. The comprehensive approach is then described in *Section 14.3*. The air quality scenario is described in *Section 14.4* along with the specific implementation. *Section 14.5* provides the discussion and conclusion of the proposed approach.

14.2 Background

This section presents the basic concepts as applied in this article. These concepts are framed by the Sensor Web, as a foundation for data and services. Other strategies for discovery of sensor data using mobile applications are for instance presented by OpenSearch[1]. However, this is outside the scope for our paper, the architecture

[1] OpenSearch website: www.opensearch.org.

is similar, only the encoding of the query between application layer and sensor layer differ.

14.2.1 Sensor Web

The Sensor Web Enablement (SWE) initiative of the Open Geospatial Consortium has established a suite of standards to realize the vision of the Sensor Web. In this context, the *Sensor Web* is defined as a Web service-based infrastructure which enables the discovery of sensor related resources, the access to sensor observations, the tasking of sensors, as well as eventing and alerting within the Sensor Web. Consequently, this functionality is provided through standardized Web service interfaces and data encodings defined by the SWE framework. Based on the SWE framework, an architecture can be created incorporating three layers: The application layer, the Sensor Web layer and the Sensor Layer. The Sensor Web layer provides standardized and uniform access to the heterogeneous sensor protocols and communication details of the underlying sensors, which are hidden to the application layer.

The SWE framework incorporates two data models and their encodings for a standardized exchange of information. First, the Sensor Model Language (SensorML) (Botts & Robin, 2007) describes the sensor‘s characteristics. Second, observed sensor data is modeled and encoded according to the Observations & Measurements (O&M) (Cox, 2007) standard. Further, the Sensor Observation Service (SOS) provides sensor observations (encoded as O&M) through spatio-temporal queries.

Based on the limitations of the initial design of the Sensor Web, a new generation of Sensor Web interfaces and encodings has been created as described by Bröring, Echterhoff et al. (2011). In this context, the Sensor Instance Registry (SIR) and Sensor Observable Registry (SOR) have been added to the SWE framework to allow clients to discover data and services (Jirka, Bröring, & Stasch, 2009). The SOR is a service providing an interface to definitions of phenomena for management and discovery. These phenomena can have semantic relationships, like equivalence or being a subtype, that a user can exploit in his searches, for example "water temperature" when searching for "air temperature", or "precipitation accumulated over 24 hours" when searching for "precipitation". A SOR can use different ontologies for calculating matches.

The SIR is a catalogue service offering operations to administrators for managing their sensors and services as well as to end-users who want to discover sensors based on spatial, temporal and thematic criteria. For these use cases SIR offers insert, update, delete and retrieve operations for specific sensor descriptions based on SensorML. For dynamic parameters the SIR provides a status handling mechanism, which allows higher updating intervals and even subscription to status events (Jirka & Nüst, 2010).

The search can be tailored to the user's needs. For spatial and temporal queries the user can apply a bounding box and a time period respectively. Identifier-based and thematic searches, the latter using a full text search, are also possible. Based on the thematic parameters, the observed property (e.g. temperature, precipitation, or ozone) can be retrieved, as typical parameter for discovering appropriate sensors. It can even be semantically enhanced by using a SOR connection. That way, a user can accept matching phenomenon types using different matching rules. Sensor searching also offers the option to request a simplified response containing only an identifier, the associated services, a description text, and a link to the full sensor description.

14.2.2 Mobile Applications Accessing Sensor Data

Several mobile applications for accessing sensor data have been developed. Examples are traffic maps, m-health (Istepanian, Jovanov, & Zhang, 2004) and weather applications such as weather underground[2]. All these applications are tailored to their specific purpose. Some mobile applications retrieve their content through web feeds encoded in for instance Atom[3] or RSS[4]. Moreover, the Hydrosys project develops both, the sensors and the mobile clients (Kruijff, Mendez, Veas, & Gruenewald, 2010). It uses a specifically deployed network of sensors for avalanche detection based on proprietary protocols.

In the context of SWE, only little research has been documented on mobile applications such as Stasch, Bröring, & Walkowski (2008), who proposed a new data model for the SOS to handle sensor data collected by mobile sensors, or Müller, Fabritius, & Mock (2011) who presented an SOS and desktop client for real-time visualization of mobile sensors. Recently, Tamayo, Viciano, Granell, & Huerta (2011) described a generic mobile client for SOS that can visualize any SOS data in a map or tabular view (no discovery included). Regarding the standardization of data (e.g. stemming from sensors) for mobile applications, the Augmented Reality community has started to define a new specification[5]. This specification aims especially at being incorporated in software for mobile Augmented Reality applications. However, a specific interface for discovery of services and data has not been developed yet and the inclusion of existing SWE specifications and applications has not been considered so far.

[2] Weather underground website: http://www.wunderground.com/.
[3] The Atom Syndication Format website: http://tools.ietf.org/html/rfc4287.
[4] RSS Specification website: http://www.rssboard.org/rss-specification/.
[5] AR standards website: http://arstandards.org/.

14.3 Discovering the Sensor Web through Mobile Applications

In this section, the basic requirements for integrating Sensor Web data into mobile applications are described. Based on the requirements the architecture is defined. The architecture applies services from the new generation of Sensor Web Enablement standards *(Section 14.2.1)*.

Discovery in the former generation of the Sensor Web has been solely based on the operations of SOS. Therefore, the particular service had to be known in advance, which is contrary to loose-coupled and distributed services. Moreover, despite websites like the Group on Earth Observations's (GEO) GeoPortal[6], service discovery is a cumbersome task.

After downloading the service metadata of a particular SOS instance (using *GetCapabilities* operation), a mobile client needs to iterate through the list of available sensors and needs to sequentially request the full sensor description, to extract a few key terms as well as spatial and temporal scope of a sensor. This search could be reduced if a user selects a topic of interest (technical term offering) out of the available ones, but the amount of transferred data is still higher and the information is more detailed than necessary for discovery purposes.

Furthermore, the first generation SOS interface does not allow clients to retrieve the latest available observation, which is a typical use case. Here a client has to "guess" a suitable interval or area to request the first data and potentially increase the extent until values are returned. This produces transmission overhead not only because of the increased number of requests, but also because a lot more data than desired might actually be returned.

Some drawbacks of the current SWE framework have been mentioned already, but shall now be put in context. Limitations of mobile devices and subsequent specific challenges are well-known and have been discussed at length in recent years (Istepanian et al., 2004; Nah, Siau, & Sheng, 2005). Among them we identified the following aspects as crucial for a successful integration of the sensor web in mobile applications: limitations in bandwidth and processing power, energy supply (i.e. battery life), and processing capabilities.

14.3.1 Requirements

Based on the previously described limitations, the following requirements for the presented architecture can be extracted. At this point, we exclude usability requirements explicitly, which are acknowledged, but are outside the scope of this analysis.

[6] http://www.geoportal.org/

REQUIREMENT 1: Resource efficiency.

The architecture must minimize energy intensive transfer and processing of data. Limiting the transfer of data is achieved through reduced communication overhead (e.g. light-weight protocols), reduced number of requests (e.g. client-side caching of retrieved information, and specific requests tailored to the respective task, e.g. discovery versus retrieving a complete description of components). Reduced communication overhead and reduced number of requests also results in reduced computational processing effort on the mobile device.

REQUIREMENT 2: Spatial context.

The spatial context of the user can be determined by the locating capabilities of the mobile device. Almost all mobile devices support some kind of detecting the current location (using GPS or Wi-Fi). From a range of everyday applications users expect this information to be included automatically while fulfilling their task at hand. Using additional sensors and other applications incorporated in mobile devices can be used to determine further aspects of context (Abowd et al., 1997). Apart from such a spatial sub-setting, further filtering can support the first requirement.

REQUIREMENT 3: Temporal context.

Temporal subsetting for both most recent data and historic data must be possible on user input. This allows users to browse real-time data, or the most recent values, as well as historic information. Even having bandwidth limitations, the user should not have to wait for first information, which can be the case without proper filtering capabilities.

REQUIREMENT 4: Thematic context.

The architecture needs to support the user for determining the thematic context. This allows the user to interact with the appropriate data

To sum up, the desired mobile application shall support the exploitation of the current location as a key component of a user's context to allow adapted discovery of desired services. Based on the generic and interoperable interfaces of the SWE framework our approach must allow to answer the apparently simple question: What information captured by sensors near to me and for a certain topic is available right now?

These requirements are applied to the mobile application and implemented in a technical architecture as described in the following section.

14.3.2 Architecture

Based on the identified requirements *(Section 14.3.1)* this section describes the developed architecture. The architecture consists of the sensor layer, the sensor

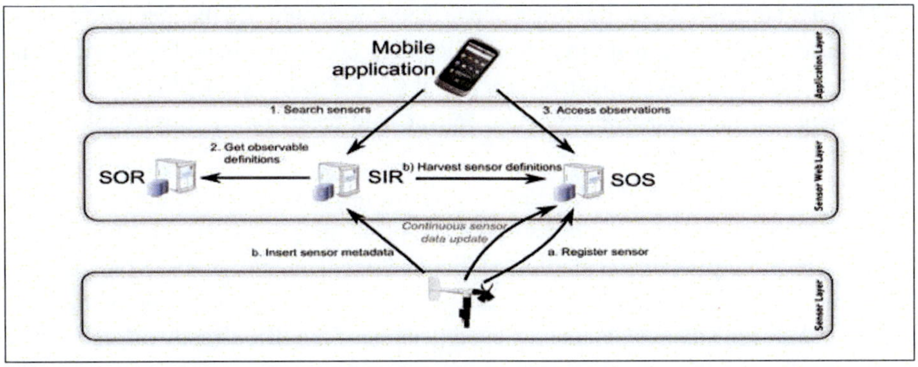

Fig. 14.1. Architecture for enabling discovering in the Sensor Web through mobile applications.

web layer and the application layer. The sensor layer can comprise any proprietary sensor or sensor network, which provides digital readings regarding a specific phenomenon such as air quality. The Sensor Web layer is based on the SWE framework *(Section 14.2.1)*. The application layer is represented by the mobile application, which retrieves context-specific sensor information from the SWE framework *(Figure 14.1)*.

The architecture supports the following workflows:

- Registration of sensors
- Discovery of sensors
- Access to sensor data and metadata.

To make the SOS aware of the specific sensor, it has to be registered accordingly (Step a. in *Figure 14.1)*. After successful registration, sensor data is continuously transmitted to the server. This can either be implemented via a push-based or pull-based communication. The transmitted sensor readings are stored in the SOS to build an archive of sensor data. Subsequently, the provider can register the sensor metadata at the SIR or the SIR harvests the data automatically. Based on the harvested information, the SIR can use this data to answer future requests of the user (Step b. *Figure 14.1)*.

The user is now able to retrieve specific sensor data, which matches his context, using the Sensor Web Layer. In principle, the user is able to retrieve any kind of sensor data. The mobile application does not access all the available sensor data and services, but can query the SIR accordingly for specific sensor information. The SIR is queried with a bounding box and a generic keyword search. The response has been extended to include a spatial reference in a search result element. This is essential for mobile applications (Requirement 2). The current SIR interface requires a retrieval of the full sensor description to obtain location or observed area of a sensor.

Based on the user input, the SIR can use the SOR to enhance the thematic context of the user query. Based on the stored metadata about the available services and data, the SIR provides the suitable service instances and the suitable observables. The client can then retrieve the actual sensor data based on the observable of the specific service instance.

Based on the requirements *(Section 14.3.1),* the presented architecture is reviewed. To fulfill requirement 1, we utilize interfaces of the incorporated web services that follow the principles of the Representational State Transfer (REST) (Richardson & Ruby, 2007). This allows an easy integration of the offered functionalities into existing Web applications. While the specification of the SIR already offers a REST interface, for the SOS, we build on previous work that defines a meaningful URI scheme for sensor data, a RESTful proxy for SOS instances (Janowicz, Bröring, Stasch, & Everding, 2010). Moreover, the architecture is resource efficient, as it allows clients to discover the relevant data and services using a single request to the SIR, instead of querying each SOS separately. Finally, the discovery ensures that all potentially available services are searched and not only those, which are known to one specific mobile application.

The requirements regarding the context of the user (requirement 2-4) is achieved through discovery. The location can be automatically determined and is attached to each search request to the SIR. The thematic context is determined based on the search query of the user. The temporal context is supported through the sensor data, as the data can be queried regarding time through the SOS directly.

14.4 Discovery of Air Quality Data in Mobile Applications through the Sensor Web

Based on the architecture *(Section 14.3.2),* this section describes the use case of accessing air quality data regarding the spatial context of the mobile application. Furthermore, this section describes the implementation, which is based on Free and Open Source Software and the Android platform by Google[7].

14.4.1 Use Case

The following use case is presented as a user story, i.e. a short text in every day's language describing a task that a user has to perform. In our user story, we have a climatologist who makes manual in-situ field measurements of air quality parameters.

[7] Android developer website: developer.android.com/.

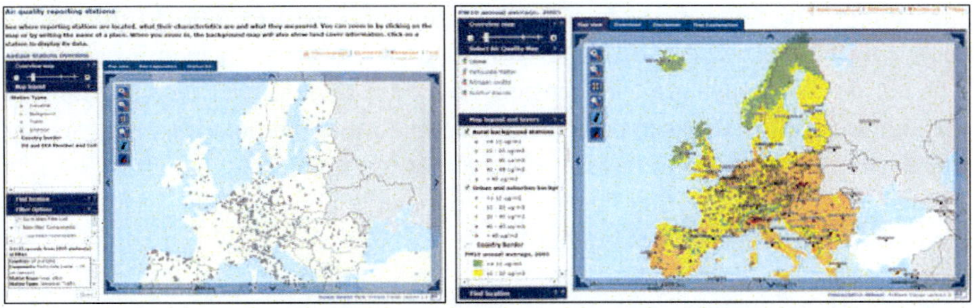

Fig. 14.2. EEA air quality data on the web. Station data (left) and interpolated data (right).

> *As a user in the field I want to access the data of the nearest long-term air quality reports collected in the AirBase dataset so that I can directly compare them to my own up-to-date local measurements and possibly adjust my observation position.*

According to the European Environment Agency (EEA) the AirBase dataset "*is the air quality information system maintained by the EEA through the European Topic Centre on Air and Climate Change. It contains air quality data delivered annually [...] establishing a reciprocal exchange of information and data from networks and individual stations measuring ambient air pollution within the Member States*"[8]. The data is downloadable in several well-defined formats completely or for country subsets, as well as viewable in specialized applications. These viewers are amongst others available for station location and metadata, and interpolated and individual values. *Figure 14.2* shows screenshots of the AirBase Viewer application for station information (left) and interpolated data (right). No public interfaces are known to query data for specific stations, spatial extends or time intervals with other applications.

However, in the UncertWeb project (Pebesma, Cornford, Nativi, & Stasch, 2010) a tool has been developed to make observations from AirBase dataset accessible via an SOS. Such a service comprising data for over 1000 stations in Germany is available online[9].

14.4.2 Implementation

To realize the use case, the presented architecture is implemented. The presented use case requires adaption of the existing SIR service interface. We based our implementation on the 52°North SIR[10] and extended both the schemata and the service imple-

[8] AirBase dataset description: http://www.eea.europa.eu/data-and-maps/data/airbase-the-european-air-quality-database-3

[9] Sensor Observation Service endpoint: http://giv-uw.uni-muenster.de:8080/AQE/sos

[10] 52°North SIR website: http://52north.org/communities/sensorweb/incubation/discovery/

mentation[11]. The schema now includes the element *ObservedBoundingBox*, in the element *SimpleSensorDescription*, from the search response document. The reason for choosing a 2-dimensional over a 1-dimensional representation of location is that the latter can be contained in the former but not vice versa. So, our approach allows sensors to have an observed area if applicable, but also a specific location if the bounding box degenerates to a single point. *Listing 14.1* shows an example of such a sensor description.

```
<sir:SearchResultElement>
  <sir:SensorIDInSIR>383</sir:SensorIDInSIR>
  <sir:SimpleSensorDescription>
  <sir:SensorDescriptionURL>http%3A%2F%2Fgiv-genesis.uni-muenster.de%3A
8080%2FSIR%2Fsir%3Fservice%3DSIR%26amp%3Bversion%3D0.3.1%26a
mp%3BREQUEST%3DDescribeSensor%26amp%3BSENSORIDINSIR%3D383</
sir:SensorDescriptionURL>
  <sir:DescriptionText>![CDATA[
Identifications: urn:ogc:def:identifier:OGC:1.0:uniqueID - urn:ogc:object:featu
re:Sensor:EEA:airbase:4.0:DEHB005; urn:ogc:def:identifier:OGC:1.0:longNam
e - Bremerhaven; urn:ogc:def:identifier:OGC:1.0:shortName - Bremerhaven;

Classifications: intendedApplication - air quality; sensorType - Background;
typeOfSensor - Background; stationOzoneType - urban; stationAreaType -
urban; stationSubCatRural - unknown;

Keywords: GERMANY; DE; AIRBASE; AIRQUALITY; EEA;
]</sir:DescriptionText>
    <sir:ObservedBoundingBox crs="-1" dimensions="2">
      <ows:LowerCorner>53.559998 8.569406</ows:LowerCorner>
      <ows:UpperCorner>53.559998 8.569406</ows:UpperCorner>
    </sir:ObservedBoundingBox>
  </sir:SimpleSensorDescription>
  <sir:ServiceReference>
    <sir:ServiceURL>http://giv-uw.uni-muenster.de:8080/AQE/sos</
sir:ServiceURL>
    <sir:ServiceType>SOS</sir:ServiceType> <sir:ServiceSpecificSe
nsorID>urn:ogc:object:feature:Sensor:EEA:airbase:4.0:DEHB005</
sir:ServiceSpecificSensorID>
   </sir:ServiceReference>
</sir:SearchResultElement>
```

Listing 14.1. A simple sensor description element encoded in XML which was extracted from a search response document.

The practical benefit of this extension can be illustrated when comparing the sizes of the search result documents. In the first SIR specification, to obtain the loca-

[11] SIR schemas are available at http://giv-genesis.uni-muenster.de/schemas/sir/, SIR instance is available at http://giv-genesis.uni-muenster.de:8080/SIR/.

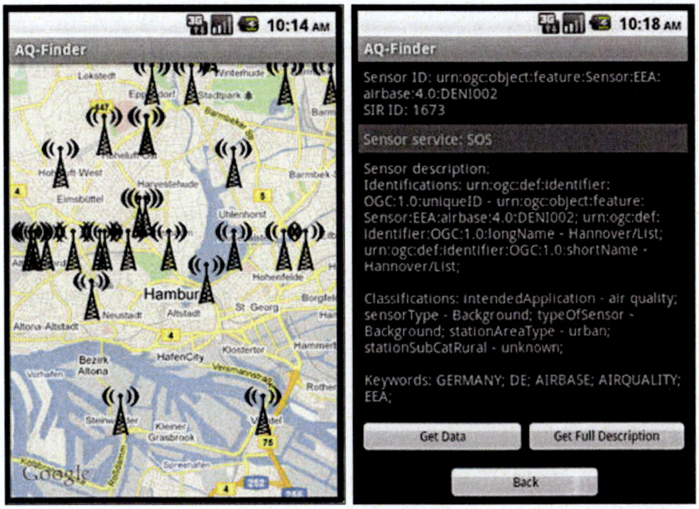

Fig. 14.3. Screenshot of the mobile application accessing the available sensors using the described architecture. Map of sensor locations (left) and exemplary sensor description (right).

tion of 376 sensors (arbitrary bounding box) a client must download and process a 2.283 KB large XML file. The simple sensor descriptions containing the information as shown in *Listing 14.1* and also the service references reduced this file almost by factor 10 to 285 KB. This improvement helps to fulfill requirement 1 *(Section 14.3.1).*

Further, we developed a client application for Android smartphones. The client application provides a map to explore all sensors provided by a SIR *(Figure 14.3),* centered at the current location of the device. The center of the sensor's observed bounding box is symbolized by a point symbol, as in our use case of in-situ sensors the bounding box is reduced to a point. Browsing the map, for example pan and zoom the map with gestures, triggers new search requests for the changed bounding box. When selecting a certain sensor station a description of the particular sensor is provided, namely the sensor identifiers and the sensor description *(Figure 14.3).* The user can now further analyze, e.g. request data using the service reference (see *Listing 14.1).*

14.5 Discussion & Conclusion

The integration and discovery of sensor data into mobile applications is required to provide contextual information to the specific user. Current sensor data is currently made available through the Sensor Web such as realized by SWE *(Section 14.2.1).* This article describes an approach to discover such data accessible through the current SWE framework (realizing Sensor Web 2.0) of standardized data formats and services. In particular, the architecture is based on SIR (for discovery) and SOS (for data access). The presented architecture *(Section 14.3.2)* is based on a requirements analysis *(Section 14.3.1)*. Furthermore, the presented architecture is applied to an air quality use case, in which a climatologist queries specific air quality sensors regarding his context (spatial, temporal, thematic). The architecture has been implemented based on Free and Open Source Software.

Our analysis has demonstrated, that existing standards such as SWE 1.0 do not support discovery and thereby Sensor Web 2.0 is required, as used for this work. In particular, current SWE standards are enhanced with specific discovery and support the requirements of mobile devices.

The current discovery uses a full text search and the specific location based the metadata in the SIR (sensor registry). In the future, this search could be more specific regarding the field of metadata (e.g. type of sensor station). The location of the sensors is portrayed currently on the map, future research will investigate the combination of mobile applications and augmented reality to discover and access the sensor data interactively.

Naturally, discovery is only the first step in an analysis for which suitable sensor metadata encoding is required. However, we imagine that future developments will integrate processing capabilities for actual data values. For example users could task web processing services (Foerster, Schaeffer, Baranski, & Brauner, 2011), to deduce, or more specifically interpolate, data with a higher spatio-temporal coverage for the user's context such as developed in the INTAMAP project (Pebesma, Cornford, Dubois, et al., 2010).

Acknowledgements

We want to acknowledge the work of Cliff Pereira student at the Institute for Geoinformatics, who developed the mobile client application. This work was supported by the European Commission through the GEOVIQUA project, contract number 265178.

References

Abowd, G., Atkeson, C., Hong, J., Long, S., Kooper, R., & Pinkerton, M. (1997). Cyberguide: A mobile context-aware tour guide. Wireless Networks, 3, 421–433.

Botts, M., & Robin, A. (2007). Sensor Model Language (Implementation specification No. 07-000). Wayland, MA, USA: OGC. Retrieved from http://www.opengeospatial.org/standards/sensorML

Botts, M., Percivall, G., Reed, C., & Davidson, J. (2008). OGC Sensor Web Enablement: Overview And High Level Architecture. GeoSensor Networks, Lecture Notes in Computer Science (pp. 175–190). Heidelberg: Springer.

Bröring, A., Echterhoff, J., Jirka, S., Simonis, I., Everding, T., Stasch, C., Liang, S., et al. (2011). New Generation Sensor Web Enablement. Sensors, 11(3), 2652–2699. doi:10.3390/s110302652

Bröring, A., Maué, P., Janowicz, K., Nüst, D., & Malewski, C. (2011). Semantically-Enabled Sensor Plug & Play for the Sensor Web. Sensors, 11, 7568–7605. doi:10.3390/s110807568

Cox, S. (2007). Observations and Measurements – Part 1 – Observation schema (Implementation specification No. 07-022r1). Wayland, MA, USA: OGC. Retrieved from http://www.opengeospatial.org/standards/om

Dey, A. K., & Abowd, G. D. (2000). Towards a better understanding of context and context-awareness. CHI 2000 workshop on the what, who, where, when, and how of context-awareness (Vol. 4, pp. 1–6).

Foerster, T., Schaeffer, B., Baranski, B., & Brauner, J. (2011). Geospatial Web Services for Distributed Processing – Applications and Scenarios. In P. Zhao & L. Di (Eds.), Geospatial Web Services: Advances in Information Interoperability (pp. 245–286). Hershey, PA: IGI Global.

Istepanian, R. S. H., Jovanov, E., & Zhang, Y. T. (2004). Guest Editorial Introduction to the Special Section on M-Health: Beyond Seamless Mobility and Global Wireless Health-Care Connectivity. IEEE Transactions on Information Technology in Biomedicine, 8(4), 405–414. doi:10.1109/TITB.2004.840019

Janowicz, K., Bröring, A., Stasch, C., & Everding, T. (2010). Towards Meaningful URIs for Linked Sensor Data. In A. Devaraju, A. Llaves, P. Maue, & C. Kessler (Eds.), Proceedings of the Workshop "Towards Digital Earth: Search, Discover and Share Geospatial Data 2010" (p. 6). Presented at the Workshop at Future Internet Symposium, Berlin, Germany. Retrieved from http://ceur-ws.org/Vol-640/paper2.pdf

Jirka, S., & Nüst, D. (2010). OGC Sensor Instance Registry (OGC Discussion Paper No. OGC 10-171) (p. 97 pp.). Wayland, MA, USA: OGC.

Jirka, S., Bröring, A., & Stasch, C. (2009). Applying OGC Sensor Web Enablement to Risk Monitoring and Disaster Management. Presented at the GSDI 11 World Conference, Rotterdam, Netherlands.

Jirka, S., Bröring, A., & Stasch, C. (2009). Discovery Mechanisms for the Sensor Web. Sensors, 9(4), 2661–2681. doi:10.3390/s90402661

Jirka, S., Bröring, A., & Walkowski, A. (2010). Sensor Web in Practice. Geoinformatics, (13), 42–45.

Klopfer, M., & Simonis, I. (Eds.). (2009). SANY – an open service architecture for sensor networks. SANY consortium.

Kruijff, E., Mendez, E., Veas, E., & Gruenewald, T. (2010). On-Site Monitoring of Environmental Processes using Mobile Augmented Reality (HYDROSYS). In A. Berre, D. Roman, & P. Maue (Eds.), Proceedings of the Workshop "Environmental Information Systems and Services – Infrastructures and Platforms" (Vol. 679, p. 7). Presented at the Environmental Information Systems and Services – Infrastructures and Platforms, Bonn, Germany: CEUR. Retrieved from http://sunsite.informatik.rwth-aachen.de/Publications/CEUR-WS/Vol-679/paper2.pdf

Müller, R., Fabritius, M., & Mock, M. (2011). An OGC compliant Sensor Observation Service for mobile sensors. In S. Geertman, W. Reinhardt, & F. Toppen (Eds.), Advancing Geoinformation Science for a Changing World (Vol. 1, pp. 163–184). Berlin, Heidelberg: Springer Berlin Heidelberg. Retrieved from http://www.springerlink.com/index/10.1007/978-3-642-19789-5_9

Nah, F. F.-H., Siau, K., & Sheng, H. (2005). The value of mobile applications: a utility company study. Commun. ACM, 48, 85–90. doi:http://doi.acm.org/10.1145/1042091.1042095

Pebesma, E., Cornford, D., Dubois, G., Heuvelink, G. B. M., Hristopoulos, D., Pilz, J., Stöhlker, U., et al. (2010). INTAMAP: The design and implementation of an interoperable automated interpolation web service. Computers & Geosciences. doi:10.1016/j.cageo.2010.03.019

Pebesma, E., Cornford, D., Nativi, S., & Stasch, C. (2010). The Uncertainty Enabled Model Web (UncertWeb). In A. Berre, D. Roman, & P. Maue (Eds.), Proceedings of the Workshop "Environmental Information Systems and Services – Infrastructures and Platforms" (Vol. 679, p. 9). Presented at the Environmental Information Systems and Services – Infrastructures and Platforms, Bonn, Germany: CEUR. Retrieved from http://ceur-ws.org/Vol-679/paper10.pdf

Raape, U., Teßmann, S., Wytzisk, A., Steinmetz, T., Wnuk, M., Hunold, M., Strobl, C., et al. (2010). Decision Support for Tsunami Early Warning in Indonesia: The Role of OGC Standards. In M. Konecny, S. Zlatanova, & T. L. Bandrova (Eds.), Geographic Information and Cartography for Risk and Crisis Management (pp. 233–247). Berlin, Heidelberg: Springer Berlin Heidelberg. Retrieved from http://www.springerlink.com/index/10.1007/978-3-642-03442-8_16

Richardson, L., & Ruby, S. (2007). RESTful web services. O'Reilly Media, Inc.

Stasch, Christoph, Bröring, A., & Walkowski, A. C. (2008). Providing Mobile Sensor Data in a Standardized Way-The SOSmobile Web Service Interface. 11th AGILE International Conference on Geographic Information Science.

Tamayo, A., Viciano, P., Granell, C., & Huerta, J. (2011). Sensor Observation Service Client for Android Mobile Phones. Proceedings of SWE 2011. Presented at the SWE 2011 workshop, Banff, Alberta, Canda.

Chapter 15

Positioning under Adverse Conditions Using High Sensitivity GPS

Klemen Kozmus Trajkovski, Dušan Petrovič, Oskar Sterle and Bojan Stopar

Faculty of Civil and Geodetic Engineering, University of Ljubljana, Slovenia
{klemen.kozmus, dusan.petrovic, oskar.sterle, bojan.stopar}@fgg.uni-lj.si

Abstract

Positioning is a key component of location-based services (LBS). An LBS user wants to acquire the current location in any possible environment. GNSS positioning, the fundamental component of a positioning system, is limited to open areas with unobstructed reception of GNSS satellites signals. In order to assure a ubiquitous positioning, other sensors have to be included in the positioning system. Inertial sensors are the most commonly used to compliment the GNSS sensors. However, the inertial sensors are very susceptible to drifts and other errors. Recently, other sensors have been considered for use in the positioning systems, e.g. UWB, Wi-Fi, pseudolites and high sensitivity GPS. High sensitivity GPS receivers have extended the use of GNSS navigation to environments which were previously believed unsuitable for satellite signal reception. However, extra care has to be considered when using high sensitivity GPS because autonomous solutions can cause major errors in the estimated position in the challenging environments. A careful and consistent treatment of the observation data can yield adequate results even under adverse conditions for signal reception.

Keywords: Positioning, GNSS, High sensitivity GPS

G. Gartner, F. Ortag (Eds.), *Advances in Location-Based Services*, Lecture Notes in Geoinformation and Cartography, DOI 10.1007/ 978-3-642-24198-7_15,

15.1 Introduction

Obtaining user's location is an essential part of the Location-Based Services (LBS). There are many positioning techniques available. However, GNSS is by far mostly used in daily practice. In general, GNSS has many advantages: it is global, 3D, always available, has constant accuracy over time and the achieved accuracy is decent. GNSS receivers are cheap, small and easy to implement in other devices, e.g. smartphones and navigators.

Unfortunately, there are also some downsides on the GNSS positioning. The main deficiency is that it is not everywhere available. Environments, such as urban canyons, forested areas and indoors, cause signal attenuation and signal reflection. As a result, low number of tracked satellites and multipath can cause large errors in the acquired position, or, in the worst case, no position can be obtained.

Due to the fact that most actual LBS requests occur in urban areas, where the satellite reception is limited, managing the positioning component of LBS is a very challenging task.

In order to provide ubiquitous positioning, other sensors beside the GNSS receivers have to be implemented in the positioning system. For many years, GPS receivers have been combined with inertial sensors to form integrated GPS/INS systems. Such systems rely heavily on the GPS part and inertial sensors are used only when the number of tracked satellites is too low to determine a GPS-only position. However, inertial sensors have time dependent biases and errors. Stable INS are very expensive, therefore, low cost instruments MEMS (Micro Electrical Mechanical Systems) are usually fused in the integrated systems. MEMS sensors usually drift very quickly, causing the position error to grow rapidly with time. Errors in the acquired position can quickly grow over 100 meters, also noted in Lachapelle (2007).

In recent years, other sensors and techniques have been considered for the implementation in the positioning systems. Wi-Fi positioning relies on the existing network of Wi-Fi access points, therefore, is not everywhere available. UWB (Ultra Wide Band) is a technology where the position of a tag is determined by TDOA (Time Difference of Arrival) and/or AOA (Angle of Arrival) to fixed sensors. Again, a network of sensors has to be installed before use.

High sensitivity GPS (HS GPS) is able to determine positions in some problematic areas. However, use of HS GPS can lead to large errors in the position. Therefore, a proper and careful processing of the observation data is required. A custom positioning procedure has been developed, using raw code and Doppler observations from HS GPS receivers. The custom procedure is based on L1-norm robust estimation method and separate determination of the receiver clock offset and the position unknowns. A so-called conditional DGPS is also introduced. The main purpose of the developed methods is achieving reasonable GPS accuracy even in harsh conditions for signal reception. Some of the representative results from Kozmus Trajkovski et al. (2010) are summarized in the following chapters.

15.2 High Sensitivity GPS

Until recently, the availability of the GNSS solutions was limited to open spaces where either few or none physical obstacles were present on signals' paths. Signals are attenuated when passing through physical obstacles, and signals can be reflected off nearby reflective surfaces. In general, urban canyons, thick vegetation and indoors are considered as challenging environments.

With introduction of high sensitivity receivers the areas of possible use of GPS technology have broaden since such receivers are capable to determine their position in difficult conditions for reception of the signals. HS GPS receivers are able to detect very weak signals by using large number of correlators and longer integration time. Regular GPS receivers incorporate up to 50 correlators, whereas the latest HS GPS chips incorporate more than one million correlators. The typical integration time for regular GPS receivers is a few ms, while HS GPS receivers perform integration over a longer interval. Integration can be extended to several hundred ms (Lachapelle, 2004).

According to Wieser (2006), the tracking performance does not entirely depend on the signal strength but also on the signal-to-noise ratio (SNR) and more specifically on the SNR in a 1 Hz bandwidth, i.e. the carrier-to-noise density ratio (C/N0). Regular GPS receivers are able to acquire and track signals with a C/N0 above 33–35 dBHz. The HS GPS receivers in our experiments were able to track signals with a C/N0 as low as 10 dBHz.

HS GPS receivers track signals from GPS satellites on the L1 frequency. The results of the signal tracking are usually code pseudoranges, carrier-phase and Doppler observables. A SNR value is also typically available. Most of the HS GPS receivers available on the open market determine the current position internally and output their results via NMEA 0183 messages. Only a few of them can be setup to output raw observation data.

15.3 Development of the Custom Positioning Procedure

Custom positioning methods can be developed only by processing raw observation data. Two u-blox evaluation kits with a raw data output option were used in the tests, namely the AEK-4T and EVK-5T. The AEK-4T employs the Antaris GPS chip and offers rates of up to 10 Hz for raw data, whilst the EVK-5T relies on the u-blox 5 Positioning Engine which supports a 2 Hz raw data update rate. Both receivers are able to determine and output the following data: code pseudoranges, carrier-phase measurements, Doppler measurements and SNR values. However, the AEK-4T receiver is able to output carrier-phase observables only for signals with a SNR over 30 dBHz. Because of this limitation, the custom positioning solution depends on code pseudoranges and Doppler observables.

Basic autonomous positioning is based on processing the code pseudoranges in the least squares adjustment. Four unknowns have to be determined in the process: the 3D position coordinates of the receiver and the receiver clock offset. A pseudorange is affected by a number of biases and possible errors: the satellite position, the satellite clock offsets, the ionosphere bias, the troposphere bias, multipath and noise. All biases have to be modelled to obtain the estimated unknowns to the highest possible degree of accuracy and reliability.

Multipath is a phenomenon whereby the receiver's antenna receives a GPS signal from different paths, caused by the reflection of the signal off reflective surfaces near the receiver. Multipath depends entirely on the local environment and is therefore nearly impossible to model. Theoretically, multipath can reach up to one half of a wavelength of a pseudorange (up to 150 m for code observations) and up to one quarter of a wavelength of the carrier phase (a few cm), according to Xu(2007). Since Doppler measurements are affected by the bias rates only, the multipath effect on Doppler observations should be on the level of the carrier phase multipath effects. Multipath is undoubtedly the most significant bias in code pseudoranges observed in "indoor" positioning.

Outliers are common in observations that take place in multipath and attenuated environments. The first option is to try to eliminate outliers from the mathematical model. However, based on our experience, it is difficult to accurately detect and eliminate outliers in such conditions. The second option is to minimise the effect of gross errors on the final solution. This can be achieved by robust estimation. Least squares adjustment methods yield a correct interpretation of the results if no outliers are present in the solution and all the systematic errors are modelled. Robust estimation methods perform considerably better than least squares when gross errors are present in the mathematical model. One of the most widely used robust estimation methods is the L1-norm. The robust estimation procedure relies on the basic least squares equations with the additional observation weighting.

15.3.1 Code-based and Doppler Positioning

If Doppler observations are available, errors in code-based observations can be mitigated by smoothing code-based observations. This smoothing technique is typically achieved using a so-called Hatch filter. In our experiments, however, this procedure has not produced promising results.

Doppler values can also be used for so-called Doppler positioning. Doppler measurements basically measure the range change between a receiver and a satellite between consecutive epochs. The following equation can be used to calculate the delta range:

$$\Delta D = \tfrac{1}{2}\left(D1_{i-1} + D1_i\right)\cdot \lambda \cdot dt \tag{15.1}$$

where $D1_{i-1}$ and $D1_i$ are Doppler observables in the previous and the current epochs, respectively, λ is the wavelength of GPS L1 frequency and *dt* is the time interval between consecutive epochs. However, this method of Doppler positioning requires reference range values. These can only be achieved using code pseudoranges. Therefore, the initial position can only be determined by code-based observations. This would ideally be determined in an environment with little or no multipath or attenuation. Differential GPS (DGPS) can also be used in such environments if the reference station or VRS (Virtual Reference Station) data is available. DGPS increases the accuracy of the initial positions and consequentially all the following positions if determined relatively by Doppler positioning.

The custom procedure for the positioning of the following epochs (after the initial epoch) determines the receiver clock offset and the position unknowns separately. The receiver clock offset is determined by Doppler observables only. After the receiver clock offset has been determined, the position unknowns are determined using code and Doppler observations. Both types of observation are equally weighted.

One additional feature of the procedure is the recalculation of the ranges after each epoch. The ranges between the receiver and the satellites are calculated using the adjusted current position. These ranges then become input data for the following epoch instead of the pseudoranges.

15.3.2 DGPS and Conditional DGPS

DGPS is a positioning technique whereby the position of the receiver is determined using code pseudorange corrections, calculated by the reference receiver. The achieved position accuracy is advanced to a scale of 1 metre, whereas the accuracy of an autonomous position is around 10 metres (Lachapelle, 2007). DGPS assumes good operational conditions in the proximity of both receivers. DGPS does not function well under adverse conditions, because the noise and multipath are amplified in the differencing process (Lachapelle, 2007).

Since DGPS does not perform well under adverse conditions where signal reception is poor, so-called conditional DGPS (cDGPS) has been developed, in which

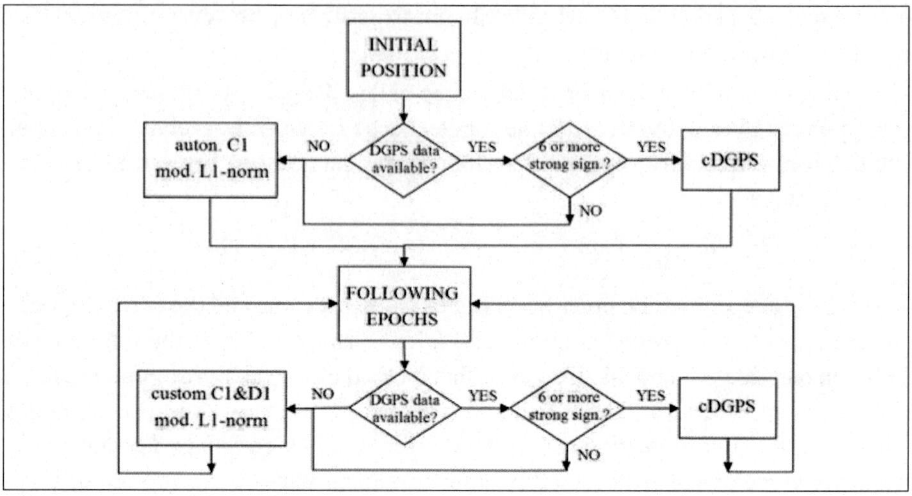

Fig. 15.1. Flowchart of the custom positioning procedures.

only strong signals are used to secure the positioning. Strong signals are defined as those with a SNR value above 35 dBHz, which is usually the limit value for the signal tracking of regular GPS receivers (Wieser, 2006). The cDGPS is performed in each epoch with 6 or more strong signals which are also tracked by the reference receiver. If the cDGPS cannot be performed, the positions are determined by the custom autonomous positioning procedure. Another useful feature of the cDGPS is its ability to calculate residuals for "no cDGPS" epochs. At the epoch the cDGPS is possible again, the position is determined by the cDGPS and by the custom procedure. The difference between both positions is time-related distributed back to "no cDGPS" epochs. This feature has no bearing on real-time surveys. Nevertheless, it is useful for a post-analysis of the calculated positions. *Figure 15.1* displays the basic data flow of the custom developed autonomous and cDGPS positioning procedures.

15.4 Experiments and Results

The experiments were carried out on-site at the Faculty of Civil and Geodetic Engineering in Ljubljana, Slovenia. The tests were performed under different conditions and at different times in static and kinematic modes. All tests were carried out and processed in a 1 Hz kinematic mode. Several autonomous solutions were determined from each survey: C1 (code-based positioning), D1 (Doppler positioning), C1&D1 (common processing of code and Doppler observations), custom C1&D1 (as described in *Section 15.3.1)*, NMEA (positions, determined by the receiver, extracted from NMEA messages). DGPS solutions are labelled DGPS, whereas the conditional DGPS is represented by the label cDGPS.

15.4.1 DGPS and Conditional DGPS

DGPS was assessed under different conditions. As expected, DGPS performs well under normal conditions. The results of a 10-minute static survey are shown in *Figure 15.2.* The DGPS positions (in red colour) are compared to the basic positioning mode using code pseudoranges only (in blue colour). The graphs illustrate discrepancies in the solutions from the reference values – known coordinates of the surveyed point. The discrepancies of DGPS positions are held within a few meters whereas the discrepancies of the basic C1 mode exceed 10 and more meters.

Under adverse conditions, the results are markedly different. *Figure 15.3* illustrates the results of DGPS (red colour) and basic positioning (blue colour) in a challenging environment – the receiver was placed underneath a roof turret. The results of DGPS are similar to or worse than those achieved by basic positioning.

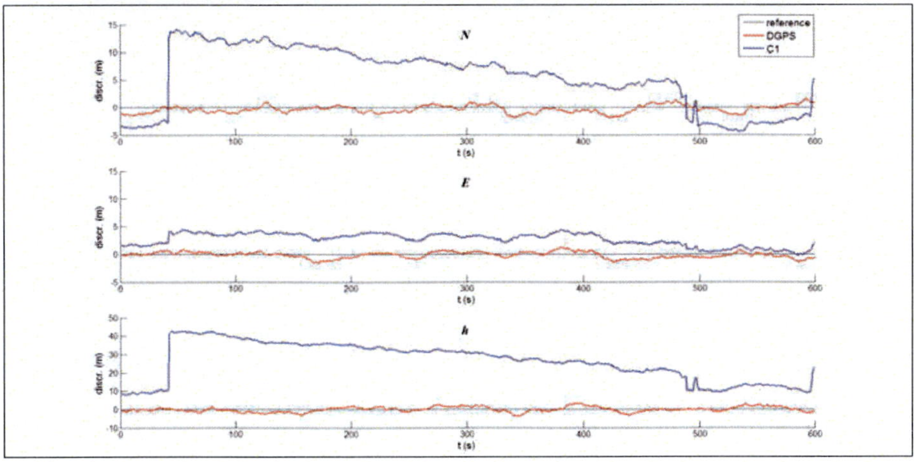

Fig. 15.2. Performance of DGPS under normal conditions.

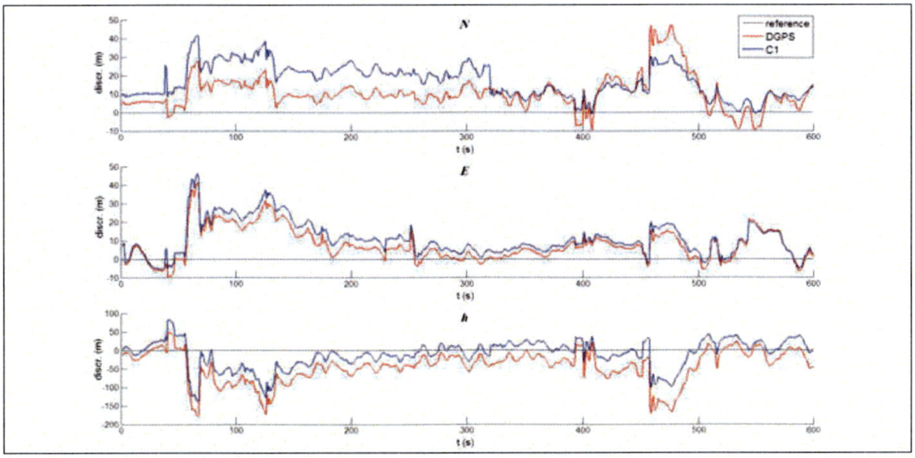

Fig. 15.3. Performance of DGPS under adverse conditions.

15.4.2 Indoor Positioning

All signals are attenuated and multipath becomes a permanent feature inside objects. This causes large errors in positions which were estimated using code-based positioning. As seen in *Figure 15.4,* the discrepancies from the reference position exceed 100 m in a single component. The best solution is also illustrated for comparison. The best solution is the custom procedure using C1 and D1 with the application of the Klobuchar ionosphere model and a troposphere model. Basic statistics (minimum deviation, maximum deviation, mean and standard deviation by coordinate components) of both solutions are listed in *Table 15.1.*

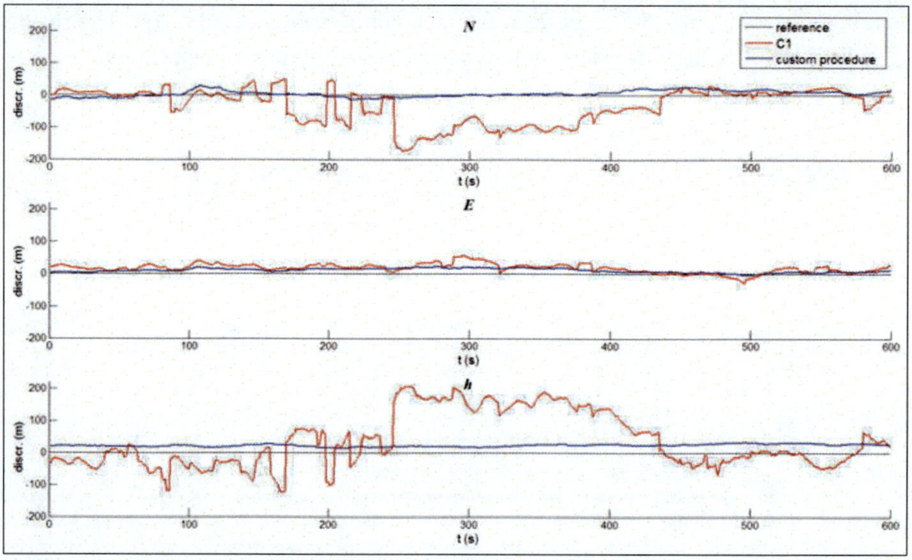

Fig. 15.4. Indoor positioning.

Table 15.1. Statistics of the worst and best indoor solutions. Units are in meters.

	C1			custom		
	N	E	h	N	E	h
min (m)	−175.8	−29.3	−124.0	−14.5	−5.4	12.9
max (m)	49.1	55.5	206.2	29.0	21.4	32.4
mean (m)	−34.0	17.0	36.4	3.5	9.3	22.7
st.dev. (m)	53.2	13.3	84.8	10.5	6.0	4.8

15.4.3 Kinematic Surveys under Adverse Conditions

The positioning procedures and the equipment were both tested in different kinematic surveys and under different conditions. Results from a mixed condition environment will be presented. The start and the end of the survey represent relatively normal conditions, however, in the middle, the survey stops underneath the overhanging roof turret for approximately 30 seconds. The whole survey takes 60 seconds. Some of the results of the developed solutions are shown in *Figure 15.5*. The black line represents the approximate trajectory and the blue outline depicts part of the overhanging roof. The first and the last determined positions are marked with larger dots. The positions are shown in the projection plane. Plotted units are in metres.

Once again, code-only positioning yields major position errors – see plot (i). The results are considerably better using the custom positioning procedure, when either the initial position is known (ii) or the ionosphere and the troposphere models are used (iii). As a comparison, the internal solution is also shown – see plot (iv). Deviations from the trajectory for the solution C1 exceeds 20 meters while deviations of other solutions are held within few meters from the trajectory.

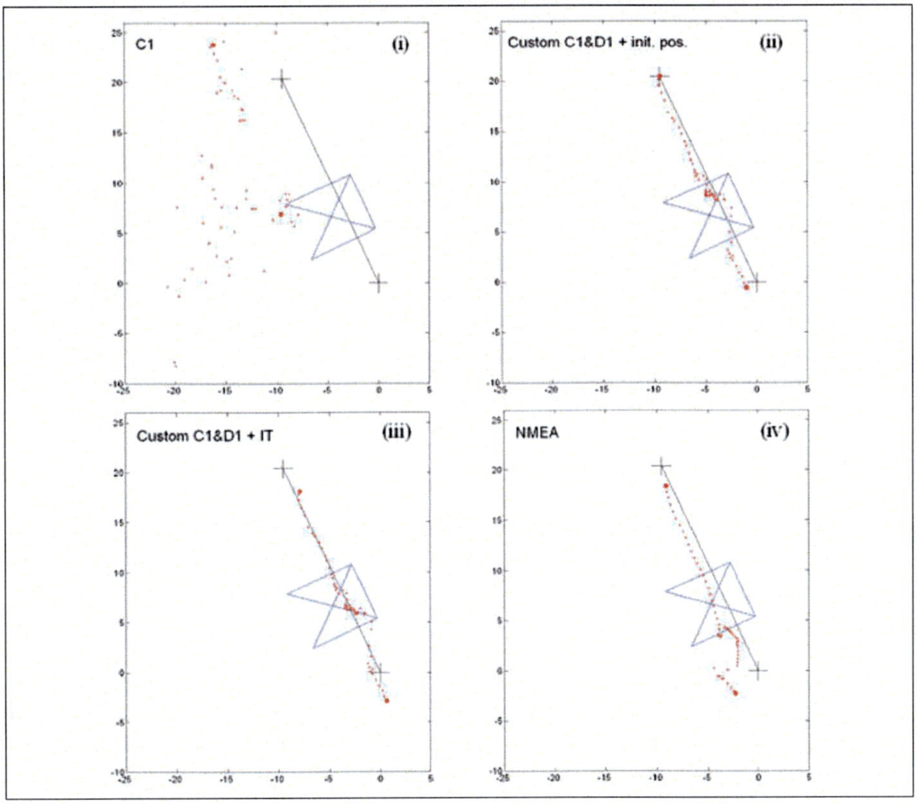

Fig. 15.5. Solutions in a mixed-condition environment: (i) autonomous C1 positioning; (ii) custom positioning procedure with C1&D1, initial position presumed known; (iii) custom positioning procedure with C1&D1, plus ionosphere and troposphere models; (iv) internal positioning of the receiver. Units are in meters.

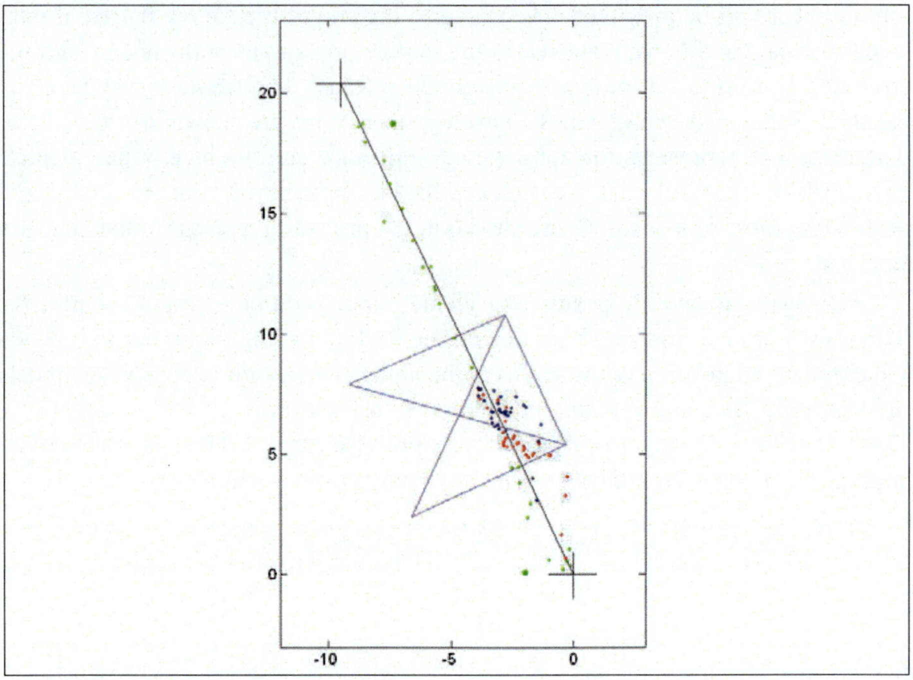

Fig. 15.6. Conditional DGPS in a mixed-condition environment.

The results of the cDGPS of the same survey are shown in *Figure 15.6.* Whenever possible, the cDGPS was performed (green dots).When this was not possible, the custom autonomous positioning was used (red dots).The recalculation of the "no cDGPS" positions was also applied (blue dots).

15.4.4 Analysis of the Results

In general, basic positioning using code pseudoranges only tend to cause major position errors beyond 100 m under adverse conditions. The application of Doppler observations in addition to code-based observations and proper processing procedure can drastically improve the accuracy of the results achieved.

The best overall solution and also the sturdiest positioning technique is the custom procedure using both types of observables, code and Doppler. As can be seen from *Figures 15.4* and *15.5,* the quality of the calculated positions is considerably higher in comparison to the basic positioning. The accuracy of the positions falls within a reasonable value range, i.e. 10–20 metres, even under severe conditions. It should be noted that the positioning mode is still autonomous. The results proved to be better than the results of the internally derived positions from the receiver (see *Figure 15.5).*

The ionosphere and the troposphere both affect the propagation of all signals, irrespective of the conditions near the receiver. Therefore, the ionosphere and the troposphere models should be used in any case to enhance the accuracy of the results.

The conditional DGPS is applicable in a mixed-condition environment of a kinematic survey. If the conditions for the reception of the signals are particularly bad, then the cDGPS cannot be used. Conditional DGPS can be used, however, under conditions with minor attenuation and multipath level. Results of the cDGPS from one case can be seen in *Figure 15.6*. The results are noticeably improved following recalculation of the positions.

15.5 Conclusions

Determining quality positions is important for the quality of LBS. Since the availability of GNSS positioning is limited, other sensors and positioning techniques have to be taken into account. Large drifts in low cost inertial sensors, MEMS, cause large errors over time. Use of MEMS is suitable only for a very short period of time. Other technologies, developed in recent years, such as Wi-Fi and UWB, have had better outlook than actual results. Also, some kind of established infrastructure is required, limiting the possibility to use the systems everywhere.

High sensitivity GPS extends the use of GNSS positioning to environments where signals are attenuated and reflected. HS GPS enables positioning even indoors, although only to a certain degree. GPS positioning inside objects is usually limited to top floors and the proximity of outer walls.

Code-based autonomous positioning can often result in major errors of the estimated position. Errors can grow beyond 100 meters. Therefore, observations under adverse conditions should be treated with caution. The presented custom positioning procedure provides results with relatively good precision. High precision in such conditions cannot be expected, especially since DGPS cannot be used. We can expect the precision within 20 m for horizontal components and within 30 m for the height component, even in very difficult conditions, when DGPS is not available. Considering an expected precision in normal conditions and actually bad conditions for performing observations, the result can be considered as good. The quality of the developed solution is proved by comparing the results to results, determined by the firmware integrated in the receiver. If the data from the reference receiver is available and if DGPS can be used at least at the beginning of a survey, the accuracy of the positions, even of those relatively determined, is much higher. The accuracy of the initial position, if determined by DGPS, is within 2–3 meter level.

References

Lachapelle G (2004) GNSS Indoor Location Technologies. Journal of Global Positioning Systems 3:2–11

Lachapelle G (2007) Pedestrian Navigation with High Sensitivity GPS Receivers and MEMS. Pers Ubiquit Comput 11:481–488

Kozmus Trajkovski K, Sterle O, Stopar B (2010) Sturdy Positioning with High Sensitivity GPS Sensors Under Adverse Conditions. Sensors 10(9):8332–8347

Wieser A (2006) High-sensitivity GNSS: The trade-off between availability and accuracy. 12th FIG Symposium, Baden

Xu G (2007) GPS Theory, In Algorithms and Applications, 2nd ed. Springer-Verlag, Berlin/ Heidelberg

Section IV

General Aspects of LBS

Chapter 16

Privacy Issues in Geospatial Visual Analytics

Gennady Andrienko and Natalia Andrienko

University of Bonn and Fraunhofer Institute IAIS, Germany
{gennady.andrienko, natalia.andrienko}@iais.fraunhofer.de

Abstract

Visual and interactive techniques can pose specific challenges to personal privacy by enabling a human analyst to link data to context, pre-existing knowledge, and additional information obtained from various sources. Unlike in computational analysis, relevant knowledge and information do not have to be represented in a structured form in order to be used effectively by a human. Furthermore, humans can note such kinds of patterns and relationships that are hard to formalize and detect by computational techniques. The privacy issues particularly related to the use of visual and interactive methods are currently studied neither in the areas of visualization and visual analytics nor in the area of data mining and computational analysis. There is a need to fill this gap, which requires concerted inter-disciplinary efforts.

Keywords: Mobility, privacy, visual analytics

16.1 Introduction

Collection and analysis of data about individuals is vital for progress in many areas such as health protection, transportation, security, to name a few. Technologies enabling collection and analysis of various kinds of personal data develop rapidly. A negative side of these developments is the growing threat to the personal privacy. This particularly applies to data containing locations of people. Analysis of such data may conflict with the individual rights to prevent disclosure of the location of one's home, workplace, activities, or trips. A number of geoinformation scien-

G. Gartner, F. Ortag (Eds.), *Advances in Location-Based Services*, Lecture Notes in Geoinformation and Cartography, DOI 10.1007/ 978-3-642-24198-7_16,

tists have been working on the privacy issues associated with the use of geospatial technologies, e.g., Armstrong 2002, Kwan et al. 2004, Armstrong & Ruggles 2005, Duckham et al. 2006, Gutmann & Stern 2007, Cho 2008.

Intensive research on protecting personal privacy in data publishing and analysis is done in the areas of statistics and data mining, which address, among others, the problems of preserving geographical privacy. The recently completed European research project GeoPKDD (Geographic Privacy-aware Knowledge Discovery and Delivery; http://www.geopkdd.eu/) had a particular focus on data about mobility (Giannotti & Pedreschi 2007) and resulted in creation of new methods for anonymization and privacy-preserving analysis of such data. The ongoing European project MODAP (Mobility, Data Mining, and Privacy; http://www.modap.org/) is a coordination action that continues the efforts of GeoPKDD by coordinating and boosting the research activities in the intersection of mobility, data mining, and privacy preservation.

Being involved in the MODAP project, we represent the visual analytics perspective on the problem of preserving personal privacy in analyzing mobility data. In this position statement, we outline our vision of the possible contribution of the geovisualization and geospatial visual analytics to the research on geographical privacy.

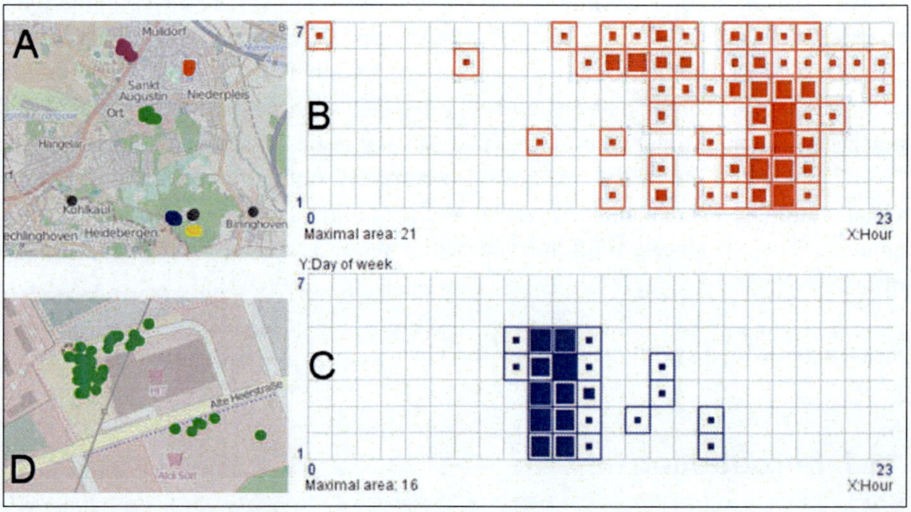

Fig. 16.1. A: Places frequently visited by a person are marked on a map by coloured dots. B, C: Two-dimensional temporal histograms with the rows corresponding to days of a week, from Monday at the bottom to Sunday at the top, and columns to hours of a day show the times when the person appeared in two of the places. The "red" place can be identified as person's home since the person came to it in the evenings of the working days. The "blue" place, at which the person arrived at 9 or 10 AM on the working day, is person's work. D: a closer look at the cluster of green dots on the map reveals a shopping centre frequently visited by the person.

16.2 Visual Analytics against Privacy

Privacy issues have not so far received close attention in the area of visual analytics research (the paper by Weaver and Gahegan 2007 could be mentioned as one of a few exceptions). While privacy is acknowledged as an issue in some papers, many researchers may be still unaware of its particular relevance to visual analytics. The following argument is meant to explain why we deem it relevant and why privacy protection needs to be considered from the visual analytics perspective in addition to the research that is done in other disciplines.

The essence of Visual Analytics is enabling synergistic work of humans and machines in analyzing large and complex data and solving complex, ill-defined problems. In other words, Visual Analytics is about creating such working conditions in which humans and computers can utilize their inherent capabilities in the best possible ways while complementing and amplifying the capabilities of the other side.

Humans have many unique capabilities that are valuable for analysis and problem solving. Among them, two inherent qualities are especially relevant to the topic of privacy protection:

- the capability to flexibly employ previous knowledge and experience, not only those related to special education and to professional activities but also those related to the everyday life and common sense intelligence;
- the capability to establish various associations among pieces of information.

Since these qualities are precious for analysis, Visual Analytics aims at enabling humans to utilize them in the most effective ways. However, the utilization of these capabilities in data analysis has the potential of increasing the threats to the privacy of individuals whose characteristics or activities are reflected in the data. This applies, in particular, to data about mobility.

For example, Andrienko et al. (2007) demonstrate the ease of identifying person's home and work places and other frequently visited places by interpreting spatial and temporal patterns of the person's moves and stops from the positions of the human common sense. The interpretation emerged from considering movement data in spatial and temporal contexts. The spatial context was provided by visualizing the data in a map display. The temporal context (particularly, days of the week and times of the day) was provided by temporal histograms. The example is partly reproduced in *Figure 16.1*.

Researchers on privacy protection in data analysis are typically concerned with the possible threats to privacy arising from computational data processing and from integration of two or more datasets. Little is done on studying the privacy issues arising from the involvement of human analysts empowered with interactive visual tools. Regarding mobility data, it appears necessary to investigate what associations can be established and what inferences can be made by a human, in particular, by considering the data in context.

16.3 Visual Analytics for Privacy

Visual Analytics can contribute to the privacy protection research in two ways. First, visual analytics researchers can identify what kinds of information can be extracted from various forms of mobility data by means of visually supported analysis and consider potential implications to personal privacy. These findings can be communicated to privacy protection researchers for developing methods to remove or decrease the detected privacy threats. Second, to allow humans to deal with large datasets, visual analytics researchers often employ techniques for data generalization and abstraction. Some of the techniques that are devised for the purposes of visualization can be adapted for protecting personal privacy (Maciejewski et al. 2008, Monreale et al. 2010). Both work directions require close inter-disciplinary collaboration. The MODAP project aims at promoting such collaborations. The following research directions are suggested for the inter-disciplinary research community.

16.3.1 Taxonomy of Movement Context

Sensitive personal information may be uncovered by linking movement data to the context, which includes, according to Andrienko et al. (2011):

- geographical space and inherent properties of different locations and parts of the space (e.g. street vs. park);
- time and inherent properties of different time moments and intervals (e.g. day vs. night);
- various objects existing or occurring in the space and time: static spatial objects (having particular constant positions in space), events (having particular positions in time), and moving objects (changing their spatial positions over time).

Human analysts are very flexible in using various kinds of context information available in various forms, e.g. as structured data, as background knowledge, or as texts or images retrieved from the Web. The research question is: What kinds of general and specific knowledge about context can enable unwanted discoveries of personal information from movement data?

Creation of a taxonomy describing various elements of movement context, their relevant properties, and possible relationships to people's activities and movement may form a basis for a systematic investigation of the potential threats to personal privacy arising from linking movement data to context. The taxonomy should include typologies of spatial locations, paths, spatial objects, time moments/intervals, events, etc. with regard to people's activities and movement. For instance, the typology of locations should contain such notions as home place, work place, shopping place, recreation place, business area, and so on. The typology of paths should include notions of high speed road, main street, minor street, footpath, crossing, etc. The taxonomy of context should also include the possible types of relationships that may occur between moving objects and elements of the context (e.g. spatial proximity, temporal proximity).

16.3.2 Taxonomy of Activities

Movement of people is connected to people's activities. There are examples demonstrating that general knowledge of the possible types of activities and their typical places, and/or typical times, and/or typical durations may allow a human analyst to extract personal information from movement data (Andrienko et al. 2007). An analyst may also use specific knowledge about the kinds of activities that are usually conducted in particular places. A taxonomy of activities and their possible relationships to elements of movement context (places, times, objects, events) would allow researchers to go beyond particular examples and derive general understanding of what can be inferred from movement data by involving general or specific knowledge about people's activities in combination with context information.

16.3.3 Taxonomy of Derivable Knowledge/Information

This taxonomy should describe the types of knowledge/information that can be extracted from movement data linked to context and activity information. Potentially sensitive types of information should be identified.

A step in this direction is the taxonomy of movement patterns suggested in (Dodge et al. 2008). This taxonomy, however, is limited to defining possible relationships between movements of two or more objects. With respect to a particular moving object, other moving objects are a part of the movement context. However, other parts of the movement context and activities of moving objects are not considered.

The theoretical work outlined in *Subsections 16.3.1–16.3.3* is useful not only for the research on preserving personal privacy. It may also provide foundations for developing new analysis methods, both in visual analytics and in data mining. In particular, visual analytics researchers may use the taxonomies in the design of the visual interfaces and interactive tools that can effectively support establishing links between movement data and other kinds of information and inferring new information.

16.3.4 Generalization and Abstraction

Methods for data generalization and abstraction are used in visualization and visual analytics for dealing with large amounts of data. A side effect of using these methods is that detailed personal information may be hidden, which is a positive feature from the perspective of preserving personal privacy. Hence, data generalization methods can potentially be adapted to the needs of preserving privacy. This refers, in particular, to methods devised for movement data. An example is presented below.

A method for spatial generalization and aggregation of trajectories (Andrienko & Andrienko 2011) has been devised for obtaining visual summaries of massive movement data. The method transforms trajectories into moves between areas, which are generated automatically based on the spatial distribution of selected points from trajectories *(Figure 16.2)*. The areas can be made larger or smaller, depending on the desired spatial scale and degree of aggregation. These areas can be used not only to aggregate the data but also to generalize the trajectories by replacing their points

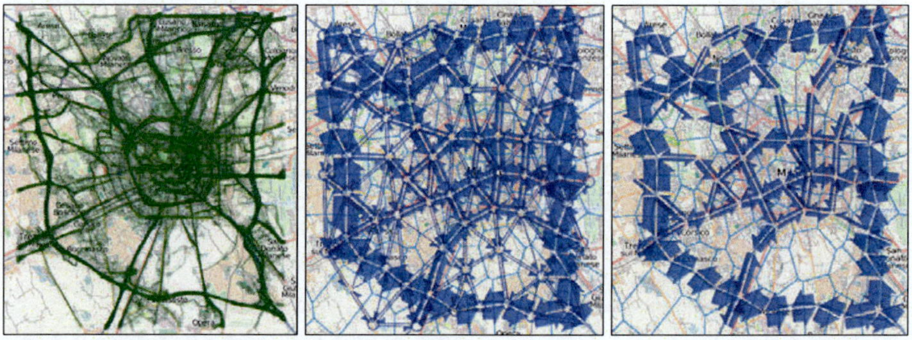

Fig. 16.2. Trajectories of individual cars (left) have been summarized into flows between areas. On the right, minor flows have been hidden for a better visibility.

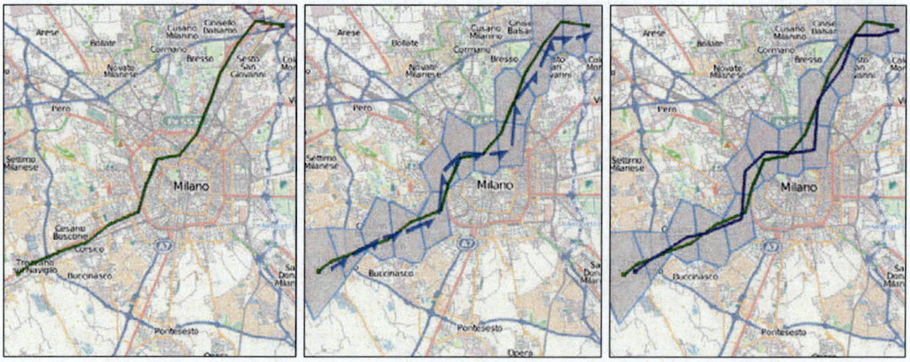

Fig. 16.3. A trajectory (left) is generalized by replacing the original points by the visited areas (centre). On the right, the original trajectory (green) is compared with the modified version (blue).

by areas *(Figure 16.3)*. This reduces the risk of disclosing sensitive private locations of people since the locations cannot be determined precisely any more.

A good feature of this transformation is that the modified trajectories can still be analyzed using various methods. Thus, it was demonstrated that the results of trajectory clustering do not significantly change (Andrienko et al. 2009). Of course, generalization alone does not necessarily guarantee data anonymity and safety. Additional work of privacy specialists was necessary for elaborating this approach into a tool for privacy protection (Monreale et al. 2010).

We suggest that one of the future research directions should be examination of existing and emerging methods for generalization and abstraction of movement data from the perspective of their possible adaptation for privacy protection. Like the other research directions, this direction requires cooperation between specialists in visual analytics, data mining, and privacy protection, as exemplified by the cross-disciplinary team of Monreale et al. (2010).

16.4 Conclusion

The problem of preserving personal privacy in publishing and analyzing data is addressed by researchers in statistics, data mining, and GIScience. However, the work on privacy protection conducted in these areas focuses on the use of computer technologies and does not consider the specific threats to privacy arising from combining the analytical capabilities of computers and humans. Visual analytics is an appropriate research field to address these issues. We suggest some research directions in which visual analytics could contribute to privacy protection.

References

Andrienko, G., Andrienko, N., and Heurich, M. An event-based conceptual model for context-aware movement analysis. International Journal of Geographical Information Science, 2011, in press

Andrienko, G., Andrienko, N., Giannotti, F., Monreale, A., and Pedreschi, D. Movement Data Anonymity through Generalization. SPRINGL 2009 – Proceedings of 2nd SIGSPATIAL ACM GIS 2009 International Workshop on Security and Privacy in GIS and LBS, November 3, 2009, Seattle, WA, USA; http://doi.acm.org/10.1145/1667502.1667510

Andrienko, G., Andrienko, N., and Wrobel, S. Visual Analytics Tools for Analysis of Movement Data, ACM SIGKDD Explorations, 9 (2), 2007, pp.38–46

Andrienko, N., and Andrienko, G. Spatial generalization and aggregation of massive movement data. IEEE Transactions on Visualization and Computer Graphics, 2011, 17(2), pp.205–219

Armstrong, M.P. Geographic Information Technologies and Their Potentially Erosive Effects on Personal Privacy, Studies in the Social Sciences, 27(1), 2002, pp. 19–28

Armstrong, M.P., and Ruggles, A.J. Geographic Information Technologies and Personal Privacy, Cartographica, 40(4), 2005, pp. 63–73

Cho, G. GIS, Personal Privacy and the Law. Wilson, J.P., and Fotheringham, A.S. (eds). Handbook of Geographic Information Science, Oxford: Blackwell Publishing, 2008, pp. 519–539.

Dodge, S., Weibel, R., and Lautenschütz, A.-K. Towards a taxonomy of movement patterns, Information Visualization, 7(3–4), 2008, pp. 240–252.

Duckham, M., Kulik, L., and Birtley, A. A spatiotemporal model of obfuscation strategies and counter strategies for location privacy. In Raubal, M., Miller, H., Frank, A., and Goodchild, M. (eds), Lecture Notes in Computer Science 4197, Springer, 2006, pp. 47–64.

Giannotti, F., and Pedreschi, D., eds. Mobility, Data Mining and Privacy – Geographic Knowledge discovery. Springer, Berlin, 2007.

Gutmann, M.P., and Stern, P.C., eds. Putting People on the Map: Protecting Confidentiality with Linked Social-Spatial Data, Panel on Confidentiality Issues Arising from the Integration of Remotely Sensed and Self-Identifying Data, National Research Council, National Academies Press, Washington, DC, USA, 2007

Kwan, M.-P., Casas, I., and Schmitz, B.C. Protection of Geoprivacy and Accuracy of Spatial Information: How Effective Are Geographical Masks?, Cartographica, 39(2), 2004, pp. 15–28

Maciejewski, R., Rudolph, S., Hafen, R., Abusalah, A., Yakout, M., Ouzzani, M., Cleveland, W.S., Grannis, S.J., Wade, M. and Ebert, D.S. Understanding Syndromic Hotspots – A Visual Analytics Approach. IEEE Symposium on Visual Analytics Science and Technology VAST 2008, Columbus, OH, pp. 35–42.

Monreale, A., Andrienko, G., Andrienko, N., Giannotti, F., Pedreschi, D., Rinzivillo, S., and Wrobel, S. Movement Data Anonymity through Generalization, Transactions on Data Privacy, 3(3), 2010, pp. 91–121

Weaver, S.D., and Gahegan, M. Constructing, visualizing, and analyzing a digital footprint. Geographical Review 97, 2007, pp. 324–350.

Chapter 17

Theoretical and Methodological Framework for Measuring Physical Co-Presence with Mobile Positioning Databases

Erki Saluveer[1], Siiri Silm[2] and Rein Ahas[3]

[1] Positium LBS, Tartu, Estonia
erki.saluveer@positium.ee
[2] Department of Geography, University of Tartu, Estonia
{siiri.silm, rein.ahas}@ut.ee

Abstract

The objective of current paper is to propose methodological steps for determining the physical co-presence of people using spatio-temporal movement data. Co-presence is the presence of people at the same place at the same time. In spite of the fast developing information and communication technologies the importance of face to face meetings remains high. We assess the probability of people to be at the same place at the same time using passive mobile positioning data with 350 000 randomly selected respondents from Estonia during 10 days in April 2011. We test criteria for finding the best spatial and temporal resolution to discover co-presence from mobile positioning database. Results show that the most suitable space unit for finding co-presence is site of mobile network and time unit is one minute.

Keywords: co-presence, face to face meeting, mobile positioning, population geography, ICT, travel behaviour

17.1 Introduction

The aim of this study is to propose a methodological steps for determining the physical co-presence of people with the help of vast databases of spatio-temporal movement. We assess the likelihood of people to be at the same place at the same

G. Gartner, F. Ortag (Eds.), *Advances in Location-Based Services*, Lecture Notes in Geoinformation and Cartography, DOI 10.1007/ 978-3-642-24198-7_17, © Springer-Verlag Berlin Heidelberg 2012

time as a co-presence – a potential interaction. We propose the best spatial and temporal resolution to discover co-presence from mobile positioning databases. For empirical study we use the passive mobile positioning data of 350 000 people in Estonia within 10 days of the April 2011 as an example.

Physical co-presence is the presence of people at the same place at the same time. Physical co-presence does not necessarily mean direct contact, i.e. meeting face to face and communication. But similar spatial behaviour is a precondition for communication. Its importance in the modern information and communication technology (ICT) based world has been stressed by many (Urry et al.2007). In spite of the existence of telephones, e-mail and the Internet, important decisions are made and important communications take place being together with partners. Trust and the meaningfulness of decisions depend on it. Development of ICT itself has however changed the forms of co-presence. The importance of virtual co-presence keeps growing; many important ways of co-presence have been solved through ICT (Miller 2005).

The subject of this article is relevant, because the growth of spatial mobility of society has increased the need to assess the existence of interaction between people or its likelihood. The idea which spread from the "Global Village" idea of a Canadian communication scientist McLuhan (1964) and its developments in the 1990s (Castells 1996), that the spread of ICT decreases the need for face to face meetings, has been refuted by now. People need physical meetings with each other; it is needed in personal life as well as in occupational communication. The methodology developed in this article is one possibility for determining co-presence.

The authors thank all contributors from the Geography Department of the University of Tartu and from Positium LBS, we also thank EMT, the largest mobile operator in Estonia, for the possibility to use the data. The work was carried through with the support of grant no. 7562 of the Estonian Science Foundation and targeted financing grant no. SF0180052s07 and with the support of the Estonian Information Technology Foundation.

17.2 The Theory

17.2.1 The Meanings of Co-Presence

Simply put, co-presence means being at the same place at the same time. The important things are seeing other people and feeling their nearness and the possibility for interaction with them (Goffman 1966, Urry 2003, Zhao 2003, Lawrence at al. 2006). Goffman (1966), one of the first to discuss co-presence, referred to co-presence as the condition when people "sense that they are close enough to be perceived in whatever they are doing, including their experiencing of others, and close enough to be perceived in this sensing of being perceived" (Goffman 1966).

Physical presence refers to the sense of being physically located in mediated space. Social presence refers to the feeling of co-presence, of social interaction with a virtual or remotely located communication partner. At the intersection of these two categories, we can then identify co-presence or a sense of being together in a shared space, combining significant characteristics of both physical and social presence (IJssekstejn & Riva 2003).

Co-presence has often also been called corporal co-presence by the corporal being together of people (Urry et al. 2007), the co-presence of close people is often spoken about in this sense, including family relationships and more intimate communication. Another concept that has formed is a face-to-face (F2F) meeting, which is a form of corporal co-presence, but applies primarily to communication in the form of meetings, negotiations and discussions; it differs semantically from communication between close people.

With the spread of ICT, the meaning of co-presence has also begun to change. While Goffman in the 1960s (Goffman 1966) spoke about physical co-presence as co-presence, then today co-presence is more and more spoken about in the context of virtual environments (Miller 2005, Urry 2002, Yu and Shaw 2007, IJsselsteijn et al. 2001). Physical presence refers to the sense of being physically located in mediated space, whereas virtual presence refers to the feeling of being together, of social interaction with a virtual or remotely located communication partner (Ijssekstejn & Riva 2003).

The formation of co-presence is influenced by several obligations of people (Urry 2002), that make people move physically or act in a virtual environment. Urry (2002) has specified 6 types of obligations, which cause the formation of physical co-presence. 1) Legal, economic and familial obligations – to have to go to work, to have to attend a family event (wedding, funeral, Christmas etc.), to have to visit a public institution (school, hospital etc.). 2) Social obligations – to see specific people face-to-face, to note their body language, to hear what they say, to meet their demands, so sense people directly, to develop extended relations of trust with others. 3) Time obligations – to spend moments of quality time with family or partner or lover or friends. 4) Place obligations – to sense a place or kind of place directly, such as walking in a city, visiting a specific building, climbing a mountain. 5) Live obligations – to experience a particular "live" and not a "mediated" event (political event, concert, theatre, match, celebration, film). 6) Object obligations – to sign contracts or to work on or to see various objects, technologies or texts that have a specific physical location. The same types of obligations can also be transferred to virtual environments.

As people go about their lives, their paths intersect through space and time to others paths, sometimes the intersects are with the paths of friends and sometimes with the paths of strangers (Crandall et al. 2010).

As a result of co-presence of people – both physical and virtual – the so-called co-presence communities are formed. These are groups of individuals who regularly share the same location at the same time i.e. they are repeatedly collocated together. There are weak relationships between the members of co-presence communities (Lawrence at al. 2006), they have a strong spatio-temporal component, but they may not be interacting with each other. Members of co-presence communities have also been described with the concept Familiar Strangers (Milgram 1977). Examples of co-presence communities are frequent commuters in a train station in a physical environment (Milgram 1977) and the visitors of online casinos at work nights in a virtual environment. Co-presence communities last for different periods of time and are of different stability (Lawrence at al. 2006). The members of a co-presence community share a certain similar interest.

17.2.2 Co-Presence and Communication

ICT enables people to communicate with each other a lot more flexibly than just face-to-face communications. Janelle (1995) classified communication modes based on their spatial and temporal constraints. Spatial constraints require either physical presence or telepresence, whereas temporal constraints require either synchronous or asynchronous communication *(Table 17.1)*. Synchronous Presence (SP), i.e. co-presence, requires co-location in space and coincidence in time, it corresponds to face-to-face interaction. Asynchronous Presence (AP) requires co-location only in space, but not coincidence in time, examples are notes left on a refrigerator or hospital charts. Synchronous Telepresence (ST) requires coincidence only in time, but not co-location in space, examples are telephones, instant messaging, radio etc. Asynchronous Telepresence (AT) does not require co-location in space or coincidence in time, examples are e-mail, printed media, webpages etc (Miller 2005).

Table 17.1. Communication modes based on their spatial and temporal constraints (Janelle 1995, Miller 2005).

Temporal/Spatial	Physical presence	Telepresence
Synchronous	Synchronous presence (SP) Face to face	Synchronous telepresence (ST) Telephone Instant messaging Television Radio Teleconferencing
Asynchronous	Asynchronous presence (AP) Refrigerator notes, hospital charts	Asynchronous telepresence (AT) Mail e-mail fax machines printed media webpages

Even though communication through the means of ICT (telephone, skype, e-mail, etc.) keeps increasing, face-to-face conversations are still in a very important place in communication between people. Social life requires moments of physical proximity (Urry 2002), even if most of the communication takes place "at a-distance", whether for business, leisure, family life, politics, pleasure or friendship (Urry 2003). In addition to maintaining social relationships, face-to-face conversations are also required to exchange very complex information (Allen 1997). Face-to-face conversations involves rich, multi-layered and dense conversations, which involves not just words, but indexical expressions, facial gestures, body language, status, voice intonation, past histories, anticipated conversations and actions, turn-taking practices and so on (Urry 2002). The eye contact achieved with face-to-face conversations enables and stabilizes intimacy and trust, as well as the perception of insincerity and fear (Urry 2002, 2003).

17.2.3 Co-Presence and Time Geography

Discussing the activities of people in space and time – which co-presence clearly also is – gets its theoretical framework from time geography, which began with the works of Hägerstrand in 1970s (Hägerstrand 1970). Hägerstrand's (1970) time geography focuses on the constraints on human activities in space and time. Time geography recognizes that human activities have spatial and temporal dimensions: activities occur at particular places for limited durations. Constraints on activity participation include the location and timing of some events such as home and work, the residual time budget outside these some events and resources available for trading time for space in physical movement or virtual interaction (Hägerstrand 1970, Miller 2007).

Two central concepts in time geography are the space-time path and prism. The space–time path traces the movement of an individual in space and time. The space–time prism is an extension of the path: this measures the ability to reach (be coincident with) locations in space and time given the location and durations of fixed activities (Miller 2005).

Finding co-presence in time geography is based on the space-time path. Two people get to interact – they have co-presence – if their space-time paths intersect each other (Miller 2005). Intersection is the condition of two or more time geographic features sharing some locations in space and time. An intersection may require coincidence of the objects for a minimum threshold time (Miller 2005).

In order to distinguish different types of co-presences (home, work, shopping, etc.), we have to proceed from the approach of space-time stations. A station is a location where paths can bundle for some activity. This usually corresponds to a designated location such as a retail outlet, office, or home (Miller 2005).

The time geography approach has been extensively written about theoretically, but is has not been used much as a framework in actual studies. Miller 2005 has developed the measurement theory to develop the time geography approach for both physical and virtual environments, developing rigorous definitions of fundamental time geographic concepts and relationships such as the space–time path, prism, stations, composite path-prisms, bundles, and intersections (Miller 2005), which could be used to analyze the data achieved with the methods location-aware technologies (LAT) and location-based services (LBS). The modern approaches proceeding from time geography have also been discussed in giving meaning to special virtual environments (Couclelis 2009) as well as for example transport and modern mobility (Delafontainea et al. 2011). Couclelis (2009) argues that because ICT the traditionally close links between activity, place, and time loose connection and "may need to be reexamined in light of the new realities".

Conceptually it is very important for this study that the flows of movements are very diverse today and co-presence may be found in very different forms, spatial and temporal proximity are not always the precondition of co-presence. The methodology developed in this study however is attempting to assess the "old-fashioned" corporal (physical) co-presence by using virtual information and communication technology-based manner of thought. At the same time, this physical co-presence may mean little interaction between those present, they may not even be aware of each other's nearness. On the other hand, people located far apart may however be in very close contact with one another through the telephone or computer. Being present at the same place at the same time in the space-time prism does however indicate certain closeness, similarity, potential for interaction. When the space-time paths are very close together, the spatial behaviour is also similar. The possible existence of spatial twins, who we are or are not aware of, has been pointed out as an example.

17.3 Methodology for Assessing Co-Presence

17.3.1 Earlier Methodologies for Assessing Co-Presence

The importance and nature of co-presence have been written about a lot theoretically, but there have been few researches in this area which have achieved practical results. Therefore the subjects of methodology for finding co-presence have also not been developed and discussed. One of the subjects that have not been discussed is the scale of space and time in measuring co-presence. Co-presence is being at the same place at the same time, but what exactly does the same place and the same time mean, i.e. what is the extent of the units of space and time, has not been specified. The following paragraphs provide an overview of on the basis of which data and to which spatio-temporal extent co-presence has been researched *(Table 17.2).*

Co-presence has been found with the help of Bluetooth tracking data (Kostakos et al. 2010, Lawrence et al. 2006), mutual call operations databases (Calabrese et al. 2011) and Flickr photos and data of social networks (Crandall et al. 2010).

Different spatio-temporal frameworks have been used in assessing co-presences with different methodologies. The spatio-temporal framework of co-presence should be set so that co-present individuals are able to sense each other directly (Goffman 1966) and it is possible for the participants to interact.

Spatial scale should be such that people would be close enough to perceive each other's existence and would be able to communicate if they wish. In reality however, depending on the used sources of data, spatial dimension has been a lot more extensive in researches. The range of determining co-presence has been the smallest in the researches based on Bluetooth, approximately 20–25m (Kostakos et al. 2010, Lawrence et al. 2006). The largest area however, where co-presence has been assessed, was a square of the size of 1° longitude and 1° latitude, which is 80x80 km on the average (Crandall et al. 2010).

Temporal scale of co-presence should be such, where people are actually together in the same spatial unit at the same time. It can be achieved on the basis of Bluetooth studies by finding the period when two Bluetooth devices were together, the accuracy of which depends on the measuring accuracy of the scanner (second, minute) (Kostakos et al. 2010, Lawrence et al. 2006). Temporal scale of one year has been used in the case of proceeding from mobile phone call operations data (Calabrese et al. 2011).

Table 17.2. Comparison of the temporal-spatial frameworks of the methodologies for assessing co-presence.

Author	Temporal scale	Spatial scale	Source of data
Calabrese et al. 2011	1 hour, 1 year	Cell tower	Anonymized call detail records
Kostakos et al. 2010	Time spent in the area of the Bluetooth scanner	Area of Bluetooth scanner	Bluetooth
Lawrence et al. 2006	Time spent in the area of the Bluetooth scanner	Bluetooth area (20–25m)	Bluetooth
Crandall et al. 2010	1 day	1° longitude and 1° latitude (the average of 80 km)	Flickr photos, social network

Co-presence can be assessed using place-based or people-based approaches (Miller 2007). Place-based approach counts all people who were located in a certain area at a certain moment in time. People-based approach however counts proceeding from each person who they were with in a certain area at a certain moment in time. The same procedure is carried though for each person who was in the area at the specified time.

All co-presence studies presented here brought as an example used people-based approach. Co-presence has been assessed through the co-presence between pairs (Calabrese et al. 2011, Crandall et al. 2010, Kostakos et al. 2010, Lawrence et al. 2006). We have used the same methodology to carry through a research on ethnical segregation in Estonia using the data of passive mobile positioning (Silm et al. 2011).

17.3.2 Our Conceptual Framework and Methodology

The aim of this article is to find appropriate spatial and temporal scales for determining co-presence, i.e. the size of the measuring "window". Solving this problem is important, because the spatial and temporal variety among co-presences is very wide. Thus choosing a "window" of appropriate size is very important for research design. It is also very important from the point of view of the amount of data, there are namely not enough observations made at the same place at the same time to enable to perform various analyses. Choosing the right scale enables to act more purposefully and shape appropriate research strategies.

Based on the communication modes specified above, the precondition of our study, which is based on the data of passive mobile positioning, is either Synchronous Telepresence or Asynchronous Telepresence, i.e. people must have made call operations the location (space) and time (time) of which have been stored in the database. It is Synchronous Telepresence when the person has made a call and Asynchronous Telepresence when the person has sent an message. The object of the research is Synchronous Presence, i.e. we are studying the co-presence of people – being at the same place at the same time – with other people at the moment the call operations were made. We have no information about the location of the person receiving the call or the message; it is not necessarily the same place. We do however know that the people who used their phones were at the same place at the same time.

The conceptual framework is based on Hägerstrand's space-time concept of time geography. Proceeding from the concept of time geography, locations of call activities are points on the space-time path. Locations of call activities do not enable to draw the whole space-time path of a person, but provide points on it in certain places and at certain times, where we can examine the appearance of co-presence. The lengths of time paths recorded by us enable to map routine moving trajectories quite well.

Passive mobile positioning data does not enable to research co-presence by the definition most mentioned in theoretical literature – people are at the same place at the same time and need to be able to directly sense each other (Goffman 1966). The minimum spatial unit in the case of using a passive mobile positioning database is a cell tower, the temporal accuracy however is sufficient, 1 second. However, many other researches (Calabrese et al. 2011, Crandall et al. 2010, Kostakos et al. 2010, Lawrence et al. 2006) finding co-presence have also used more extensive spatial areas, which do not enable people to sense each other's existence directly.

We are using space-based approach to assess the spatio-temporal scale of co-presence. In the case of space-based approach we count the number of people in the space-time window. Passive mobile positioning is good data for researching co-presence, because it enables to find co-presence without performing separate inquiries or installing special devices (such as Bluetooth scanners). Researchers have shown that people meeting each other physically does not mean that they are not using mobile means of communication (Ito 2003).

17.4 Data and Methods

17.4.1 Data

Mobile operators' core networks have to know the locations of mobile phones, so they could deliver calls and messages fast and efficiently through the antennae with the best signal strength to end-user. This location information is frequently stored in the databases of the mobile operator for monitoring the network. In the case of call detail records (CDR), the data is stored and the history of a network subscriber is maintained. Passive mobile positioning is a method to analyze the spatio-temporal data, which is stored in the memory files and logs of mobile operators (Ahas et al. 2008).

The area of Estonia, the area involved in the research, is 45 215 km^2, the population is 1.39 million people, the population is located unevenly with 550 000 people gathered in the region of the capital, Tallinn, in the north of the country. Mobile network covers almost the whole territory, with the sparsely populated areas of woods and wetlands covered with a less dense network of base stations. There are 3 mobile operators in Estonia and the passive mobile positioning data in this study is from the CDR data of EMT, the largest mobile operator in Estonia. EMT has a market share of 47% and according to the European Commission Report, the penetration rate of mobile phones in Estonia is 116.1 percent (European Commission Report 2010). 95% of Estonians are using mobile telephones as their everyday communication channel.

In the current study, the CDR data is from the GSM and WCDMA network technologies of Ericsson's platform for telecom operators. The data is anonymized and handed for processing to a mobile positioning solution provider Positium LBS using software of Positium Data Mediator.

Each CDR record represents the call activity of a mobile subscriber, who is for example calling or sending instant/multimedia messages. The spatial resolution is a network cell (Cell ID) and temporal resolution is a second. Exact dimensions of network cells depend on the density of cells. The density of cells is higher in places, where there is more phone use, such as urban areas, and so the spatial accuracy of positioning is much more precise in urban areas (~100m – 1km) than in rural areas (~1km – 30km). Each subscriber of the mobile network has been assigned an anonymous ID number, which is consistent over time. In the case of some subscribers, there is also information about their gender and age available.

In this study we used a randomly generated sample of 350 000 EMT network subscribers. The CDR data was taken from a period of 10 days from April 11, 2011 to April 20, 2011, and the call operations could have been performed in any place in Estonia. The total number of positionings within this period of 10 days was around 13 million data records. The average number of calls per user per day was 5.25 and it was slightly higher for men (5.63) and for the age group 20–39 *(Figure 17.1)*.

17.4.2 Methods

There were five spatial resolutions and seven time resolutions chosen for measuring different space and time windows for the co-presence of people. Simple methods were used to determine the possibilities for the existence of co-presence, the number of call activities in different spatial and temporal divisions and the number of people at the same place at the same time were determined. The time resolution was divided into the intervals of 1 second, 1 minute, 1 hour, 3 hours, 6 hours, 12 hours and 1 day. The spatial resolution was divided as follows *(Figure 17.2)*:

- cell – network antennae coverage area, the number of cells in Estonia is 4632, the average size of a cells is 380 people and 10 km^2;
- site – network base station coverage area, there are usually 3 or 4 cell's per one site, the number of sites in Estonia is 1013, the average size of a sites is 1380 people and 40 km^2;
- local settlement unit – the smallest settlement unit in Estonia, which in addition to the densely populated areas called villages includes the uninhabited areas surrounding villages, the number of settlement units in Estonia is 4687, the average size of a settlement unit is 2480 people and 9 km^2;
- municipality – the smallest administrative unit in Estonia, this includes both towns and less populated rural areas, the number of municipalities is 226; the average size of a municipality is 6140 people and 190 km^2;
- county – local administrative centre in Estonia, the number of counties is 15, the centre is a town, the average size of a county is 89 350 people and 2900 km^2.

The total of 35 different space and time windows were used and for each of the windows co-presence was calculated based on the data of 350 000 subscribers within 10 days. We used the place-based calculation method to calculate co-presences, which means that the minimum number of co-presences was two. If there are three people in four different space and time windows with 2 people together in one window and one in the other 3 windows, then the co-presences are also 2 and

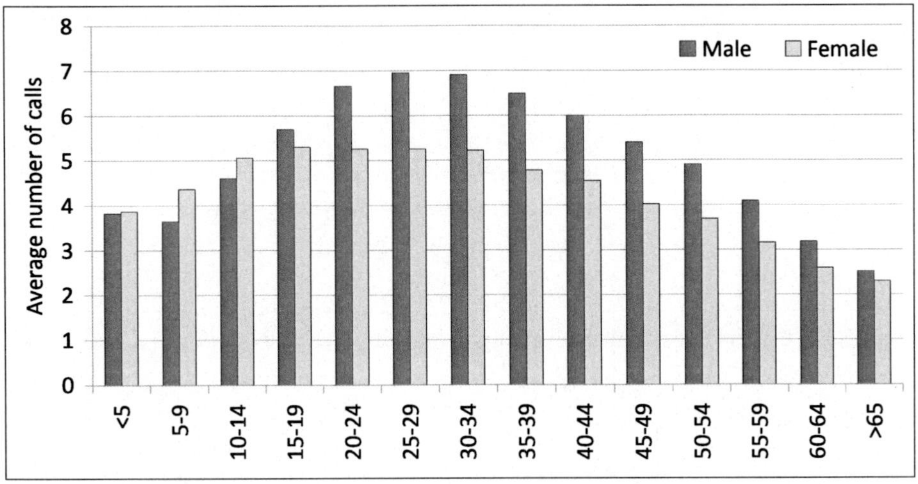

Fig. 17.1. Average number of calls per user per day by age groups.

Fig. 17.2. Example of distribution network cells and sites in city of Tartu (100 000 inh.).

3. If these four windows were aggregated into one window, we would count it as 3 co-presences as there are three different individuals *(Figure 17.3)*.

The spatial measuring unit of cells and sites used in the case of the spatial units used in the analysis was a geometrical area unit found with Voronoi tessellation. The actual administrative borders were used in the case of local settlement units, municipalities and counties.

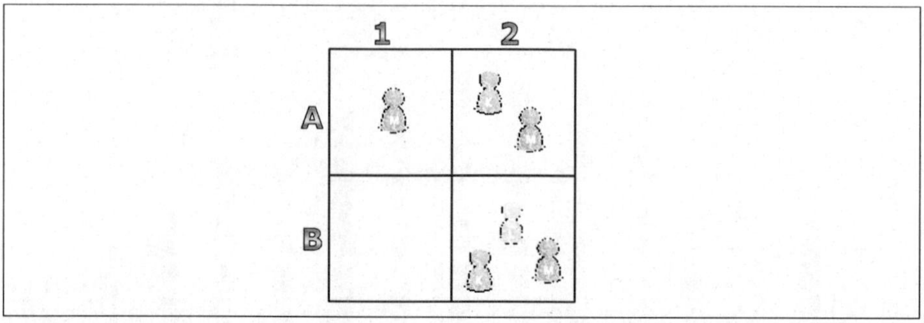

Fig. 17.3. Example of calculation of co-presence for three individuals (K, L, M) in four space and time windows.

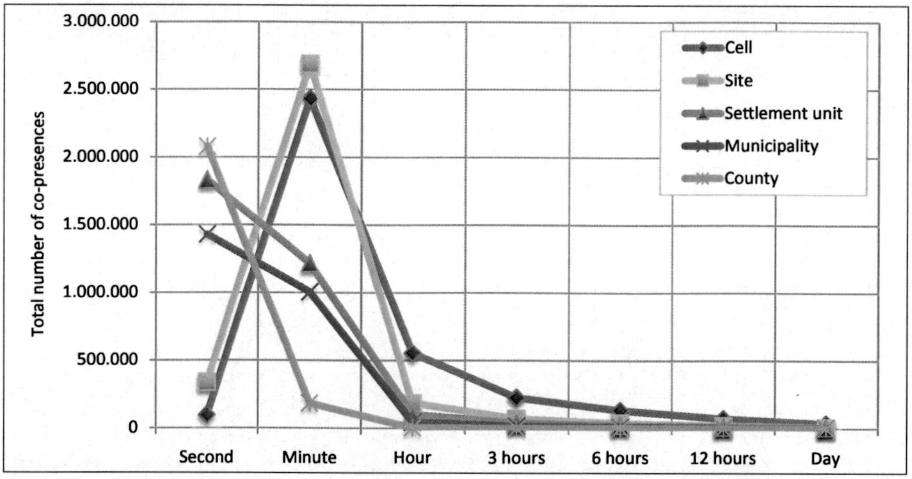

Fig. 17.4. Total number of co-presences in a specific space and time window.

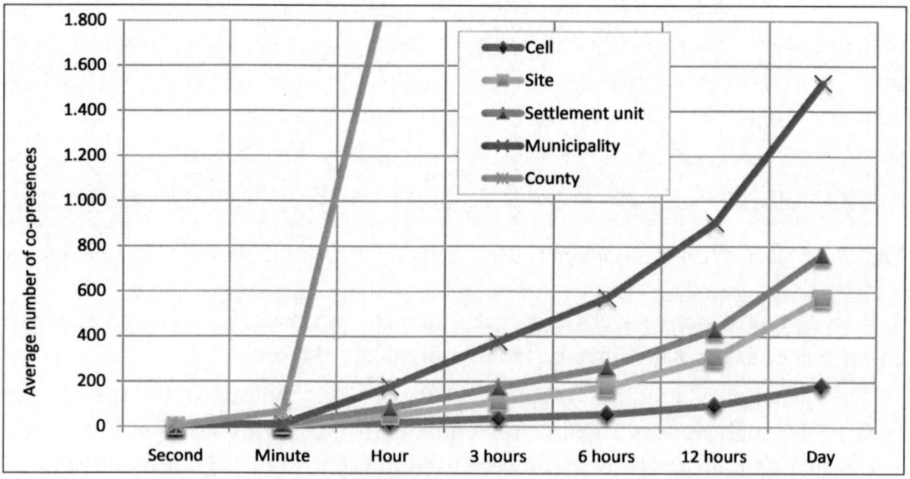

Fig. 17.5. Average number of co-presences in a specific space and time window.

17.5 Results

17.5.1 The Frequency of Cases of Co-Presence

The results show that the amount of spatial co-presences depends to a significant extent on the chosen spatial and time scale. In the case of the total number of 2.5 co-presences for 350 000 analyzed persons, the total number of co-presences is the highest for one minute temporally and for a cell and a site spatially *(Figure 17.4)*. In a longer period of time than a minute and a site as spatial unit, the amount of possible co-presences decreases steadily. The reason is the decrease in the number of possible space and time units. In the case of the smallest time unit however, a second, the frequency of performing call activities together in the smallest space units (cell and site) is also lower.

In the case of content analyses performed on the basis of co-presence however we are much more interested in the average number of co-presences per one space and time window. The number of co-presences in a space and time window is influenced by the size of the space and time window, i.e. the larger the spatial area and the longer the period of time of determining co-presences, the more people happen to be together (perform call operations) in the same space and time unit *(Figure 17.5)*.

The number of co-presences is the highest in the case of the largest space and time window – county-day – the average of 16 160, which forms 4.6% of all people involved in the research. In the case of a county, the number of co-presences exceeds the result of the next level (municipality) of spatial unit even in the case of a 1-hour time step, with the average of 2032 people per county and hour and the average of 1524 people per municipality and day respectively. In the case of smaller space and time windows than county-hour and municipality-day, the average number of people in a space and time window remains below 1000 people *(Table 17.3)*. In addition to those mentioned above, there are more than 500 people in the same space and time window in the case of municipality-6 hours and municipality-12 hours and in the case of the time step of 1 day, settlement unit-day and site-day. There have been less than 10 people together in the same space and time window on all spatial levels in the case of the time step of one second, and on the smaller spatial levels (cell, site, settlement unit) in the case of one minute.

Table 17.3. Numbers of co-presences per space and time units.

Space	Secund	Minute	Hour	3 hours	6 hours	12 hours	Day
Cell	2	3	17	37	57	95	181
Site	2	4	49	111	178	306	573
Settlement unit	3	9	83	176	264	429	761
Municipality	5	12	177	376	573	904	1524
County	5	67	2032	4498	6833	10307	16160

Choosing a space and time window, a version should be preferred where the number of co-present people is higher. A larger area and a longer period of time however lead to a situation where the two people may not actually meet.

The selection of the size of the space and time window of co-presence is significantly influenced by the aim of the study. If the aim is to analyze co-presence on the basis of certain attributes (gender, nationality, age), it must be considered that the number of co-presences per each attribute category must be sufficiently high.

17.5.2 Dependence on the Hierarchy of Settlement System

On the level of municipalities, the number of co-presences depends on the level in the hierarchy of settlement system; there are more co-presences in the centres of higher levels than in smaller centres and in the commuting area of centres *(Figure 17.6)*. The capital city Tallinn, where there is the average of more than 11 000 people in the space and time window municipality-hour, holds the first place with respect to the number of co-present people, it is followed by regional towns with a bit more than 1000 people and county seats with a bit less than 300 people. In the commuting area of centres, including in the near commuting area of the capital city, the number of co-present people per hour remains below 250 people. The number of co-present people is the lowest in rural municipalities, which are not the commuting area of any centres (44), and in the far commuting area of county seats and regional towns, 48 and 49 people respectively.

The dependence of co-presences on the level in the hierarchy of settlement system results from the good correlation between co-presences and the number of residents of municipalities. Division of the number of co-present people by municipalities reveals a similar structure of settlements as the number of residents; the larger and smaller centres and their commuting areas correspond to the size of the centre. The largest centre, the capital city Tallinn, forms 33.3% of the number of co-presences and, according to Statistics Estonia, 30.7% of the number of residents, approximately 11 000 co-presences and 400 000 residents respectively *(Figure 17.7)*. The share of the second largest town, Tartu, amounts to 7.3% with respect to the number of residents and 9.1% with respect to co-present people. The share of the remaining regional towns is 2–5% according to both databases. With respect to the share of municipalities, the numbers of co-present people in the primary city Tallinn and the second largest centre Tartu dominate; according to Statistics Estonia, the centres of the Ida-Viru County (Narva, Kohtla-Järve, Sillamäe) however dominate in the numbers of residents.

The average number of co-presences on the level of a site does not follow the level in the hierarchy of settlement system as does the level of municipalities. The number of co-presences per site is influenced by the density of mobile network. The average number of co-present people per site is lower in the areas, where there are more sites.

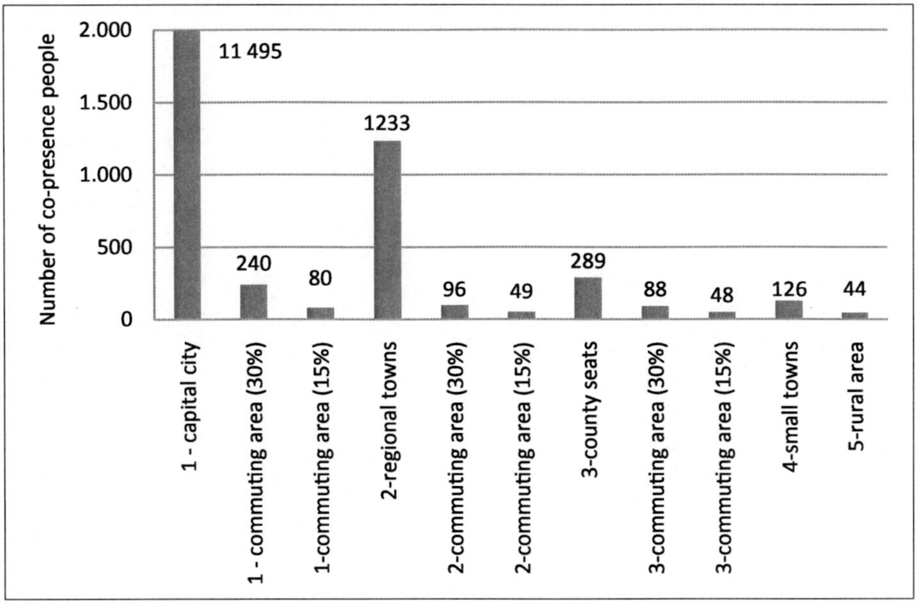

Fig. 17.6. Average number of co-presences in a space and time window municipality-hour by levels in the hierarchy of settlement system.

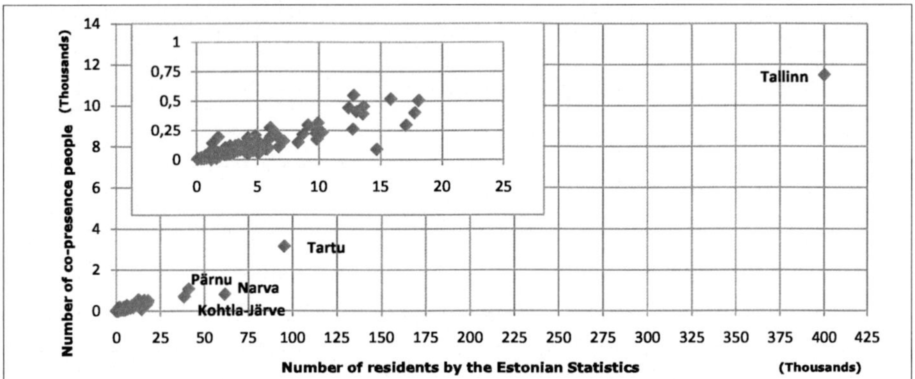

Fig. 17.7. Correlation between the number of co-present people modelled home anchor points and the number of residents according to Statistics Estonia in municipalities in Estonia.

There are 274 sites in Tallinn and the average number of co-present people is 63, regional towns however have the average of 19 sites and the number of co-present people per site is 85 *(Figure 17.8)*. The number of co-present people per site is the highest in county seats – 89 people per hour (the average number of sites in county seats is 4). In the commuting area of centres, the number of co-presences per site and hour decreases moving away from the centre. The number of co-present people is the lowest in rural municipalities and in the commuting area of county seats.

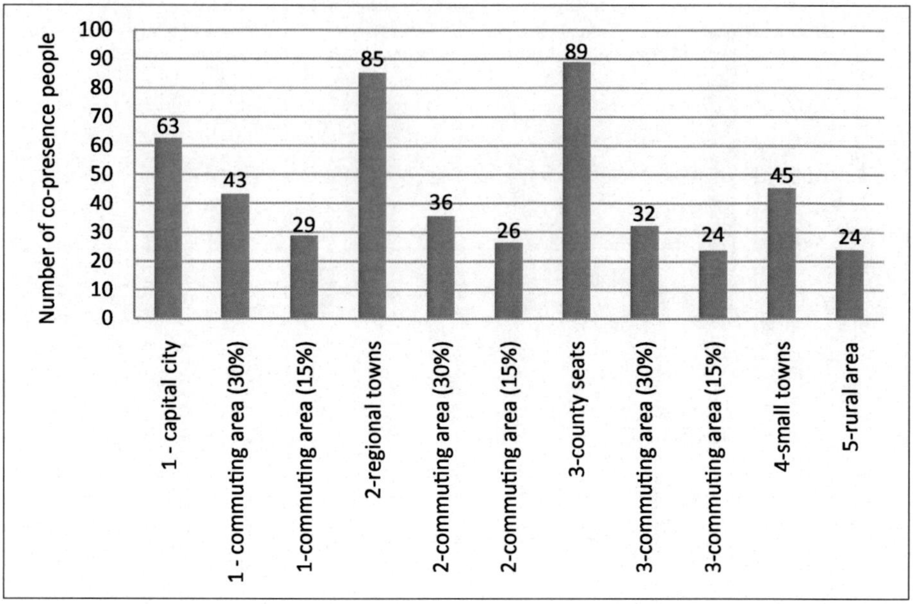

Fig. 17.8. Average number of co-presences in a space and time window site-hour by levels in the hierarchy of settlement system.

17.6 Conclusion and Discussion

The aim of this study was to find a spatial and temporal scale for determining co-presence on the basis of mobile positioning data. Several scientists have worked on determining co-presence theoretically (Miller 2005, Lawrence et al. 2006, Urry 2002), the variety of these approaches shows that there are very different possibilities for defining co-presence or "being together". In addition to the physical being together, communication through means of communication or visiting the same place at different times have also been discussed as being together in space and time (Miller 2005). The size of the spatial and temporal framework of co-presence largely depends on the aims of the study. If the aim is to find face-to-face meetings, for example in studying communication or social networks, the space and time framework is "strict", i.e. it has to reflect the physical closeness of the locations of people and being together at the same time, to enable communication between people. However, when the objects of interest are communication and spread of information, the space and time window of co-presence may be more flexible; in addition to physical co-presence virtual contacts or communication through message boards are also acceptable in such case (Miller 2005). In the case of certain implementations of space-time behaviours, such as spread of diseases or organization of public transport however, the object of interest is the potential crossing or overlapping of paths.

In addition to selecting the scale based on the aim, the nature of data is also important. Even though there are many databases and lots of possibilities for searching from them today, the amount of databases suitable for studying co-presence is limited. Such data has been gathered through censuses, registers and inquiries about places of residence, places of work and use of space during free time. Based on these, co-presence can only be determined for long-term places of stay (e.g. place of residence or work or holidays). Most of the problems brought up on the basis of the subject of co-presence however proceed from everyday spatial behaviour: tourism, business meetings, transport, communication between people, spread of innovations, etc. The number of databases suitable for studying such subjects is however considerably lower: interviews, travel diaries and ICT databases (Calabrese et al. 2011). Passive mobile positioning (Ahas et al. 2008) is one of the databases, which enables to study the everyday spatio-temporal behaviour of people. This study based on the data of passive mobile positioning and on the need to find a suitable spatial and temporal window for determining co-presence grew out of studies of ethnical segregation in Estonia (Silm et al. 2011).

Our results show that in the case of place-based approach, co-presences in the case of different space and time windows vary. Using the data of passive mobile positioning, the smallest possible spatial unit of co-presence is a cell and the smallest possible time unit is a second. The maximum space unit used in this study is a county and maximum time unit is a day. Initially it could be presumed that the effect of the time unit on the number of co-presences has a certain direction *(Figure 17.4)*, but in reality the number of co-presences determined with place-based approach is influenced by the amount of possible space and time units. In the case of units of larger surface area (county, municipality, settlement unit), even the number of co-presences measured with the accuracy of a second is sufficient, prolonging the period of time however decreases the number of potential co-presences, because the number of time units is smaller in such case. The number of possible co-presences is the largest for a space scale of one mobile cell and time scale of one minute. The average number of co-presences however increases by prolonging the period of time and increasing the extent of spatial area in the case of all possible time and space windows *(Figure 17.5)*.

This result shows that setting the correct space and time window, determining its geographical and temporal division, are very important. It is of course primarily based on the data and aims, but knowing the dimensions of the scale of correct size for finding co-presences is also important. This is influenced by the sizes of population and the sample, the locations of settlements and surface area of the area involved in the research. The correlation of co-presences with the number of residents according to Statistics Estonia revealed a linear relation between density of population and co-presences, a higher number of co-presences corresponds to a greater density of population *(Figure 17.7)*. But the number of co-presences is also influenced by the size of a settlement unit, i.e. the level in hierarchy of settlement system. The correlation with the level in hierarchy is high on the level of municipality-hour *(Figure 17.6)*, on the level of site-hour *(Figure 17.8)* however the results are different. There is namely a large number of sites in bigger cities (high density of mobile cells), due to which the number of potential co-presences per one site decreases. This is one of the aspects which has to be considered when choosing space and time window for research projects.

It is also important to be acquainted with the technological solutions of different operators and networks. In this paper we have used the data based on Ericsson's 2G and 3G technologies in Estonia. We know that the Network areas, cells and sites have different dimensions and meanings in the case of Nokia-Siemens and other network systems. It is still thought that based on standards, the peculiarities of broadcasting and geography and population, the main structures of mobile networks are similar and the results of this study enable to generalize co-presence more widely.

Conclusively it can be said that in the case of determining co-presence with the data of passive mobile positioning, the largest number of possible co-presences is achieved using the accuracy of one minute and network cell. The levels of a county and municipality also provide a comparable number of co-presences at the accuracy of one second, but the spatial unit is too large here (max 60–80 km) for it actually to be named as the physical co-presence of people. The average number of co-presences in a space and time window is however the highest in the case of the area of the largest spatial extent and the longest period of time, thus a county and a day out of the analyzed versions. However, for assessing actual co-presence, this area is too large and the period of time is too long. Thus, based on the results of this study in Estonia, we have found that the most suitable space unit for finding co-presence is site of mobile network and time unit is one minute. In some study cases, time unit could be extended to an hour for increased average number of co-presences. On this basis, the number of possible co-presences in the population of Estonia is the highest possible and the spatial unit is also small enough to study physical co-presence.

In the future it is necessary to study the same spatio-temporal division on the basis of personal co-presences and compare the similarities and differences of spatial and personal co-presences. It is also necessary to assess the effect of implementing different samples; studies are usually not carried though with such massive databases, in the case of smaller samples however the frequencies of being at the same place at the same time may be very different. The different logics of mobile networks GPS-based data also require attention, the more precise the georeference of collected data, the lower the frequency of co-presence determined on the bases of the data.

References

Ahas R, Aasa A, Roose A, Mark Ü, Silm S (2008a) Evaluating passive mobile positioning data for tourism surveys: An Estonian case study. Tourism Management 29(3):469–486

Allen T (1997) Architecture and Communication Among Product Development Engineers. Sloan School of Management, MIT, Cambridge. 1–35

Calabrese F, Smoreda Z, Blondel VD, Ratti C (2011) The interplay between telecommunications and face-to-face interactions – an initial study using mobile phone data. Presented at CoRR

Castells M (1996) The Rise of the Network Society, The Information Age: Economy, Society and Culture Vol. I. Blackwell, Cambridge, Oxford

Couclelis H (2009) Rethinking time geography in the information age. Environment and Planning A 41(7):1556–1575

Crandall DJ, Backstrom L, Cosley D, Suri S, Huttenlocher D, Kleinberg J (2010) Inferring social ties from geographic coincidences. PNAS 107(52):22436–22441

Delafontainea M, Neutensa T, Schwanenb T, van de Weghea N (2011) The impact of opening hours on the equity of individual space–time accessibility. Computers, Environment and Urban Systems 35(4):276–288

European Commission Report (2010) Progress report on single European electronic communications market 2009. http://ec.europa.eu/information_society/policy/ecomm/doc/implementation_enforcement/annualreports/15threport/comm_en.pdf

Goffman E (1966) Behaviour in Public Places. Free Press, New York

Hägerstrand T (1970) What about people in regional science?. Papers of the Regional Science Association 24:7–21

IJsselsteijn W, Freeman J, Ridder H (2001) Presence: Where Are We?. CyberPsychology & Behavior 4(2):179–182

IJsselsteijn W & Riva G (2003) Being There: The experience of presence in mediated environments. In Riva G, Davide F, IJsselsteijn (Eds) Being There: Concepts, effects and measurement of user presence in synthetic environments. Ios Press, Amsterdam. 3–16

Ito M (2003) Mobiles and the appropriation of place. Receiver. 1–3

Janelle D (1995) Metropolitan expansion, telecommuting and transportation. In Hanson S (Eds) The Geography of Urban Transportation. Guilford Press, New York. 407–434

Kostakos V, O'Neill E, Penn A, Roussos G, Papadongonas D (2010) Brief Encounters: Sensing, Modeling and Visualizing Urban Mobility and Copresence Networks. ACM Transactions on Computer-Human Interaction 17(1):1–38

Lawrence J, Payne TR, Roure DD (2006) Co-Presence Communities: Using Pervasive Computing to Support Weak Social Networks. In WETICE 149–156

McLuhan M (1964) Understanding Media: The Extensions of Man. The MIT Press, Cambridge, Massachusetts, London, England

Milgram S (1977) The Familiar Stranger: An Aspect of Urban Anonymity. In Milgram S (Eds) The Individual in a Social World. Addison-Wesley. 51–53

Miller HJ (2005) Necessary space – time conditions for human interaction. Environment and Planning B 32:381–401

Miller HJ (2007) Place-Based versus People-Based Geographic Information Science. Geography Compass 1/3:503–535

Urry J (2002) Mobility and Proximity. Sociology 36(2):255–274

Urry J (2003) Social networks, travel and talk. British Journal of Sociology 54(2):155–175

Urry J, Larsen J, Axhausen K (2007) Networks and tourism. Mobile social life. Annals of Tourism Research 34(1):244–262

Silm S, Toomet O, Ahas R (2011) Co-presence as a measure of ethnic segregation. Presented at AAG

Yu H & Shaw S-L (2007) Revisiting Hägerstrand`s time-geographic framework for individual activities in the age of instant access. In Miller HJ (Eds) Societies and Cities in the Age of Instatnt Access. Springer. 103–118

Zhao S (2003) Toward a taxonomy of copresence. Presence 12(5):445–455

Chapter 18

Investigation and Development of Mobile Touristic Applications

Eva Hauthal and Dirk Burghardt

Institute for Cartography, Dresden University of Technology, Germany
{eva.hauthal, dirk.burghardt}@tu-dresden.de

Abstract

With mobile devices a challenging media for cartographic information presentation became available. On the example of transferring static hiking maps into a mobile framework, research questions are identified on cartographic communication in general and cartographic design in particular. A high level communication model is proposed to identify and illustrate tasks and roles of cartographer and (prod-)user around the creation and usage of location based applications. The conception of a mobile application for hiking in Saxon Switzerland is based on an analysis of touristic mobile apps. The comparison presented within this paper provides an overview of existing functional requirements and interaction possibilities. The mobile hiking application is implemented prototypically containing user generated spatial content from *Hike & Bike Map*. The developed app serves as a platform for integrating further user generated content, such as specific thematic and regional contents (e.g. climbing paths). Future research and development will investigate possibilities on the adaptive personalized information presentation enhancing the current adaptable version.

Keywords: mobile application, user generated content, adaptation, communication model

G. Gartner, F. Ortag (Eds.), *Advances in Location-Based Services*, Lecture Notes in Geoinformation and Cartography, DOI 10.1007/ 978-3-642-24198-7_18,

18.1 Introduction

The ongoing technological development in the domain of geoinformation processing has a huge impact on the handling of and interaction with cartographic information. This becomes obvious with the capturing of spatial data, as well as with the presentation of maps at a variety of output media. Cartographic applications on mobile devices have enormous research and development potential, because they provide the combination of interactivity and mobility in extension to paper maps, thus they allow an adaptive, personalised information exchange. A forerunner for the era of mobile cartography was on the one hand the emerging of mobile internet in the end of the last century and on the other hand the remove of Selective Availability in the GPS signal. Thereby information and communication opportunities were not only available at the computer 'at home' but everywhere and any time. This causes an increasing number of invocations of web maps and user created maps which are often produced 'quick and dirty' ignoring fundamental cartographic concepts and design principles (Hoffmann 2011). The fast production of maps has its reasons in simple and intuitive user interfaces, open APIs and mashups. Subsequently an ambivalence emerged between traditional cartography and the popularity of new technological applications (Gartner & Schmidt 2010).

With the civil use of GPS it became possible to integrate the current location of the user into maps. New challenges in cartography emerged, e.g. the compensation of technical limits of the mobile devices through the utilisation of intuitive interaction methods, consideration of context as well as adaptation and personalisation of mobile maps. GPS on the one hand and mobile internet on the other hand enabled the user to collect and to publish own spatial data. User generated content became an important source of cartographic information. Thus the complete cartographic communication process evolved from a unidirectional into a multi user communication. Unfortunately cartographic theories did not reflect the new technological opportunities adequate so far. Kraak and Ormeling (2010) illustrated the increasing gap between technological opportunities and cartographic theory *(Figure 18.1)*.

Thinking about the perceived gap between cartographic theory and technological development a number of arguments can be made. Traditionally, structural diagrams are used in cartography to reflect the cartographic communication process, the interplay between cartographer and user, as well as for the illustration of the influencing factors. In fact the role of the cartographer is changing and traditional cartographic communication models of theoretic cartography that are based on the sender-channel-receiver-model are no longer valid for the reason that a strict separation into map maker and map user can not be accomplished anymore (Hoffmann 2011). New structural models have to be derived for the identification and illustration of current and future research trends in cartography. Further missing is the theoretical foundation of map design rules and cartographic guidelines on recent technological possibilities.

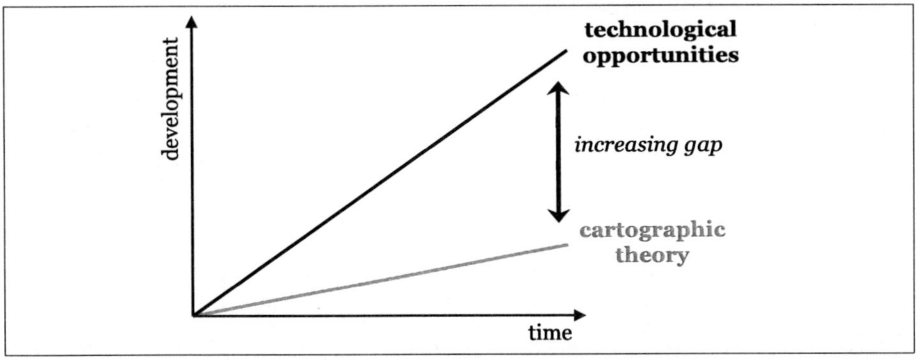

Fig. 18.1. Increasing gap between the development of cartographic theory and technological opportunities (Kraak & Ormeling 2010).

Selected design guidelines exist, e.g. android developer guidelines (http://developer.android.com/index.html) or experiences at numerous developer forums. Nevertheless much more scientific research is required for the presentation of spatial information on mobile devices, such as about the usage of spatial and semantic filters, the integration of spatial and non-spatial information or the utilisation of interactive legends. A final remark on the perceived gap between cartographic theory and technological development has to qualify the statement. There are also a number of scientific publications available which illustrate future location-based mobile applications not available today, e.g. in the domain of context consideration, adaptivity or augmented reality (Jiang & Yao 2006, Raper et al. 2007).

An application field of mobile cartography are mobile map applications for smartphones in tourism. These touristic applications are an up-to-date kind of travelling service in the century of mobile technology and information. A mobile map application for the pilot region Saxon Switzerland was conceived and implemented prototypically. This application was inspired by a cartographic hiking calendar for Saxony consisting of twelve hiking trips about an annually varying topic and appropriate static maps. The newly developed app for Android transferred the hiking calendar into an interactive, mobile cartographic environment. The presented investigation and development of a mobile cartographic application aims in identifying theoretical and practical research questions.

Section 18.2 presents current research in mobile cartography regarding context consideration, adaptation and utilisation of volunteered geographic information. In *Section 18.3* a theoretical model for mobile cartographic communication is proposed. A comparison of existing touristic applications in *Section 18.4* was the basis for the conception of the hiking application. In *Section 18.5* details of the prototype implementation are described. The final section discusses conception and implementation of the mobile hiking application in the context of the derived communication model.

18.2 State of the Art

The realisation of wireless internet access brought web maps back to mobile environment thus to the environment where they are most needed (Meng & Reichenbacher 2005). From this union of real and virtual worlds numerous challenges for cartographers are arising. Not only technical parameters like the limited display size or battery life but also not-technical parameters such as time-critical user tasks or the permanently changing environment of the user force cartographers to embed only really needed information and interactions in mobile maps (Sarjakoski & Nivala 2005, Cartwright et al 2007, Millonig & Gartner 2011). The minimalist displays make mobile maps more personal than their predecessors and require an adaption of contents and representation to the present demands and cognitive abilities of the user (Setlur et al 2010). For adapting an application to different needs arising from mobile using situations the context of the mobile user has to be understood.

Context in mobile cartography has physical, semantic and social facets as well as aspects regarding the system of the mobile device. This way specific context dimensions can be defined: situation, user, activity, information and system. These dimensions are not independent at all but rather connected by complex relations. A challenging task is to find the most important parameters out of this large quantity of context that are relevant for mobile cartography and that have a significant influence on mobile geospatial information use (Reichenbacher 2003).

Adaption can be seen as the answer to a changing context. Thereby a distinction between adaptive and adaptable can be made. A system is adaptable when the adaption is carried out by the user in an interactive way whereas a system is adaptive when the diagnosis of requirement for user support and the execution of adaption are carried out by the system itself.

The implementation of the mechanism of adaption in cartography has the aim to help the user by accessing geographic information more efficiently, to adapt their visualisation to the user requirements and the limited resources as well as to improve the entire relevance. The result is an adaptive map that can be defined as a map-like presentation which was adapted extensively by the user or the system (Reichenbacher 2005).

Not only the information in a mobile map, also the interactions provided to the user have to be adapted which means they have to be reduced to the required extend (Bereuter et al 2009). The adapted interactions on a mobile device are definitely an extension of a static paper map, but an interaction decrease is recognizable in comparison to the interactive maps of a desktop PC.

As already mentioned user generated content is a part of recent cartography as well. The standard example for user generated content in cartography is *OpenStreetMap*, but also *Wikipedia*, *YouTube* or *Flickr* are further indicators for the increasing importance of user generated content in general. Goodchild (2007) describes this phenomenon that thousands of people are willing to spend a lot of time for publishing geospatial data in the internet without a chance for financial reward and without any certainty that these contents are needed, as Volunteered Geographic Information (VGI) or Volunteered Geography.

The most convincing advantages of user generated content are little costs for the provider and up-to-dateness. However virtual vandalism is the biggest worry of users. They fear that their contents could be manipulated, falsified or even deleted (Coleman et al. 2009). Another negative aspect is the suspected lack of quality in user generated content caused by missing editorial works, anonymous publishing and doubtful trustworthiness of the authors. In the form of rating mechanisms (e.g. *ebay)* the doubtful trustworthiness of the authors is taken into account. Another example is Wikipedia where doubtful contents can be marked or in *facebook* where users can publicise the inapposite behaviour of other users. A similar mechanism does not exist for cartographic user generated content yet (Gartner & Schmidt 2010).

The aspects of context, adaption and user generated content described above are not or merely hardly taken into consideration in existing cartographic communication models of theoretic cartography. Although in some recent models the user is noticed as a map maker (e.g. Kelnhofer 2003, Lechthaler et al 2007), the adaption of a mobile map to the user and the current using context cannot be found. Nevertheless the determining factors originating from the first important cartographic communication model of Kolacny (1969) can be seen as context. Only Kelnhofer (2003) states a 'user defined thematic map' in his model, but doubts the necessary geographic and cartographic competence of the user.

This clarifies that adaption has to be taken into consideration in a newly derived communication model as well as the fact that also the user undertakes the role of cartographer. So the cartographer backs out and provides expert knowledge in the form of base maps, software or interaction techniques that can be used by the map user as a user interface for integrating own data. So it is not appropriate to act on the assumption of a huge data bucket that is supplied by the cartographer and the map user, but also of an amount of interaction techniques that are provided by the cartographer to the map user. Out of this total quantity a selection has to be made according to the user and the current using context which is presented on the mobile device.

18.3 Theoretical Model for Mobile Cartographic Communication

The derived cartographic communication model presented in *Figure 18.2* is an attempt of further developing Kolacny's (1969) cartography related sender-channel-receiver-model. On the left side the cartographer is placed as sender and coder who is subdivided into designer and developer. The developer supplies algorithms, software and interaction techniques for the user, the cartographer provides base maps and style files which serve for the design and the coding of geospatial information. In doing so the result of the coding and sending process are influenced by the cartographic-artistic and informatic-technical abilities and experiences of the cartographer. Both are forming the cartographer's context (with grey font in *Figure 18.2)*.

On the right side stands the map user who is now called ProdUser (Budhathoki et al. 2008). That person is both producer and user in the sense of consumer of geospatial information. Of course the consumer and producer do not necessarily have to be united in one person. That is why in the model the ProdUser is split into data producer and data user. So the ProdUser acts as a sender but also as a receiver and decoder. As a sender the person can integrate raw data into the base maps provided by the cartographer and can equip them with qualitative information and/or pictures. The coding of these raw data is carried out by applying the style files created by the cartographer. The context influencing the ProdUser is formed by personal experiences, abilities, goals and interests. The interests are determining what the contents are about. The ProdUser as data producer differentiates from the educated cartographer by abilities and experiences. The ones of the cartographer guarantee a better cartographic quality whereby the differences between several cartographers are not as big as in the heterogeneous crowd of ProdUsers.

In the centre of the model stands the mobile application subdivided in software components and required hardware. The hardware represents the mobile device its properties matter the context of the user. The software is formed by the contents such as maps, texts and pictures collected by the cartographer and the ProdUser as well as by the interaction techniques developed by the cartographer. According to Wolodtschenko (2010) maps, texts and pictures are semiotic meta-variables. Out of these components of software a selection is made according to the context of the user and subsequently visualised on the mobile device.

The sender-channel-receiver-model is realised by the three components cartographer, mobile application and ProdUser but the original one-way aligned form is not present anymore for the reason that two senders are existing.

The real world is located in the model above the modified sender-channel-receiver-model. Information are gathered from the real world. The real world contains two overlapping subsets: the world of the cartographer and the world of the ProdUser.

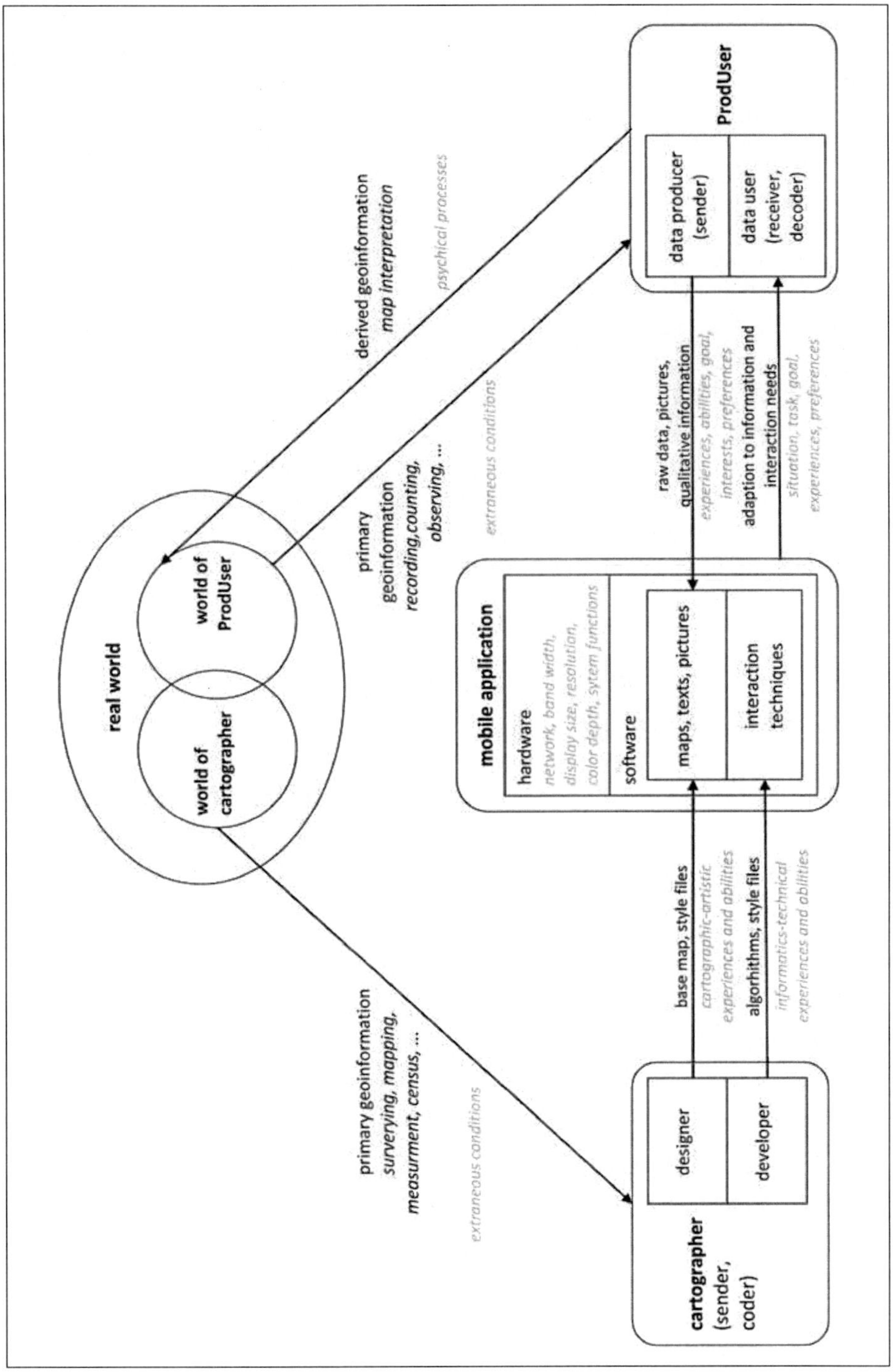

Fig. 18.2. Cartographic communication model for interactive mobile maps.

The overlapping part represents on the one hand the common character repertoire which is necessary for a successful communication and on the other hand the common abilities to work with shared software and the interaction techniques. From the real world extraneous influences take effect on the context of the cartographer and the ProdUser. Gathering information from the real world is done by the cartographer in the form of surveying, mapping, measurement and census; the ProdUser proceeds by recording, counting and observing.

One example – actually the paramount example per se – that reflects the presented communication model is *OpenStreetMap*. In this project users are building up collaboratively a world map with tools provided by the developers. This map is used in mobile application such as *Skobbler*, a costless navigation service users can integrate own local knowledge as 'local heroes'. Another mobile application that represents the communication model is *The North Face Trailhead* for iPhone analysed in *Section 18.4*. With this app user generated trails for sports activities such as hiking, biking or canoeing can be retrieved and are displayed on top of *Google Maps*. The user can also create and upload own trails equipped with pictures and information. The platform in the background of this application is *EveryTrail*.

18.4 Investigation on Touristic Applications and Derived Conception

As a preparation for the conception of a mobile application for Saxon Switzerland nine mobile touristic applications were examined: five applications for iPhone (*Lonely Planet Berlin*, *Lonely Planet Paris*, *Ostseeküste – ADAC Wanderführer*, *Allgäu*, *The North Face Trailhead*) and four applications for Smartphones with an Android operating system (*outdooractive*, *TravelBook Berlin*, *Vienna Travel Guide*, *Qype*). Doing this, twelve questions characterising the functionality extent of these applications were emerging:

1. Is the current location of the user considered?
2. Are the map and other contents linked with each other?
3. Is the used map produced by the developer?
4. Is it possible to change the map design?
5. Is it possible to show/hide certain map contents?
6. Are further interactions than panning and zooming possible with the base map?
7. Is it possible to personalise contents?
8. Is a data input possible?
9. Does the application have a search function?
10. Does the application have an intern routing function?

11. Is an augmented reality available?
12. Is it possible to use the application unrestrictedly without an internet connection?

The answers to these questions for each examined app are summarised in *Table 18.1*. The numbers in the heading are corresponding to the numbers of the questions listed above. The last column of the table contains the number of satisfied criteria in percent.

For the reason that these criteria are not equivalent in every case, not meaningful with regard to usefulness and not representative for the entire number of existing touristic applications, a more general classification concerning purpose and territorial validity of an application was carried out. The examined applications can be separated in city guides and leisure guides. Leisure guides suggest trips for exploring a certain region in the form of sporty activities such as hiking or biking. On this occasion it can't be eliminated that leisure guides contain sightseeing tours as well. However city guides introduce a city by listing interesting venues for tourists such as sights, museums, restaurants etc but mostly without combining them as a trip. The territorial validity can be distinguished in regional and global, i.e. if an application is meant for a specific place or region or significantly more wide-ranging. Four kinds of classes result from this classification: regional city guides, global city guides, regional leisure guides and global leisure guides. An assignment of the examined applications can be seen in *Table 18.2*. Applications with a global territorial validity are based exclusively on user generated contents and derive those contents from internet. The one with a regional territorial validity reveal this characteristic already by their name.

Before starting with the conception of the mobile application of Saxon Switzerland for Android, it should be summarised what information are provided by the already mentioned hiking calendar that can be used for the app:

- information box with properties of the hiking trip (route, journey, distance, duration, altitude difference)
- detailed description of the hiking trip
- map
- information on the hiking region
- pictures

These information are spread on three calendar sheets: description of the hiking trip, map and information on the hiking region. The fourth calendar sheet with the monthly picture and the calendar table will be disregarded for the conception. It is obvious to retain the previously described trisection of the calendar contents in the application.

Figure 18.3 depicts the conceived application as a flow chart. Each box represents a screen page. The title explains the purpose of the screen page. Inside of the

Table 18.1. Examined mobile applications and their extent of functionality.

Android	1.	2.	3.	4.	5.	6.	7.	8.	9.	10.	11.	12.	
outdooractive	✓	✓	–	✓	–	–	–	–	✓	–	–	–	33%
Qype	✓	✓	–	–	–	–	✓	✓	✓	–	–	–	42%
TravelBook Berlin	✓	✓	✓	✓	✓	✓	✓	–	✓	✓	–	–	81%
Vienna Travel Guide	–	✓	✓	–	✓	–	–	–	✓	–	–	✓	42%

iPhone	1.	2.	3.	4.	5.	6.	7.	8.	9.	10.	11.	12.	
Allgäu	✓	✓	✓	✓	–	–	–	–	–	–	–	–	33%
Lonely Planet Berlin	✓	✓	✓	–	✓	–	–	–	–	–	–	✓	42%
Lonely Planet Paris	✓	✓	✓	–	✓	–	✓	–	✓	–	–	✓	58%
Ostseeküste – ADAC Wanderführer	✓	✓	✓	–	–	–	–	–	–	–	✓	–	33%
The North Face Trailhead	✓	✓	–	✓	–	–	✓	✓	✓	✓	–	–	58%

Table 18.2. Classification of the examined mobile applications regarding territorial validity and purpose.

Android	territorial validity	purpose
outdooractive	global	leisure guide
Qype	global	city guide
TravelBook Berlin	regional	city guide
Vienna Travel Guide	regional	city guide

iPhone	territorial validity	purpose
Allgäu	regional	leisure guide
Lonely Planet Berlin	regional	city guide
Lonely Planet Paris	regional	city guide
Ostseeküste – ADAC Wanderführer	regional	leisure guide
The North Face Trailhead	global	leisure guide

box all of the possible interactions within this screen page are listed, subdivided into *Click*, *Options Menu* and *Map*. *Click* involves interactions that are triggering an action or invoking a new screen page by a click on a button or the like. The menu items of the screen pages Options Menu are pooled under *Options Menu* and if the screen page contains a map, all the possible interactions with this map are listed under *Map*. Furthermore *Figure 18.3* points out the paths the user can reach the screen pages on, but only in one direction because it is always possible to go back with the *Back*-button of the smartphone. In the last line of each box the Android layout view of the screen page is named.

The examined touristic applications introducing hiking trips or the like presented these trips in the majority of cases in a list, more seldom in a map. The entry of the mobile application for Saxon Switzerland shall be a choice between a list of all the trips and a general map of the trips. In both of these versions one trip can be selected

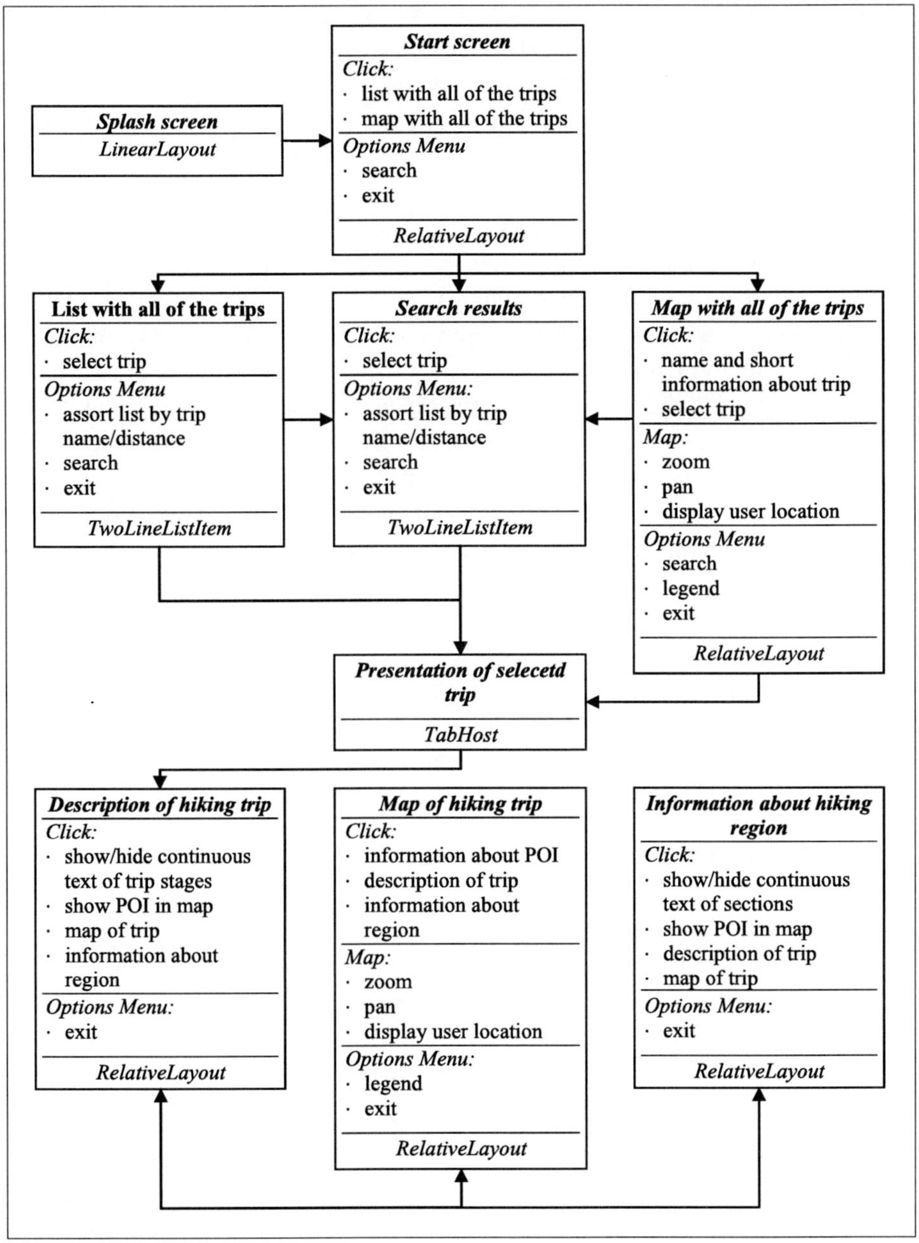

Fig. 18.3. Flow chart of conception of mobile application for hiking in Saxon Switzerland.

and will be presented in the mentioned trisection. In the list each list item consists of two lines. In the first line the name of the hiking trip is written, the second line provides short information about the trip such as distance and duration. Through the Options Menu the list can be assorted by trip name or by distance. Furthermore the search function can be invoked which scans all contents of the app for the user defined search keyword. The search result is a two-line list as well. Moreover the search function shall be available at the starting page and at the general map. A click on a trip in this map displays the same information like the list of all the trips (trip name, distance, duration). A second click invokes the entire tripartite information about a trip that are displayed in three tabs (route, map, information). The page structures of the trip description and of the information on the hiking region are similar. The continuous text is subdivided into several stages or sections. At first only the headings are visible for getting an overall view. With an alongside button the appropriate continuous text can be shown or be hidden. Within these continuous text pictures are located that are a link to the related POIs in the map of the second tab. This map can be panned and zoomed (as well as the general map), POIs can be selected and information about them can be invoked by doing so. Furthermore a legend of the base map can be displayed via the Options Menu.

For the prototypical implementation of the application for Saxon Switzerland *Hike & Bike Map* which is based on *OpenStreetMap* serves as a base map. The map that can be used by Androids *MapView* by default is *Google Maps* but *Google Maps* is slightly appropriate as a base map for a hiking application because it contains only streets whereas *Hike & Bike Map* contains also smaller ways and paths and moreover climbing peaks, caves, rest places, parking sides etc, i.e. objects typical for hiking maps.

18.5 Prototype and Implementation

The result of the prototypical implementation of the mobile hiking application for Saxon Switzerland is most widely adequate to the conception described above. The only not implemented items are the links between pictures within the continuous texts and POIs in the map as well as the legend. Until now only one trip completely filled with content is available. All of the other trips are merely placeholders.

According to the flow chart in *Figure 18.3,* the first screen page that allows an interaction is the start screen (see *Figure 18.4(a))* where the user can choose between a presentation of all available hiking trips in a list (see *Figure 18.4(b))* or in a map. In each of these three *activities* the search function can be invoked. In the list of all available hiking trips as well as in the general map one trip can be selected and is subsequently presented tripartite. This trisection of content of a hiking trip is realised by a *TabHost*-View with three *TabWidgets* for the description of the trip (see *Figure 18.4(c)),* the map (see *Figure 18.4(d))* and the information about the hiking region. The first *TabWidget* with the description of the hiking trip is the default tab, which means it is the initially focussed one when a trip is invoked.

The data storage of all the contents of the hiking application for Saxon Switzerland is realised with a SQLite database. Data queries are made by SQL commands. For data concerning the description of the trip and the information about the hiking region, a layout is generated dynamically according to the amount of data in the database. The coordinates of the trip routes are stored in the database in the form of a KML file. For displaying a route in the map, all the coordinates are read and connected by lines whereby the starting and ending point of the route is represented by a circle. For each POI along the trip route, a marker is displayed in the map at the appropriate location. A click on the marker changes it into a bubble containing the name of the POI object and a short description of it. These markers do also appear in the general map but here at the location of the trips starting point. They show the trip name, distance and duration of the appropriate trip after a click on it. In the general map, the POIs of each trip are not displayed, only the trip routes. When the GPS function is started by the user, the current user location is displayed, updated frequently and taken as the centre point of the map.

Instead of the default Android design a customised design was applied for the entire hiking application. The look of some Android *views* can easily be changed by *styles* and *themes* which can be described as the 'CSS of Android'. But how to customize other *views* or how to realize other design aspects is occasionally hard to find out and to apply. Sometimes it is necessary only to define an attribute, sometimes an extra function has to be implemented or even own *NinePatch*-files have to be created. For the customisation of the application, design guidelines of the official Android Developer website were considered as well as the results of a research paper of the Institute of Cartography (Alexander 2010) that examines the design of signatures for screen output. Especially the aspects of modern design described in this research paper were followed. These aspects are: highlights (e.g. highlight line), gloss, textures, rounded corners, colour gradients and plasticity. Further suggestions for the design of egocentric maps were made by Meng (2005). More guidelines concerning the structure of mobile applications ascertained during the evaluation of a travel planning tool for Finland (Schneider et al. 2010) were taken into account for the hiking app for Saxon Switzerland.

Fig. 18.4. Screenshots of the mobile hiking application for Saxon Switzerland (a, b, c, d).

In addition to the completion of the implementation according to the conception and the upgrading of further hiking trips, the mobile hiking application for Saxon Switzerland could be extended in future with regard to several aspects: own maps could be created for the regions of the hiking trips, more user generated content could be integrated in the application as well as an augmented reality is conceivable for the navigation of the user, for defining the names of surrounding mountaintops or for displaying climbing paths on rocks.

18.6 Conclusion

The presented communication model for mobile cartography tries to identify and to illustrate research areas, which can close the gap between modern technological opportunities and cartographic theory. Examples are the usage of user generated content for information delivery, the adaptation and personalisation of both content and presentation style as well as further developed cartographic design guidelines.

A comparison of theoretical cartographic approaches and the state of the examined mobile applications above shows contrary cases than the one detected by Kraak and Ormeling (2010). Indeed the processes of adaptation and personalisation were conceived and particularly implemented prototypically in projects such as *CRUMPET* (Poslad et al. 2001), *USHER* (Titkov & Poslad 2004) or in the formal model for mobile map adaption of Raubal and Panov (2009), but these implementations are not applied in more ordinary applications like the examined ones. Concerning this matter there are scientific approaches and theories existing which gives indications for future technological developments.

The conception, implementation and use of the mobile hiking application for Saxon Switzerland are reflected on an abstract level in the derived communication model for mobile cartography. Furthermore the model illustrates, that modern cartography is influenced and fostered by technological developments and incentives more than ever. During the development of cartographic communication theory focus was set on semantic and syntax whereas nowadays even more pragmatic aspects considered. This is shown in the model by terms like style files, display size etc. Maps are no longer artefacts anymore and fundamental cartographic rules for producing a readable graphic figure are hardly taken into account by numerous amateur users who produce cartographic representations in a 'playful' and 'entertaining' way. Thus two parallel worlds are arising: the traditional cartography with its long history, approved methods and fundamental knowledge and the 'quick and dirty' produced maps visualising private and individual geospatial data of amateur users (Gartner & Schmidt 2010). For these reasons, on the one hand traditional cartography should found its established standards into efficient tools for giving these 'quick and dirty' products a new quality and for converging or maybe even for harmonising the two parallel worlds. On the other hand by increasing interdisciplinarity it is inevitable to think about the key issues of cartography. This process could be supported by cartographic communication models. Generally those models have the aim to structure a subject area, to abstract from unimportant aspects and to shape research questions. But they can also be used for identification and differentiation.

References

Alexander N (2010) Gestaltung von Positionssignaturen für die Bildschirmausgabe. Institute for Cartography, Dresden University of Technology

Bereuter P, Venkateswaran R, Weibel R (2009) The Use of Filters for Adaptive Mobile Mapping Scenarios. In : Proceedings of AGILE 2009 Workshop on Adaptation in Spatial Communication, Hannover, Germany

Budhathoki N R, Bertram B, Zorica N-B (2008) Reconceptualizing the role of the user of spatial data infrastructure. GeoJournal 72(3–4): 149–160

Cartwright W, Peterson M P, Gartner G (2007) Multimedia Cartography. Second Edition. Springer Verlag, Heidelberg

Coleman D J, Georgiadou Y, Labonte J (2009) Volunteered Geographic Information: The Nature and Motivation of Produsers. International Journal of Spatial Data Infrastructures Research 4: 332–358

Gartner G, Schmidt M (2010) Moderne Kartographie – Technologische Entwicklungen und Implikationen. Kartographische Nachrichten 6: 299–305

Goodchild M. (2007) Citizens as Voluntary Sensors: Spatial Data Infrastructure in the World of Web 2.0. International Journal of Spatial Data Infrastructures Research 2: 24–32

Hoffmann K (2011) Nutzergenerierte Karten und kartographische Kommunikation im Web 2.0. Kartographische Nachrichten 2: 72–78

Jiang B, Yao X (2006) Location-based Services and GIS in Perspective. Computers, Environment and Urban Systems, 30(6): 712–725

Kelnhofer F (2003) Thematische Kartographie zwischen Persistenz und Innovation. Unpublished lecture at colloquium of honour for Wolf Günther Koch on 15 January 2003, Institute for Cartography, Dresden University of Technology

Kolacny A (1969) Cartographic Information – A Fundamental Concept and Term in Modern Cartography. The Cartographic Journal 6(1): 47–49

Kraak M-J, Ormeling F (2010) Cartography. Visualization of Geospatial Data. 3rd edition. Pearson Education Limited, Harlow

Lechthaler M, Strasser M, Todor R (2007) Die Rolle der Kartographie in räumlichen Entscheidungsprozessen. The role of cartography in spatial decision processes. In: Mitteilungen der Österreichischen Geographischen Gesellschaft, Band 149

Meng L (2005) Ego centres of mobile users and egocentric map design. In: Map-based Mobile Services. Theories, Methods and Implementations. Springer Verlag, Heidelberg

Meng L, Reichenbacher T (2005) Map-based Mobile Services. In: Map-based Mobile Services. Theories, Methods and Implementations. Springer Verlag, Heidelberg

Millonig, A, Gartner G (2011) Identifying motion and interest patterns of shoppers for developing personalised wayfinding tools. Journal of Location Based Services 5(1): 3–21

Poslad S, Laamanen H, Malaka R, Nick A, Buckle P, Zipf A (2001) CRUMPET: Creation of user-friendly mobile services personalized for tourism. In: Proceedings of 3G 2001 – Second international conference on 3G mobile communication technologies, London, UK

Raper J, Gartner G, Karimi H, Rizos C. (2007) A Critical Evaluation of Location Based Services and Their Potential. Journal of Location Based Services 1(1): 5–45

Raubal M, Panov I (2009) A Formal Model for Mobile Map Adaptation. In: Location Based Services and TeleCartography II. From Sensor Fusion to Context Models. Springer Verlag, Heidelberg

Reichenbacher T, De Sabbata S, Crease P (2009) The Concept of Geographic Relevance. In: Proceedings 6th International Symposium on LBS & TeleCartography CGS, University of Nottingham

Reichenbacher T (2005) Adaptive egocentric maps for mobile users. In: Map-based Mobile Services. Theories, Methods and Implementations. Springer Verlag, Heidelberg

Reichenbacher T (2003) Mobile Cartography – Adaptive Visualisation of Geographic Information on Mobile Devices. Technische Universität München

Sarjakoski L T, Nivala A-M (2005) Adaptation to Context – A Way to Improve the Usability of Mobile Maps. In: Map-based Mobile Services. Theories, Methods and Implementations. Springer Verlag, Heidelberg

Schneider S, Ricci F, Venturini A, Not E (2010) Usability Guidelines for WAP-based Travel Planning Tools. In: Information and Communication Technologies in Tourism 2010. Proceedings of the International Conference in Lugano. Springer Verlag, Wien

Setlur V, Cynthia K, Mikelson P (2010) Towards designing better map interfaces for the mobile: experiences from example. In: Proceedings of the 1st International Conference and Exhibition on Computing for Geospatial Research & Application, New York, USA

Titkov L, Poslad S (2004) Supporting privacy for U-Commerce tourism services. UbiAgents Workshop on AAMAS-2004 Conference, New York

Wolodtschenko A (2010) Atlaskartosemiotik: neue Möglichkeiten und Lösungen. meta – carto – semiotics, Journal for Theoretical Cartography 3

Chapter 19

Combining Indoor and Outdoor Navigation: The Current Approach of Route Planners

Ann Vanclooster and Philippe De Maeyer

Department of Geography, Ghent University, Belgium
{ann.vanclooster, philippe.demaeyer}@ugent.be

Abstract

This paper studies the use of indoor infrastructures for navigation in several currently available route planners. In the context of an increasing dependence on positioning and navigation tools, a shift has taken place from solely outdoor applications to the indoor environment. Although location based services and indoor positioning techniques may have gotten increasing attention from research and commercial point of view, ubiquitous indoor navigation systems are not yet available on the market. With people moving seamlessly from indoor to outdoor, systems that integrate navigation in both will be the next challenge in navigational research. This paper contributes to this integration of the notion of indoor and outdoor space by studying its impact on route planners. A review of various case studies in multiple route planners has been carried out which reveal different aspects and requirements for the indoor-outdoor connection in way finding. Currently, mostly data constraints prevent the optimal use of all navigation routes. Additional problems were discovered with address matching methodologies influencing the exit choice of buildings (leading in some cases to sub optimal routing). Recommendations are made for future enhancements based on the product to market implications to come to a better integration of indoor with outdoor infrastructures.

Keywords: Navigation, Indoor, Outdoor, Route planner

G. Gartner, F. Ortag (Eds.), *Advances in Location-Based Services*, Lecture Notes in Geoinformation and Cartography, DOI 10.1007/ 978-3-642-24198-7_19,

19.1 Introduction

Over the last decade, navigational tools have become more and more prevalent as a resource for reliable route planning and way finding. Generally, navigation requires tracking and guidance by a technical localization infrastructure, support of multiple navigation contexts (navigable and non-navigable space description based on user and environmental constraints) and an appropriate (for the application level) and accurate topographic representation of space (Nagel *et al.* 2010). For outdoor navigational systems, these requirements have been achieved over the years by the development of the global positioning system for tracking and guidance, a more efficient and abundant data collection using mobile mapping technology and improvements in modeling and data storage (e.g. GDF standard). However, this effort has been solely centered on pure outdoor car navigation systems.

Although pedestrian navigation systems hold similar demands for route planning, their interpretation and specification to the pedestrian context calls for a specific and individual adaptation. This is induced by differences in context, environment, mode of locomotion, scale level and technology (Walton & Worboys 2009). For example, pedestrians walk more freely in the available space. Modeling this by using the available outdoor transport networks doesn't completely reflect this freedom (Bogdahn & Coors 2009). Second, pedestrians have access to both indoor and outdoor environments requiring route guidance in both. This implies availability of both indoor and outdoor data, technological support in indoor environments and a communal space model. Third, the seamless movement from pedestrians from indoor to outdoor has to come to light again in the developed navigational models and route finding applications. Fourth, a more constrained environment makes route guidance more arduous due to a change in scale level and a more challenging landmark recognition. Current and future indoor and combined indoor-outdoor navigation systems should be able to implement these specific requirements.

Literature shows that over the last decade various researchers have begun developing systems based on situation awareness and smart environments using location based services (Gartner et al. 2007, Huang et al. 2009). A recent boost in technological advancements for tracking people in indoor environments has lead to increasing possibilities for the development of indoor navigational models. However, this research has focused solely on the technological aspects of indoor positioning and navigation (Mautz et al. 2010). From the multiple techniques available for indoor positioning, no standard has developed yet because none of them fulfill all positioning requirements. Alternatively, several researchers have developed a wide variety of indoor navigational models ranging from abstract space models (Becker et al. 2009) and 3D models (Coors 2003, Li & He 2008) to pure network models (Jensen et al. 2009, Karas et al. 2006, Lee 2001, Lee 2004) and ontological models (Anagnostopoulos et al. 2005, Lyardet et al. 2008, Meijers et al. 2005). While these models might be useful in specific situations, a general framework for indoor navigation modeling has still to reach full maturity (Nagel *et al.* 2010). At issue is that all the previously mentioned attempts remain solely applicable to indoor situations. In order to fully accommodate navigation, a connection with outdoor applications has to be made.

Most current endeavors to combine indoor with outdoor navigation are focused on tracking techniques; in particular the transition of positioning tools from indoor to outdoor environments. The majority of these efforts originated from robotic research (Pfaff et al 2008) and navigation of the visually impaired persons (Ran et al. 2004, Scooter & Helal 2005). The NAVIO project (Retscher & Thienelt 2004) is one of the few attempts focused on pedestrian indoor and outdoor navigation. It aims at developing a route modeling ontology, which provides both outdoor and indoor routing instructions by identifying and formally defining the criteria, actions and reference objects used by pedestrians in their reasoning for navigation routes (Tsetsos et al. 2007). However, the project focuses solely on location fusion (i.e. the aggregation of location information from multiple sensing elements) and user interfaces, again making the approach too narrow. In the modeling field, the most notable work is of Slingsby and Raper (2007) who model a part of the built environment with its immediate surroundings. However, their model is quite complex and not suitable for navigational applications. It is also confined to describing small scale areas. The above research overview shows has up until now no fully integrative approach for combined indoor-outdoor navigation has yet been thoroughly developed.

Apart from the theoretical research efforts, some LBS applications have already been developed as practical pedestrian navigation applications. Makkamappa (www.makkamappa.com) is a smart phone based mapping system which can be used for GPS tracking after uploading maps and making it GPS linked. Photomap (http://ifgi.uni-muenster.de/archives/photomap/Home.html) uses a technique of photographing public maps for pedestrian outdoor navigation. Both applications are focused on outdoor pedestrian routing using continuous GPS tracking. PinWi (Löchtefeld et al. 2010) is a LBS system for pedestrian indoor navigation which uses photos of an indoor YAH-map as navigation model and dead reckoning for positioning. As this may be a worthwhile approach, it is only locally applicable and not comprehensive enough for being a general indoor routing application. It is also less accurate and disregards problems of availability and indoor-outdoor integration. With above practical implementations having their merit, they still are mainly restricted to the application goal. Before developing more models for combined routing, an evaluation has to be made of the practical implementation issues with the integration of indoor and outdoor routing.

The key purpose of this paper is to evaluate the current use of indoor infrastructures for way finding in common route planners. This is done to make an evaluation of the next necessary steps and current problems in indoor and combined indoor-outdoor routing applications. Route planners are one of the first applications to acknowledge the data requirements for indoor and combined indoor-outdoor navigation since they do not require the technological advancements indispensable for full navigation applications. They focus mainly on the data and the presentation of the data in a certain data model used for traditional route calculations. Their implementation of indoor navigation requirements can serve as a base for practically improving current indoor and combined indoor-outdoor routing endeavors and for bringing theory closer to practice.

In this paper, first a review has been carried out of various case studies in multiple route planners, which reveals different aspects and requirements for the appropriate indoor-outdoor connection in way finding. The case studies each examine a current problem in the indoor-outdoor connection by comparing the results of the most commonly used route planners. Secondly, results of this review and their mutual comparison are employed in the discussion to reflect on recommendations for a better future use and integration of indoor infrastructures in route planning applications.

19.2 Route Planner Review

The objective of this review is to grasp the current state of the art on the integration of indoor infrastructures for navigation in common route planners. Without a proper connection of indoor with outdoor environments for navigation, route planners may calculate non accurate and sub-optimal routes. In this review, indoor infrastructures are considered buildings with multiple entrances above and underground, underground walkways, underground shopping centers and underground transportation systems. Since the indoor built environment can only be accessed by pedestrians, only pedestrian navigation is taken into account with a possible connection to public transport options. The used route planners are common for way finding within the geographical area of the query. For queries in Belgium, the following route planners are used:

- Bing: www.bing.com/maps
- Google Maps: www.googlemaps.com
- Mappy: www.mappy.com
- Via Michelin: www.viamichelin.com
- RouteNet: www.routenet.com
- OpenRouteService: http://openrouteservice.org

Queries in Korea are performed with the use of Google Maps and Naver (maps.naver.com). In the different case studies, multiple aspects of the indoor-outdoor connection in routing will be investigated using various route planners. A comparison of the quality of the current route planners is assessed recording their approach of handling data.

19.2.1 Indoor Data Availability

Following examples all make use of an internal network structure. However, usage is not always straightforward or optimal.

19.2.1.1 Indoor Infrastructure as Part of the Shortest Path

To test whether a route planner utilizes the indoor network structure in the shortest path calculations, a first query has been executed to navigate from Cantersteen to Ravensteinstreet in Brussels (Belgium). The optimal pedestrian and shortest path route uses the Ravenstein gallery with aboveground entrances in both streets.

Differences over the multiple route planners can be detected. Both Bing and Google Maps don't make use of the gallery, while Mappy, Via Michelin, RouteNet and OpenRouteService on the other hand do. It can be noted that Bing doesn't even recognize the gallery as part of the spatial dataset. In Google Maps the gallery is mapped with a text label, but is not part of the vector data available for routing. The other route planners map the optimal and shortest pedestrian route between departure and destination point. This query shows that in some cases both the indoor network structure and the aboveground entrances are mapped and used in the calculation of the shortest path.

A second example studies the use of an underground structure as part of a shortest path calculation in Myondong underground shopping centre (Seoul, Korea). The route planner was asked to perform a route calculation from the Lotte Department Store in Myongdong to the Ibis Hotel across the street. This street is not directly crossable by pedestrians due to heavy traffic. Instead, across the hotel entrance is an underground passage way and shopping centre which leads to the other side of the road.

With this query the usability of 3D underground structures in route planners (both the location of entrance points and network usage) is tested. For this query, local data for the city centre of Seoul was only available through Google Maps and Naver (a Korean route planner), while other route planners lacked detailed street network data.

This example shows that there is a huge difference in navigational instructions for both route planners. While Google Maps doesn't provide routing information for pedestrians in Seoul, Naver on the other hand has very detailed information of the available pedestrian roads. It recognizes the underground passage way with

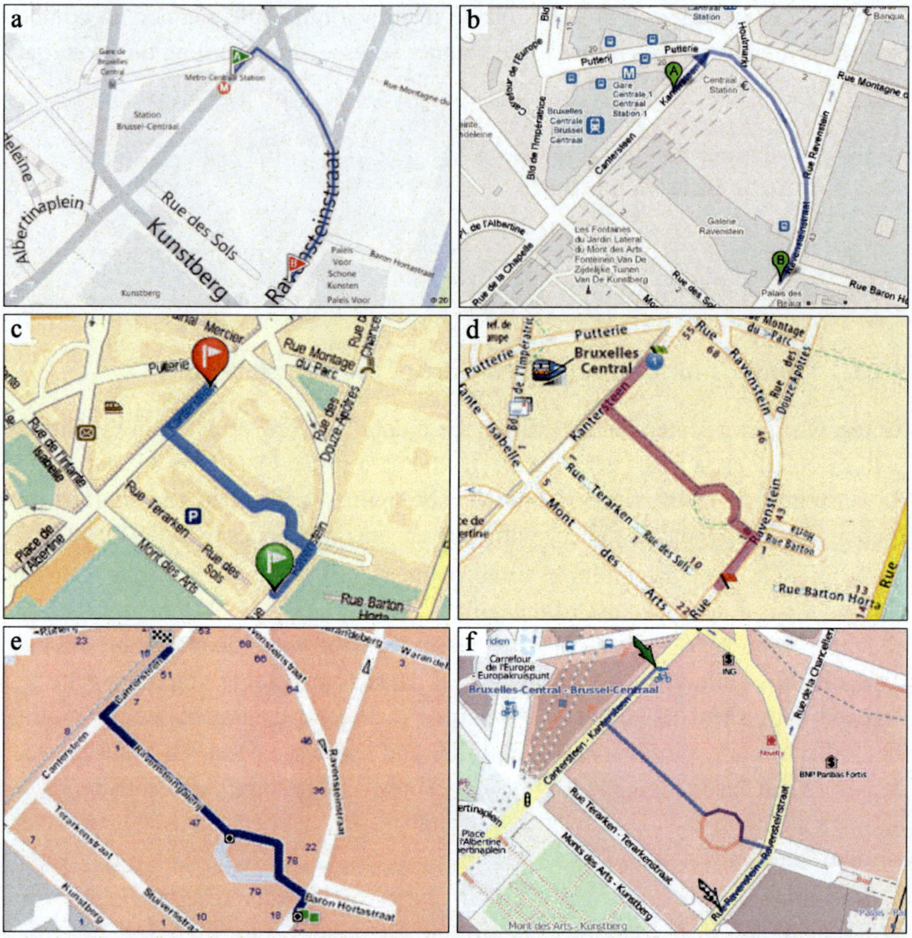

Fig. 19.1. Navigation from Cantersteen to Ravensteinstraat (Brussels, Belgium) using route planner (a) Bing, (b) Google Maps, (c) Mappy, (d) Via Michelin, (e) RouteNet and (f) Open-RouteService.

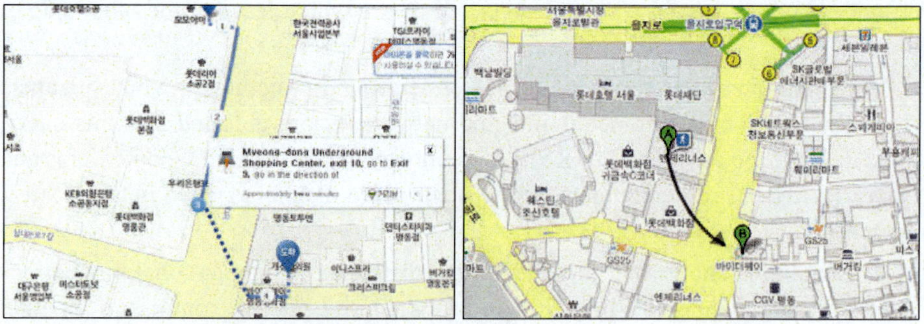

Fig. 19.2. Navigation from Myongdong Lotte Department Store to Ibis Hotel (Seoul) using Naver (left) and Google Maps (right).

Fig. 19.3. Street view of road in Myongdong (Source: Naver). The red arrows show the entrances of the underground passage way.

the corresponding entrance points and exit numbers. Consequently, the navigation instruction is described incorpora-ting all possible details.

19.2.1.2 Availability of Entrance Information

Apart from checking the use of internal network structures, it is also interesting to verify the data completeness of the route planners for navigation. Interior data can be considered complete if it can solve all queries, has the appropriate interior network edges, semantic information and ability to connect the indoor with the outdoor networks via the entrance/exit points of buildings. As is shown in the first example of *Section 19.2.1.1,* Mappy, Via Michelin, RouteNet and OpenRouteService use all aboveground entrances in the calculation of the shortest path. However, the gallery also has one underground connection with the main railway station in Brussels. The following query tests the use of this underground entrance with a query from the railway station to the Ravenstein gallery. The query is executed in all six available route planners.

It can be concluded that only OpenRouteService provides all the entrances to the indoor gallery, even the underground passage way. The spatial data sets of the other route planners are incomplete resulting in sub-optimal routing instructions. It has to be pointed out that the address matching (discussed in *Section 19.2.2)* influences the ability to calculate the routes. For the query in OpenRouteService, the start position has been manually pointed out, since this route planner does not incorporate

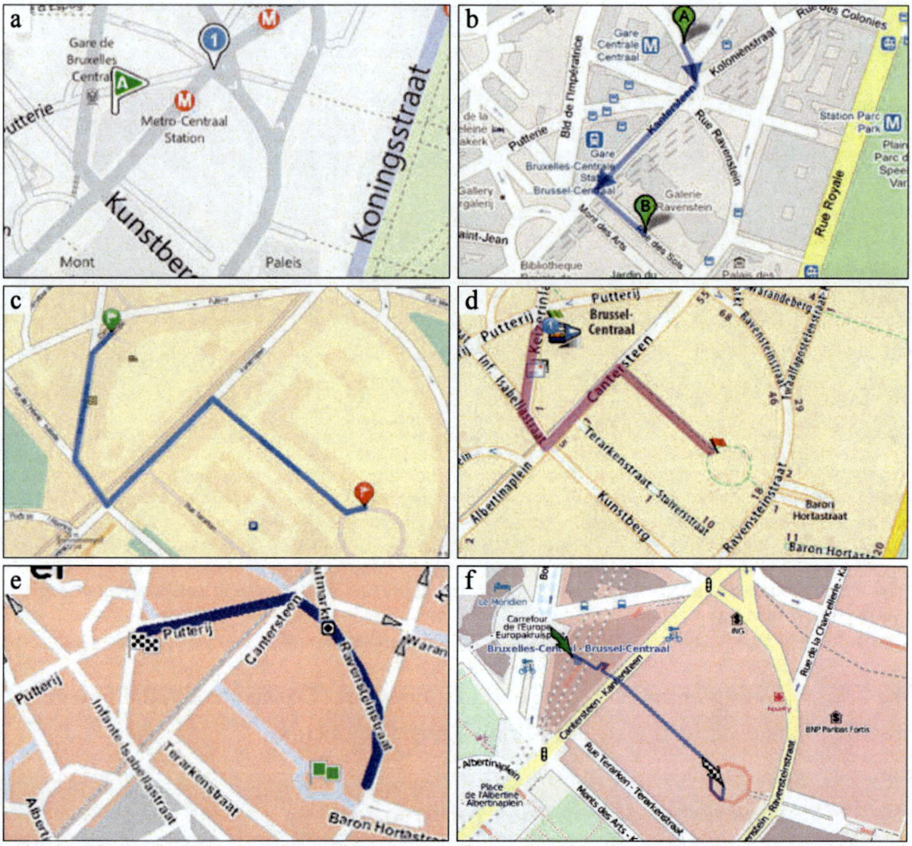

Fig. 19.4. Navigation from Brussels Central Station to Ravensteingallerij using route planner (a) Bing, (b) Google Maps, (c) Mappy, (d) Via Michelin, (e) RouteNet and (f) OpenRouteService.

appropriate address matching. In the Bing route planner, accurate data is lacking of the building itself (attribute is not found in the dataset), making it impossible to even calculate a route. Google Maps has the attribute information but the address is not linked to the network. Instead, the endpoint is linked to the closest available network data with respect to the central point of the gallery. Also, Google Maps links the attribute information for the Central Station to a different geographical location compared to the other route planners. Mappy and Via Michelin, on the other hand, both have network data inside the building complex. However, the underground passage way from the station to the gallery is not digitized. RouteNet maps the location of the gallery on the same postion. However, despite having the internal network structure, the calculated route leads to the back entrance which is the closest to the mapped location (i.e. the location of the address).

19.2.2 Address Matching

In the following examples the query requires appropriate linking between the users input and geographical coordinates.

19.2.2.1 Address Matching Within Indoor Infrastructures

As shown in *Section 19.2.1.1,* in some cases indoor network data is available. However, the availability of an indoor network is no guarantee for appropriate linking of indoor features with indoor address localization. In the following example this is tested through navigating within a certain indoor infrastructure which requires indoor addresses linked to the network structure. Note that we don't take into account the indoor tracking methods necessary for an indoor positioning system and solely focus on the navigational instructions of route planners. This case study is again carried out in the Ravenstein gallery in Brussels. As was concluded from the example above, only Mappy, Via Michelin, RouteNet and OpenRouteService were able to visualize and use the indoor network in its route calculations. Therefore only those are used in the current example.

These similar queries lead to different results over the various route planners. With the navigation instructions in the left column, both destination and departure points are situated on the same network edge which requires a linear interpolation technique for appropriate address matching. Open-RouteService completely lacks a link between addresses and spatial location. Even for outdoor environments, specific addresses in the same street are linked to one point on the network. For this query, the position of start and destination were added manually. The calculation of the shortest route makes use of the internal network. OpenRouteservice can as a consequence not be used for accurate address matching.

As can be seen, only Mappy and RouteNet are able to visualize the correct end points. However, none of them are able to actually calculate the shortest route between them. They both use a different mapping method to project the end points to the correct position on the network. Mappy maps the correct internal location, but cannot connect them through the indoor network. RouteNet searches for the closest available network edge to map the address and connects them using the outdoor network.

The second query also requires internal navigation in the same gallery, but the end point is located on a different part of the internal network. As can be seen from *Figure 19.5* (right column), in this case all route planners are able to perform a correct address matching with a proper connection to the interior network. Via Michelin and RouteNet calculate the shortest path between both points, while Mappy uses a part of the network twice in its calculations resulting in a sub-optimal navigation solution.

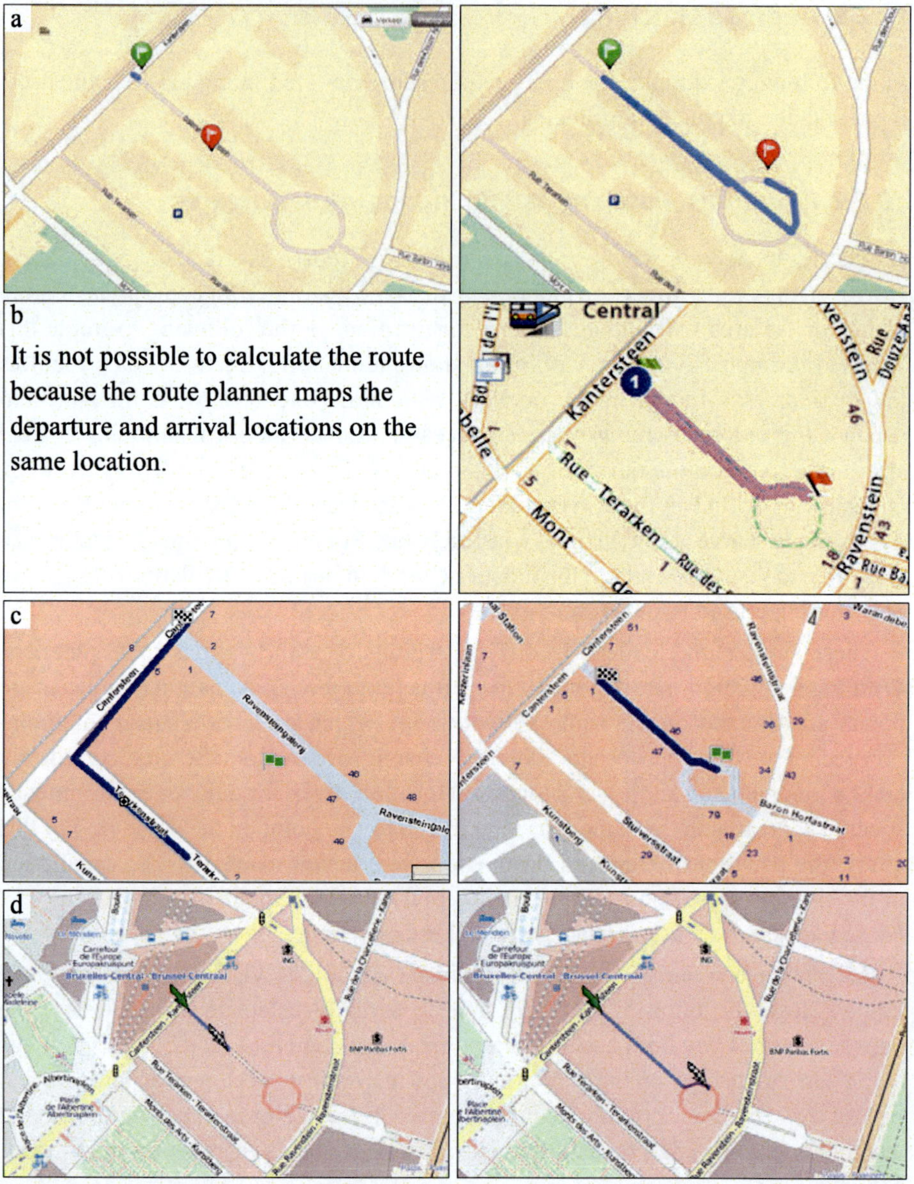

Fig. 19.5. Navigation from Ravenstein gallery 2 to Ravenstein gallery 27 (left column) and from Ravenstein gallery 12 to Ravenstein gallery 60 (right column) using (a) Mappy, (b) Via Michelin, (c) RouteNet and (d) OpenRouteService.

19.2.2.2 Address Matching Influences the Exit Choice

Another aspect of the challenges involved with the indoor-outdoor connection is the way in which exit points and address matching methods are related to each other. The next two case studies test whether route planners make use of different exit points of indoor infrastructures when calculating routes to different locations and in what way the exit choice influences the final route calculation.

This first example uses the main station in Ghent (Belgium) as starting point for two queries. The first query (left column in *Figure 19.6)* asks the route to the centre of town, north of the station. The second query (center and right column in *Figure 19.6)* requires the route to the hospital in the south of the city. The station has two main entrances, one at the front (north side) and one at the back (south side) of the station.

From this example it can be concluded that all five route planners only use one entrance/address point for route planning, no matter what the destination of the query is. Both Bing and Google Maps have the station located at the back entrance, making the route to the city centre not optimal. Interestingly enough, in this case they even use different solutions to get to the north side of the station, due to different routing algorithms used in the calculation. For the second query, the departure points with respect to the geographical location of the station remain the same over all route planners. When looking at the destination, the different route planners use multiple locations depending on the availability of the spatial data.

A second case study takes place in the Waasland shopping centre in Sint-Niklaas (Belgium). It is not so much focused on indoor networks, however the results can have major importance for future indoor-outdoor connections. The query inquires about driving directions to the shopping centre. The shopping centre has multiple entrances and parking spaces which makes driving rather complex. One of the problems here is the question of where to park your car when you want to go to a certain shop. A certain optimization can take place which requires the connection of the several entrances, the internal building layout and the immediate outdoor environment.

It can be seen that the geographic location of the endpoint differs over the various route planners. The digitalization of the outdoor parking area varies from quite rough (Bing) to very detailed (Google Maps). However, none of the route planners make use of entrance point information, making a future indoor-outdoor connection at the moment rather difficult. The algorithm for linking the address information with the spatial network information differs for every application, but is of major importance for results of the route calculations.

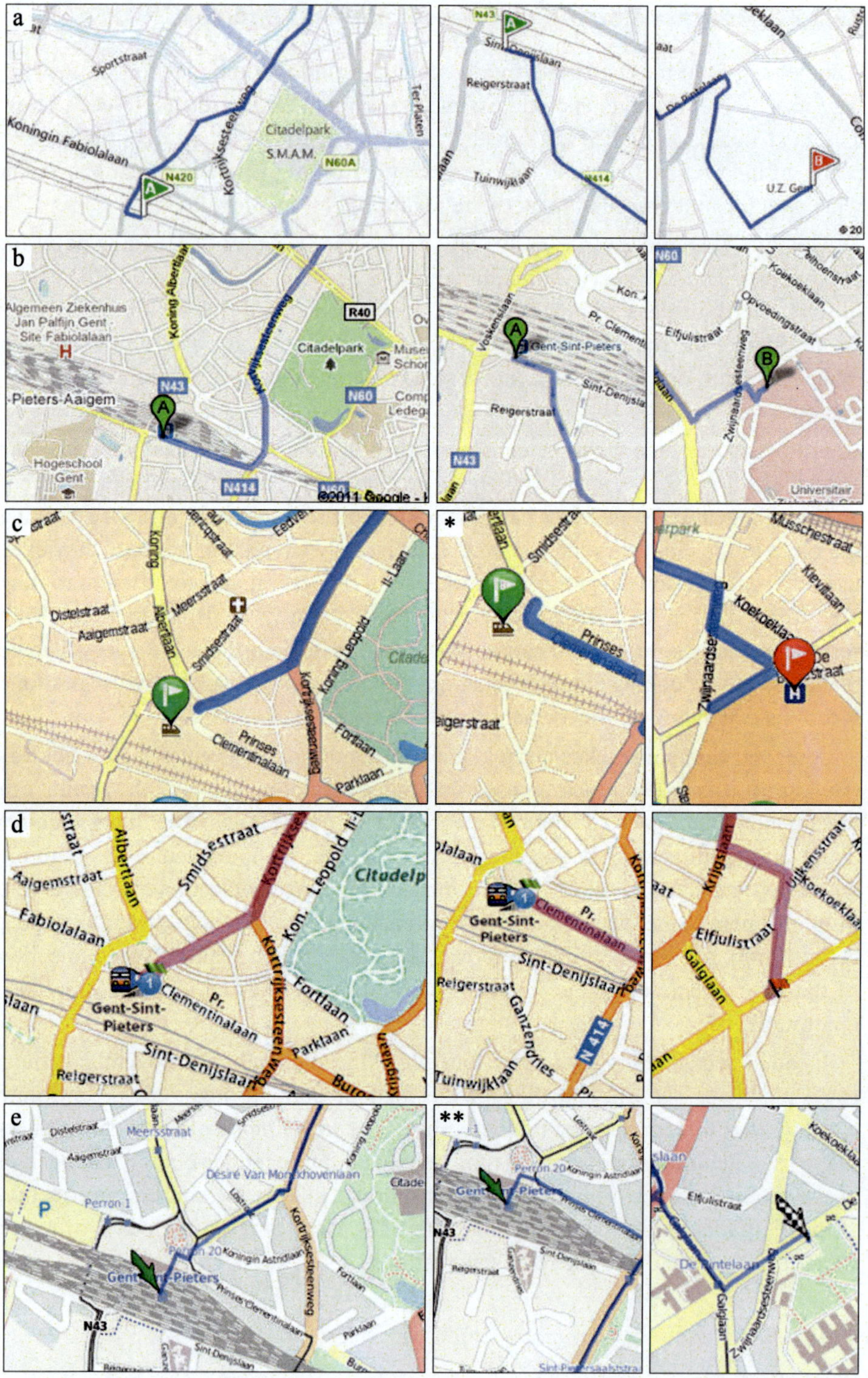
a
b
c
*
d
e
**

Fig. 19.6 on the left page. Navigation from railway station Gent-Sint-Pieters to Korenmarkt (left column) and University Hospital (center and right column) using multiple route planners (a) Bing, (b) Google Maps, (c) Mappy, (d) Via Michelin and (e) OpenRouteService.

* Via Michelin didn't recognize the name 'UZ Gent' or 'Universitair Ziekenhuis Gent'. Instead the address given by the website of the hospital (De Pintelaan 185) is used as end point of the query.

** OpenRouteService doesn't incorporate appropriate address matching capabilities. The start and end points of the queries are added manually.

Fig. 19.7. Driving instructions to Waasland Shopping Centre using (a) Bing, (b) Google Maps, (c) Mappy, (d) Via Michelin, (e) RouteNet and (f) OpenRouteService.

19.2.3 Multimodal Routing Application

One of the applications where the indoor-outdoor connection in navigation is really important is when changing mode of locomotion and this mostly related to the public transportation system. In the following case study a multimodal path using public transportion is calculated from Donuidong 30 to the University of Seoul

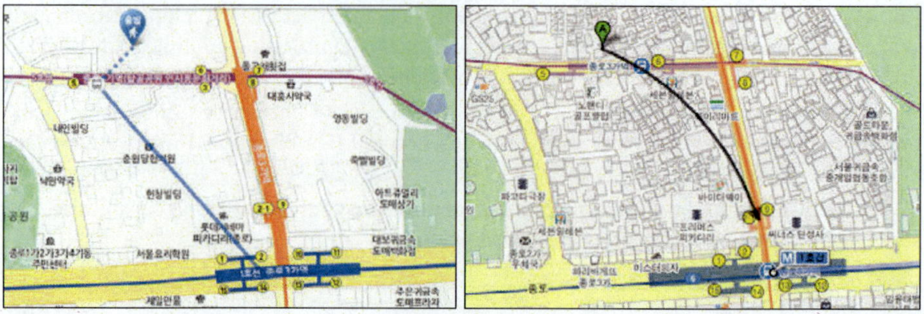

Fig. 19.8. Navigation from Donuidong 30, Seoul to University of Seoul in Naver (left) and Google Maps (right). Zoom of part 1 from Donuidong 30 to Jongno 3-ga subway line 1.

(Seoul, Korea). The calculated route involves changes from pedestrian movement to subway and bus. The first part of the route consists of the movement from the address to the subway entrance. Both route planners make use of the same subway line.

With above routing navigation, we can make the following conclusion: Google Maps doesn't support detailed and accurate navigational instructions, only the information to go to subway line 1 with stop Jongno 3-ga. Naver on the other hand is more detailed and connects the walkway from the given address to the entrance of subway line 5 (Jongno 3-ga). The route is continued using the underground subway infrastructure until line 1 is reached. However, details from within this underground infrastructure are not provided.

With the above example, it is shown that Naver knows the available underground structures and entrances. However, the entrance choice is solely based on the shortest route aboveground. In reality, when entering the subway of Jongno 3-ga at entrance 4, the route requires descending over multiple floors and is much longer and more exhausting to walk than walking directly towards entrance 6. As is shown here, knowledge of 3D underground obstacles and structures does affect the optimal route choice but is currently not taken into account.

19.3 Discussion

In the following paragraphs, we will first discuss some more general conclusions with regard to the previously described case studies. We follow the same structure of the examples given. Subsequently, some of the implications and difficulties for immediate development of indoor routing are being discussed.

19.3.1 Problems with Current Indoor Navigation Applications

From the above case studies, several conclusions can be drawn.

First, with regard to the data availability and completeness of the data we can conclude that most route planners do not incorporate indoor infrastructures in route calculations. This is most likely given by a lack of available indoor data (e.g. Bing in *Section 19.2.1.1)*. Reasons for this are likely related to the fact that indoor data gathering has only just begun over the last few years. Also, the geographical area of the query could account for the unavailability of data in some areas, since companies developing route planners will put most effort into areas with the highest commercial value (e.g. European route planners have no detailed data available from the city centre of Seoul). Among route planners which do have some indoor data available, there is a dramatic difference in their level of detail. Data ranges from very rough (e.g. Google Maps in *Section 19.2.1.1)* to quite detailed (Naver in *Section 19.2.1.1* and *Section 19.2.3* and Mappy and Via Michelin in *Section 19.2.1.1)*. When this indoor data is available, the disparate route planners mostly use it integrated with their outdoor networks in the shortest path calculations *(Section 19.2.1.1)*.

The data problem is more pronounced with regard to underground structures. Usually both the entrance points and the underground network are not available *(Section 19.2.1.2)*. Even with the most accurate information available, there are issues in calculating the optimal routes. Although the entrance location and attributes are used as connectors between outdoor and indoor network data, the actual underground network structure is not mapped or known. This results in a lack of knowledge about the 3D infrastructure which can have a detrimental effect on navigation instructions (no indication of how to move in the underground area requires the user to rely on the available exit signs or other information) and calculation of shortest path (the result is mostly not the shortest path because of the movement in 3 dimensions with entrance choice based on the shortest aboveground path). In that case, the route planner uses the knowledge of the various entrances of an underground system and the time needed to move from one to another to calculate the shortest routes.

Secondly, the discussion from *Section 19.2.2* implies that address matching is a problem for both outdoor as well as for indoor navigation. Outdoor address matching links the address to a single entrance/exit point, no matter what the destination of the query is. Not differentiating between the start point of the query with respect to the destination leads to inaccurate routing. Indoor address matching is done through linear interpolation of the indoor network structure (if available). When no indoor infrastructure is available, addresses are matched through projecting the central point on the closest outdoor network edge (see *Sections 19.2.2.1* and *19.2.1.2)*. The accuracy of the storage and location of the addresses is thus of major importance for routing in general and can highly influence optimal routing calculations.

Third, the connection of indoor and outdoor networks is mostly guaranteed when the travel mode remains the same and the entrance data is available (see *Sections 19.2.1.1* and *19.2.3)*. However, changing of mode of locomotion influences the route calculation making the calculations more complex (see *Section 19.2.2.2* shopping centre example). This depends on both the data quality of the indoor-outdoor connection as well as the general accuracy of the outdoor network. This will be an issue for the future expansion of indoor-outdoor navigation applications with optimizations of route calculations.

19.3.2 Indoor Navigation: Product-to-Market Implications

19.3.2.1 Data Acquisition, Standards and Accuracy

Data is the main ingredient for navigation and route planning. Within the area of outdoor navigation applications, a wide variety of data sources is already available from a mix of local and global data providers. The main spatial data providers are Navteq, TeleAtlas and Google. Historically, Holland-based TeleAtlas and American Navteq were interwovenly used in many navigation applications. However, purchases lately of the main data providers by commercially independent navigation producers (Navteq by Nokia and TeleAtlas by TomTom) resulted in individual vouching for your own data set. As a result, Google (who had just signed a deal for using TeleAtlas data) switched to individually conducted data gathering for their US dataset. Additional reasons for this move, were said to be the lack of accuracy and coverage in the United States from the TeleAtlas data (http://blumenthals.com/blog/2009/10/12/google-replaces-tele-atlas-data-in-us-with-google-data/). Google increased with this step its intention as one of the main contenders for spatial data information. From these data providers, no comprehensive efforts have currently been made to expand their spatial data set with ubiquitous indoor data.

As seen in the examples above, data is also crucial in the incorporation of indoor infrastructures in analysis and route calculation. The feasibility of indoor data acquisition is in this regard challenged and unseen. Nowadays, the available spatial datasets are mainly being updated and created using aerial images an mobile mapping vans. These methods are however not suitable for indoor mapping. Technically, a consensus is still lacking on a universal indoor tracking method as solution for the unavailability of GPS signals in buildings. One of the results is that the currently used user input from GPS tracks for updating and editing OpenStreetMap data cannot be applied here unless a ubiquitous indoor tracking system has been developed. Other options for indoor data gathering include photo modeling and laser scanning of individual buildings (Biber et al. 2004); but this is work intensive, expensive and not a comprehensive way of solving the data problem. Currently, many indoor data already exists in the form of for example YAH maps, CAD plans, CityGML or IFC models. These data represent the topographic building structure developed from certain application fields (e.g. structural building development, orientation purpose, evacuation maps). The problem with these indoor data sources is the huge diversity in data structure, completeness, availability, data coverage and level of detail. The area and institutional rules of the country also influence the specificity of the data source. As long as no generally accepted indoor standard is developed or a method to incorporate every possible indoor data source, comprehensive indoor data inclusion will remain challenged (Nagel et al 2010).

In either way, from these data sources correct networks have to be deducted. Since there is still no consensus on an appropriate and mathematically sound relation between data source and network creation for indoor environments, this is an additional problem needed to be solved before real indoor navigation can happen (Nagel et al. 2010). From the OGC and research environment attempts are currently made to develop a general framework and data standard (similar to GDF) for indoor navigation (Nagel et al. 2010). This is a promising step towards creating a background data model which can be used independently of the data input source.

19.3.2.2 Indoor Geocoding Challenges

A second major challenge in indoor navigation and route planning, is the geocoding of the users input to a geographical location or spatial unit. The term geocoding refers to assigning a geographic code based on a certain input information. Mostly geocoding is synonymous with address matching, arising from the prevalent use of transforming postal addresses into geographic coordinates (Goldberg et al. 2007). However, the input source can contain any other type of locational data (e.g. named buildings). Apart from the input, the fundamental components of the geocoding methodology include the processing algorithm, the reference dataset and the requested output (Goldberg et al. 2007). The challenges with the processing algorithm include identification of the separate parts of the input consistent with the reference data set (i.e. standardization and normalization process), matching of the best candidate with reference to the input data and determination of the appropriate geocode for output (Goldberg et al. 2007). The reference dataset consists of the data with which the input data will have to be matched. The output can be any geographically referenced object matching with the input data (Goldberg et al. 2007).

Goldberg et al. (2007) mention frequently induced errors in the outdoor geocoding methodology. With the most commonly used linear interpolation techniques, several assumptions are already made that affect the resulting geocoding accuracy (e.g addresses are assumed to all exists with equal parcel width). This methodology is also only restricted to outdoor address location finding, mostly on street level. However, other methodologies (e.g. area based or hybrid address matching) have similar problems and disadvantages. The reliance of 2D GIS data sources precludes the ability for highly precise geocoding of 3D structures with multiple addresses (Goldberg et al. 2007).

Indoor geocoding is susceptible for even more difficulties. First and foremost, the existing semi-uniformity in outdoor addressing is completely non-existing indoors due to country-related differences and a less rule based structure. For example, a 3D address consist of a 2D building address and a 3D subunit address, describing the location of a building's interior room (Lee, 2009). Lee (2009) suggests a 3D address geocoding methodology. It is based on a two-step process with first determination of the building within the geographical area (following the outdoor geocoding methodologies), followed by a street-like linear interpolation technique applied on an internal network of the building. This approach disregards the problems of discontinuous room numbering, for which transition tables can be a solution. Secondly, a reference dataset for indoor environments is not available. Outdoor geocoding methods mostly use existing street network data set (e.g. TIGER) with the range of house numbers linked to the street intersection or spatial street feature in the database. As long as no standard for indoor data exist, reference datasets will not be available for address matching.

19.3.2.3 General Feasibility Issues

Concluding, we are still far apart from incorporating indoor environments in routing applications. Challenges remain in data availability, storage, network completeness, linkage to the outdoor networks and geocoding. Technical innovations, research and creativity in the routing with less data might improve the feasibility for success in the next years. It is shown that the availability and quality of outdoor and indoor data and their connection is of high importance for the resulting route calculations. It appears that it is not feasible to gather and maintain all indoor data accurately from all buildings in the next years, since this would require a huge amount of data collection and maintenance. However, such a complete data gathering is not always necessary. Even small enhancements in indoor data can have a huge influence on routing (e.g. pointing out all connection points between indoor and outdoor environments, even without the actual indoor network would make the address matching more accurate and would also provide possibilities to have more optimal routes as for example shown in *Section 19.2.3)*. More accurate information will of course result in optimal route calculations.

With all the above mentioned challenges, it is not possible to do a complete data acquisition for a combined indoor-outdoor navigation. We should seek to focus on large infrastructures and transportation networks with more specific navigational directions. The benefit of accommodating navigation in those infrastructures is bigger since a lot of people daily use and rely on those. These structures are also quite often fixed and stable over long periods of time, making the indoor data gathering and maintenance also more feasible. As is shown in the examples, the 3 dimensional network aspect is here of major importance to enhance routing for everyone.

An important role in data acquisition and address matching will be for the public. Over the last year, an increase has been seen in the public participation for outdoor data following the success of the data acquisition in OpenStreetMap (i.e. Wikipedia style updating and editing of data). This was noticed and built upon by other internet based applications and could also be a solution for indoor routing applications. Already at this moment users can change addresses and location of addresses for outdoor routing. Once the technology is ready for continuous indoor tracking and more user input is allowed, this could open up the indoor world too.

19.4 Conclusion

With this comparison of how current route planners use indoor infrastructures in the calculation of pedestrian routes, several active problems with this indoor-outdoor connection are identified. The most stringent limitation of current route planners in this realm is the availability of accurate data of indoor infrastructures. This data should consist of network information, additional semantic enrichments and all entrance points. As can be seen from the examples above, nonexistent or inaccurate information can lead to sub optimal routing, and even to a lack of routing in many cases. However, when the appropriate data is available, very precise routing information is proven to be calculated. It is pointed out that even small data additions, such as entrance and exit points of major infrastructure projects, can have a huge influence for pedestrian routing. Secondly, outdoor address matching techniques cannot directly be applied to indoor datasets. Immediate indoor-outdoor connection for navigation applications still have a long way to go. This research fits in with the ongoing awareness of indoor and outdoor navigation and more specifically it gives an overview of the data requirements for navigational applications. Future applications will more often focus on this indoor-outdoor connection, not only in navigation but also for wider analyses and applications.

Acknowledgements

Financial support from the Flanders Research Foundation (FWO-Vlaanderen) is gratefully acknowledged.

References

Anagnostopoulos C, Tsetsos V, Kikiras P, Hadjiefthymiades SP (2005) OntoNav: A Semantic Indoor Navigation System. Department of Informatics and Telecommunications, University of Athens. http://sunsite.informatik.rwth-aachen.de/Publications/CEUR-WS/Vol-165/paper14.pdf. Accessed 13 May 2011

Becker T, Nagel C, Kolbe TH (2009) Supporting contexts for indoor navigation using a multilayered space model. Proceedings of the 2009 Tenth International Conference on Mobile Data Management: Systems, Services and Middleware: 680–685

Biber, P., Andreasson, H., Duckett, T., & Schilling, A. (2004, 28 Sept.-2 Oct. 2004). 3D modeling of indoor environments by a mobile robot with a laser scanner and panoramic camera. Proceedings of the 2004 IEEE/RSJ International Conference on Intelligent Robots and Systems (IROS 2004).

Bogdahn, J., & Coors, V. (2009). Using 3D Urban Models for Pedestrian Navigation Support. Paper presented at the GeoWeb 2009 Academic Track – Cityscapes, Vancouver, BC, Canada.

Coors V (2003) 3D-GIS in Networking Environments. Computers, Environment and Urban Systems 27(4): 345–357

Gartner G, Cartwright W, Peterson MP (2007) Location Based Services and Telecartography. Springer-Verlag, Berlin Heidelberg

Goldberg, D. W., Wilson, J. P., & Knoblock, C. A. (2007). From text to geographic coordinates: the current state of geocoding. Journal of the Urban and Regional Information Systems Association, 19(1), 33–46.

Huang H, Gartner G, Schmidt M, Li Y (2009) Smart Environment for Ubiquitous Indoor Navigation. International Conference on New Trends in Information and Service Science, NISS '09: 176 –180

Jensen CS, Lu H, Yang B (2009) Graph Model Based Indoor Tracking. Proceedings of the 2009 Tenth International Conference on Mobile Data Management: Systems, Services and Middleware: 122–131

Karas I, Batuk F, Akay A, Baz I (2006) Automatically Extracting 3D Models and Network Analysis for Indoors. In Abdul-Rahman A., Zlatanova S., Coors V (Ed.): Innovation in 3D-Geo Information System. Springer, Berlin Heidelberg: 395–404

Lee J (2001) 3D Data Model for Representing Topological Relations in Urban Features. Delaware County Regional Planning Commission, Delaware. http://gis.esri.com/library/userconf/proc01/professional/papers/pap565/p565.htm. Accessed 13 May 2011

Lee J (2004) A spatial access-oriented implementation of a 3-D GIS topological data model for urban entities. GeoInformatica 8 (3): 237–264

Lee, J. (2009). GIS-based geocoding methods for area-based addresses and 3 D addresses in urban areas. Environment and planning B, Planning and design, 36(1), 86–106.

Li Y, He Z (2008) 3D Indoor Navigation: a Framework of Combining BIM with 3D GIS. Proceedings of the 44th ISOCARP Congress 2008

Löchtefeld, M., Gehring, S., Schöning, J., & Krüger, A. (2010). PINwI: pedestrian indoor navigation without infrastructure. Proceedings of the 6th Nordic Conference on Human-Computer Interaction: Extending Boundaries.

Lyardet F, Szeto DW, Aitenbichler E (2008) Context-Aware Indoor Navigation. In: Hutchison D, Kanade T, Kittler J, Kleinberg JM, Kobsa A, Mattern F, Mitchell JC, Naor M, Nierstrasz

O, Rangan CP, Steffen B, Sudan M, Terzopoulos D, Tygar D, Weikum G (Ed.) Ambient Intelligence. Springer, Berlin Heidelberg 5355/2008:290–307

Mautz R, Kunz M, Ingensand H (2010) Proceedings of the 2010 International Conference on Indoor Positioning and Indoor Navigation (IPIN). IEEE Xplore: CFPI009J-ART. doi: 10.1109/IPIN.2010.5648185

Meijers M, Zlatanova S, Pfeifer N (2005) 3D Geo-Information Indoors: Structuring for Evacuation. Proceedings of Next generation 3D City Models.

Nagel, C., Becker, T., Kaden, R., Li, K.-J., Lee, J., & Kolbe, T. H. (2010). Requirements and Space-Event Modeling for Indoor Navigation, OGC Discussion Paper (pp. 54). OGC: OGC.

Pfaff P, Kümmerle R, Joho D, Stachniss C, Triebel R, Burgard W (2008) Navigation in Combined Outdoor and Indoor Environments using Multi-Level Surface Maps. Department of Computer Science, University of Freiburg. http://citeseerx.ist.psu.edu/viewdoc/summary?doi=10.1.1.69.9146 Accessed 13 May 2011

Ran L, Helal S, Moore S (2004) Drishti: An Integrated Indoor/Outdoor Blind Navigation System and Service. Second IEEE International Conference on Pervasive Computing and Communications (PerCom'04)

Retscher G, Thienelt M (2004) NAVIO – A Navigation and Guidance Service for Pedestrians. Journal of Global Positioning Systems 3(1–2): 208–217

Scooter W, Helal S (2005) RFID Information Grid for Blind Navigation and Wayfinding. Proceedings of the Ninth IEEE International Symposium on Wearable Computers: 34–37

Slingsby A, Raper J (2007) Navigable Space in 3D City Models for Pedestrian. In van Oosterom P, Zlatanova S, Penninga F, Fendel E (Ed.) Advances in 3D Geoinformation Systems. Springer, Berlin Heidelberg: 49–64

Tsetsos V, Anagnostopoulos C, Hadjiefthymiades S (2007) Semantic Enrichment of Location-Based Services. In Taniar D (Ed.) Encyclopedia of Mobile Computing and Commerce: 856–862

Walton, L. A., & Worboys, M. (2009). Indoor Spatial Theory. 6 pp. Retrieved from www.spatial.maine.edu/ISAmodel/documents/IST_ISA09.pdf

http://blumenthals.com/blog/2009/10/12/google-replaces-tele-atlas-data-in-us-with-google-data/

http://ifgi.uni-muenster.de/archives/photomap/Home.html

www.makkamappa.com

Chapter 20

LatYourLife: A Geo-Temporal Task Planning Application

Amin Abdalla

Institute of Geoinformation and Cartography,
Vienna University of Technology, Austria
amin.abdalla@tuwien.ac.at

Abstract

The aim of this research was to build a prototype application which uses state-of-the-art web-mapping and mobile technologies in order to connect people's daily tasks with geography and therefore serving them with a new perspective on task planning. This was achieved by implementing a spatial calendar, hence a scheduling application which not only stores the temporal information about each event, but also a spatial component. The major purpose was to explore the potential functionalities geo-temporal data could provide to improve task management. The application was field tested by users who responded positively, but simultaneously highlighted issues that an application like this would need to tackle, in order to be successful.

20.1 Introduction

A *Location-based-service* (LBS) is defined by Koeppel as: "*any service or application that extends spatial information processing, or GIS capabilities, to end users via the Internet and/or wireless network*" (Koeppel, 2000). Jiang and Yao argue that "*a true LBS application aims to provide personalised services to mobile users whose locations are in change*" (Jiang andYao, 2007).

Location-based services are often praised as the most promising application for the near future. Especially after the emergence of so called 'smart phones', which in many cases incorporate GPS sensors and considerable computing power, mobile users are equipped with technology which was recently exclusive to professionals.

G. Gartner, F. Ortag (Eds.), *Advances in Location-Based Services*, Lecture Notes in Geoinformation and Cartography, DOI 10.1007/ 978-3-642-24198-7_20,

The number of potential applications built upon these new possibilities of location aware devices is huge, and in fact there are already hundreds of *apps*[1] available for purchase in mobile stores, using location awareness as a tool to provide the user with context relevant information.

The range of topics covered by already existing or prototyped services include restaurant, cash machine or real estate search queries (e.g. Aloqa[2]), as well as social services (e.g. Hong *et al.*, 2004; Sohn *et al.*, 2005; Miluzzo *et al.*, 2008) or even location based gaming (Lonthoff *et al.*, 2007).

According to Jiang and Yao (2007) the questions a LBS-user is concerned about are: Where am I currently? What and where are the nearest Locations of interest? How to get there?

Or as summarised by Frentzos *et al.* in 2007, the current LBS solutions consist of three fundamental services: *What-is-around, Routing and Find-the-Nearest.*

Nevertheless, it was stated that current services "*...are rather naïve, not exploiting current software capabilities and the recent advances in the research fields of spatial and spatio-temporal databases.*"

Research fields like predictive modelling and the exploration of data collected from LBS (e.g. Karimi and Xiong, 2003; Ashbrook and Starner, 2003; Thi Hong *et al.*, 2009; Nanni and Pedreschi, 2006) could significantly benefit from these advances.

Frentzos *et al.* proposed three extensions of the current services in order to ensure the development of next-generation LBS. Amongst those the *Get-together* proposal, in which a meeting point for several users is calculated based on the future projection of the calling user's trajectory, is probably the most relevant to this research. It points to the integration of location predictions into mobile computing.

The following pages talk about the LatYourLife-application (LYL), a LBS which uses web-mapping and mobile technology to assist people in their daily life task-planning and tackling. It connects tasks and events with geographical information and attempts to make innovative use of the geo-temporal data. Therefore the application not only deals with the three services listed above, but also attempts to create a *When-to-act*-service. This is achieved by integrating route time predictions as a basis for a pro-active alert feature. The main challenges of this application are (1) the design of a user interface that is focused around a map rather than a calendar (2) the constant monitoring of the users location and adaption of the system to it.

[1] Mobile applications

[2] http://www.aloqa.com

20.2 Background

The idea of using location as an assisting factor for managing tasks is not new and was pioneered by Lamming and Flynn in 1994 with the "forget-me-not" application, a tool which enables the user to store a "biography" consisting of information about personal location, encounters with others, workstation activities, file exchange, telephone calls and other information. Based on psychological theories, which say that physical context could be helpful for recalling information out of memory (Barsalou,1988) the application tries to mimic the so called episodic or autobiographical memory, which therefore should help people remember specific things.

More recent examples of such reminders are functionalities known as "*location based alerts*", which is basically an attempt to remind people of errands by invoking alarms according to the position rather then time.

ComMotion (Marmasse and Schmandt, 2000) for example, uses learning algorithms to determine frequent locations and enables the user to apply reminders to these points (e.g. a shopping list could be attached to a grocery store, which would then be alerted when passing by it).

CybreReminder (Dey and Abowd, 2000) tries to create complex situation reminders based on a combination of time, location and other criteria. Very explicit situations could be specified, such as a reminder which is triggered only in case you are at a particular point at a particular time with or without particular weather conditions applied for example.

A user study conducted by Sohn et al. in 2005 revealed that, peoples reactions to the concept of location based reminders are quite positive. The study supplied 10 participants with mobile phones and a "Place-Its" application which "...is designed around the post-it note usage metaphor, and named for its ability to "place" a reminder message at a physical location (i.e. a place)."(Sohn et al.,2005)

Sohn et al. conclude that "*...the prevalence of mobile phones and the pervasiveness of their networks makes them a promising platform of personal ubiquitous computing. ...On the whole it appears that the convenience and unbiquity of location-sensing provided by mobile phones outweights some of their current weaknesses as sensing platform.*"

By now plenty of such applications are available for purchase on the market. (e.g.: reQall[3]; Geominder[4]; Proxido[5]; Synchrospot[6])

In terms of personal task planning or scheduling, geographic information has

[3] http://www.reqall.com
[4] http://www.ludimate.com/products/geominder
[5] http://www.hollowire.com/proxido.
[6] http://synchrospot.com/android/home.do

not entered the mainstream applications yet. Typical calendar application like the Microsoft Outlook[7] or Google Calendar[8] are based on traditional time interfaces, with Google Calendar being able to show event locations on a map, given a provided address. The lack of creativity in incorporating GIS and LBS in such applications seems baffling, considering the benefit such integrations could bear. The mobile version of Google Calendar resembles a ported version of the web-app and does not make use of the potential so-called smart phones provide.

Potential which is probably best illustrated by the CenceMe-application (Miluzzo *et al.*, 2007; Miluzzo *et al.*, 2008), an app which makes comprehensive use of facilities offered by the devices and enables the users of social networks to share personal information sensed by the mobile phone, with their friends.

Besides the ability to share the current location, it uses algorithms to classify the sensed data and recognises whether the user is alone, attending a party or walking on the street, driving or running, etc. The integration of such a service into a task planning application is interesting in terms of group or office management.

A pioneer diary study in 2000 (Colbert, 2000) examined the *rendezvousing* behaviour of students for a month and concluded that 21% of unsuccessful rendezvous were due to the lack of information about the other attendees and suggested that location information about them could enhance the ability to plan events.

An application derived from these conclusions and possibly the most similar work preceding the LYL-application, was designed and evaluated by Fithian et. al in 2003. The mobile Location-aware handheld event planner provided the user with basic event information and location details about users attending the event. The software already implemented *passive* route time predictions, such that users are able to see the estimated travel time to the event. The study though did not evaluate the benefit users could gain from such an application.

Talking about location tracking and awareness of mobile devices, privacy issues have to be part of the discussion. The idea of knowing where a person is at any time of the day raises justified concerns amongst academia and society.(Dobson and Fisher, 2003; Halliday, 2010)

The application built in this research holds very private and detailed information about the user's activities, but since there was a lot of work undertaken on the topic (Hong and Landay, 2004; Chow and Mokbel, 2009; Ghinita et. Al, 2009), this paper will not go into discussing the issue, although it is crucial to be aware of it when developing such applications.

[7] http://office.microsoft.com/en-us/outlook.
[8] http://www.google.com/calendar/.

20.3 User Requirements

The diary study undertaken by Colbert identified eight possible causes for rendezvous problems (Colbert, 2000). 37% of them were caused by the mode of travel, 25% were caused by the over-run of previous activities and 21% by the lack of information of other attendees. 10% were contributed by the lack of geographic information and 7% by the lack of travel information.

The statistics show that 17% of the problems with rendezvous were related to geography and could therefore be tackled by supplying the user with this kind of information. Thus the core purpose of the research was to explore and probably implement some of the new functionalities made possible by the geo-temporal information supplied by the users and address the 17% mentioned above.

Many studies have focused on the user requirements for LBS in terms of navigation in an urban area (e.g. Baus *et al.* 2002; Borntraeger *et al.* 2003) or the requirements for specific user groups (elderly people: Osman *et al.* 2003; blind people: Klante *et al.*, 2004). But a study particularly relevant for the LYL-application was conducted in 2008 by Nivala *et al.* and analysed the potential users and their tasks during a hike. The resulting user requirements were grouped into three different phases of the hike: *before*, *during* and *after*.

Thus the users need information to plan the hike in the *before*-phase, require services *during* the hike (e.g.: location of other hikers, navigation) and finally demand adequate functionalities *after* the hike (e.g.: sharing experiences).

Using the analogy of hiking, planning a day could be divided into the same three phases, which call for similar services. In the *before*-phase, the functionalities should help the user in his process of decision making, by providing useful (geographical) information incorporated into the planning interface.

The features in the second stage should assist users in their ambitions to achieve the aims mapped out for the day. In the last stage the user should be able to look back at past events and review information about it.

Dependent on the type of stage and its requirements, either a web- or mobile platform is more suited for implementing such services. A desktop-computer is not be the best mean to remind the user of something, as its static nature limits the occasions such a reminder could achieve its goal. A mobile device in comparison would be more appropriate, as it is in close proximity to the user and operates almost 24 hours a day. Hence the built System consists of three components:

1. A Web Interface (used in the *before*- and *after*-phase)
2. A Spatial Database (important for the *after*-phase)
3. A Mobile Application (mainly for the *during*-phase)

The decision to include a web-based interface was made in order to separate the mapping and planning of the events from the mobile application, based on the fact that "...*simplicity of use is a key issue. Currently, users rather refrain from using LBS if it requires manual data entries. Usage must be intuitive, results should be reachable within 3 clicks*"(Uhlirz, 2007)

Together the parts build a structure enabling the user to conveniently plan and map upcoming tasks on a desktop computer and then navigate through the day having the information carried away on the mobile device.

Since "...*users are central to LBS ... LBS applications should be designed based on a user-centred view*." (Jiang and Yao, 2007)

In order to comply with the above statement and investigate which functionalities are demanded for the *before*- and *during*-phase, a brief user study including 10 smart phone owners with different backgrounds and ages was conducted.

A questionnaire containing a list of three proposed features for the web interface and three for the mobile application, as well as a section for suggestions, was handed out. The participants were asked to assess the proposed functionalities according to what they thought is useful or not and urged to come up with their own ideas of what services could be useful for such an application.

Table 20.1. The sorted list of functionalities according to the user assessment (1= very useful; 5 = not useful).

Functionality	Average
one click navigation to next appointment	1.4
geographic planning aid	1.5
punctuality alert	1.7
group calendar/map	2.1
share your location	2.1
location based reminder	2.4

The outcome of the study showed that the participants were quite generous with their assessment, as no functionality exhibits an average rating higher than three, which leads to the conclusion that all of them are perceived as relatively useful.

Nevertheless the ability to share their location and event information, as well as location based alerts, were seen as less valuable.

The following points were suggested by the users:

- weather integration
- ability to view the event location (e.g. a picture of the location)
- telephone number of the venue
- navigation based on Google Street View

The Routing, Geographic Planning Aid and Punctuality Alert were implemented due to their high ranking in the list (see *Table 20.1)*, whereas Weather Integration was chosen to be important because it was the only user suggestion which occurred twice.

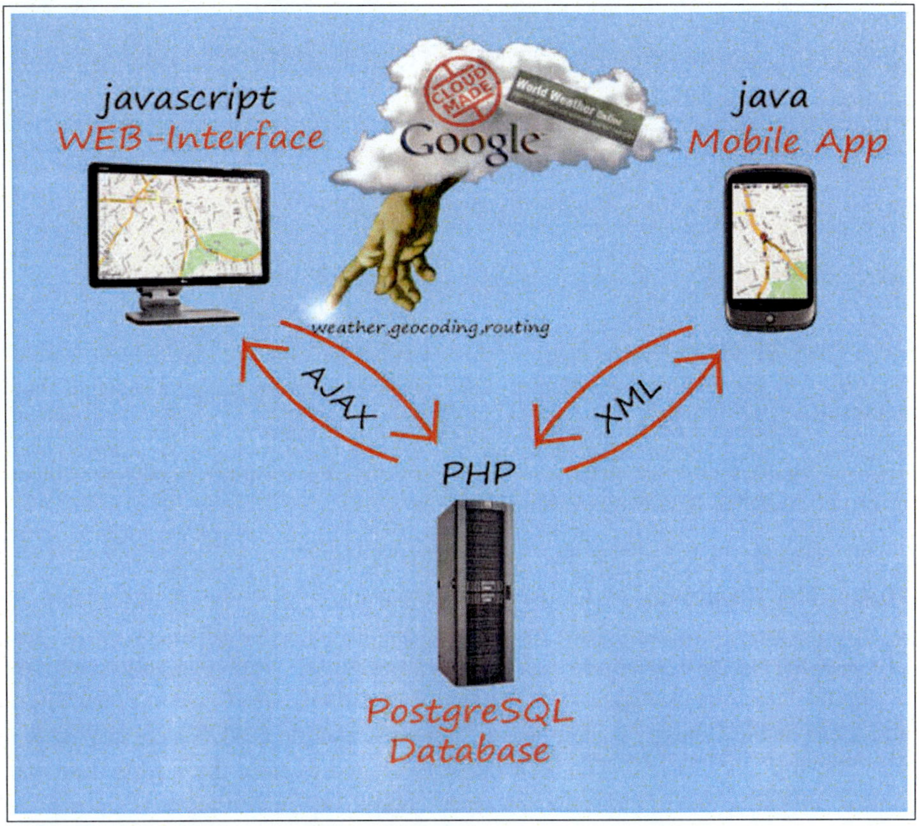

Fig. 20.1. An abstracted graphical illustration of the System Structure.

Location Viewing, another user suggestion, was chosen to be incorporated due to the simplicity of its implementation.

20.4 Application Architecture

The application architecture was designed to provide the user with a web interface as well as a mobile component which are tied together through a database.

The communication between the web interface and the database is achieved by using AJAX[9] requests, to keep the interface responsive and avoid loading times when updating the event information. PHP[10]-files are handling the interaction with the database and are chosen due to the open nature of the scripting language. Since the acquisition and management of routing or weather data is a rather complicated

[9] Asynchronous Javascript and XML
[10] Hypertext Pre Processing (PHP)

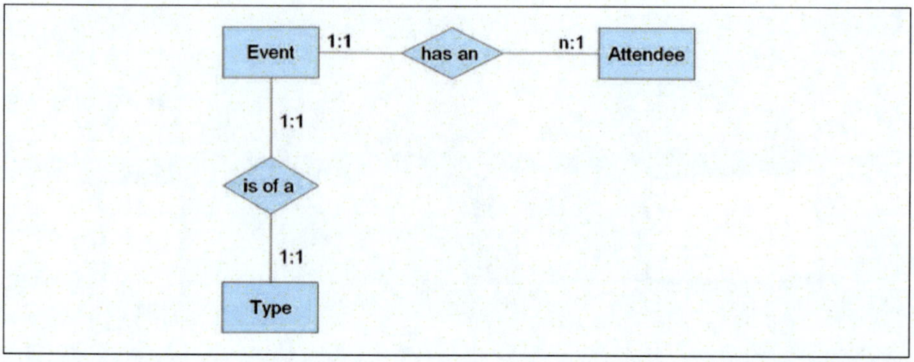

Fig. 20.2. A very simple, but satisfactory ER model to store the basic information about the events.

undertaking, Web APIs[11] are a convenient and elegant solution to acquire related information based on the geographic location.

20.4.1 The Database

The underlying database was build with PostgreSQL, an open-source object-relational Database Management System (PostgreSQL, 2010). The open nature and the possible storage of geographic components in the DBMS were key factors for choosing the software. The data model designed to store the information was kept simple in order to avoid unnecessary complexity in further development of the application. An ER-diagram[12] was used to describe and design the model (see *Figure 20.2)*. For the sake of simplicity, it is assumed that an Event could only be related to one User/Attendee, but the design could be expanded, so that users are able to share events with others.

20.4.2 The Web Interface

The core purpose of the interface is to plan and set events on the basis of a map as opposed to the traditional time based calendar interface. The display is divided into three different parts: a Sidebar, a Map and a Calendar. The Sidebar shows general information such as weather, date, time and details of the next upcoming Event. The Calendar is initially hidden and can be opened by the user.(see *Figure 20.3)*.

Events can be added onto the map and the week calendar on the right allows the user to edit time and information about them. An input form stores, updates or deletes the events in the database.

[11] Application Programming Interface

[12] Entity Relationship

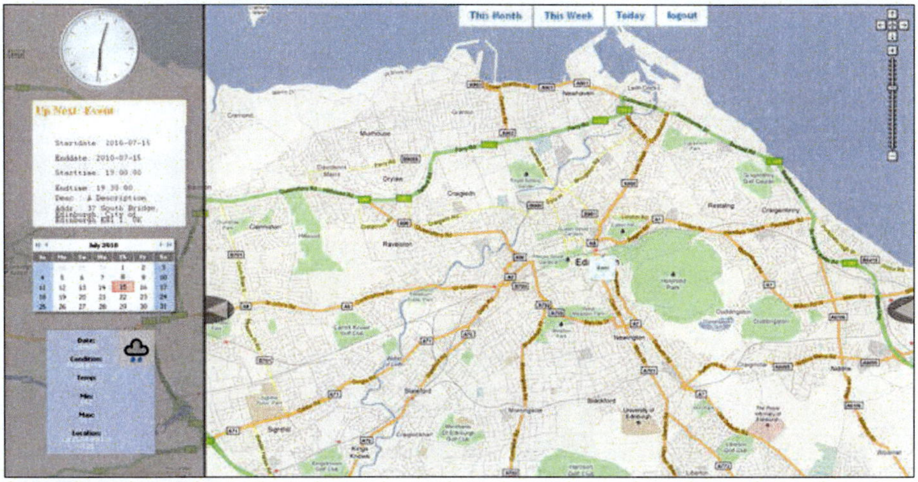

Fig. 20.3. The Web-Interface as it looks in its initial state. The Sidebar on the left contains general Information, the map in the centre shows the Events scheduled.

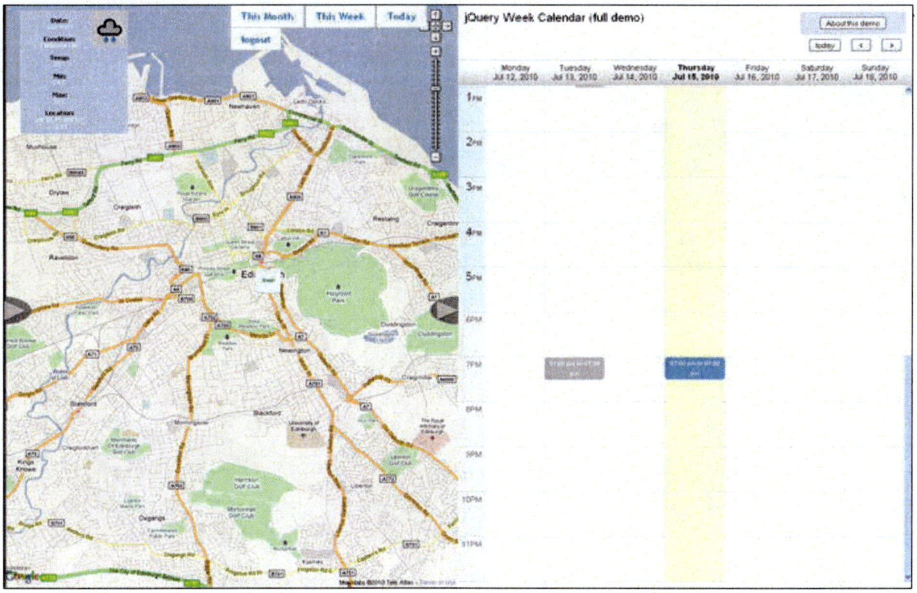

Fig. 20.4. The Web Interface with the calendar opened on the right side. Events are visualised geographically and temporally.

Fig. 20.5. The input form is used to store the event details once it was located at the map

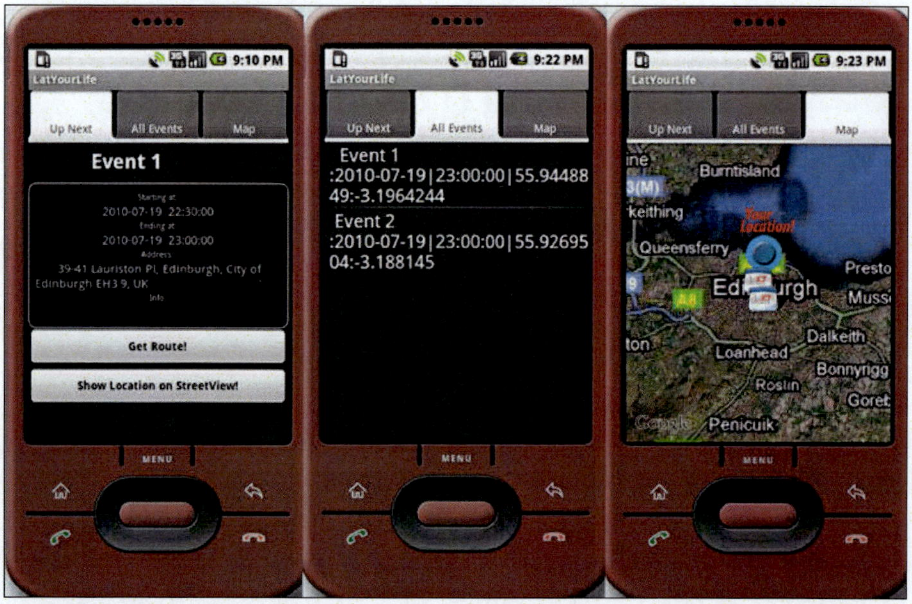

Fig. 20.6. Three of the four different views from the mobile application interface

20.4.3 The Mobile Application

The mobile application was developed for the Android Platform[13], an open source software stack created for mobile phones and other devices. The platform was chosen

[13] http://developer.android.com/index.html

due to its accessibility and coding in the java programming language. The mobile application is built out of four different user interfaces and a "service" running in the background evaluating the user's position, downloading route information and setting the time for notifications.

The application takes its information directly from the database, utilising php-generated xml-files.

Figure 20.6 shows three of the four views of the mobile application, from left to right the first view provides information about the next scheduled event, the middle view shows a list of all events scheduled for the current day and the last window contains a map with all today's events and the users current location mapped. The fourth view is simply an interface which enables the user to edit some of the event details.

Since the application is very much a prototype implementation, interface design and information display remains a matter of improvement.

20.5 Implemented Functionalities

The features implemented in the application were chosen upon their ranking in the user study and considerations in terms of technical complexity.

20.5.1 Geographical Planning Aid

This particular functionality was meant to support the user when taking decisions about setting events at different locations and time. So if it takes 30 minutes for the user to get from Event A to Event B, but the time available to be there is less than that, the application will signal the user about possible problems.

To execute such an algorithm specific information is needed:

- The time when Event A is ending
- The time when Event B is starting
- The time it will take to travel from A to B (this implies the intended mean of travel)

Once the information is provided an algorithm for evaluating the decision taken by the user can be executed.

$$\text{If } (TT < AT) \{ \text{ TRUE } \} \text{ else } \{ \text{ FALSE } \}$$

$AT_{AB} = ST_B - ET_A$

TT_{AB} = Estimated minimum travel time between Event A&B

ET_A = End time of Event A

ST_B = Start time of Event B

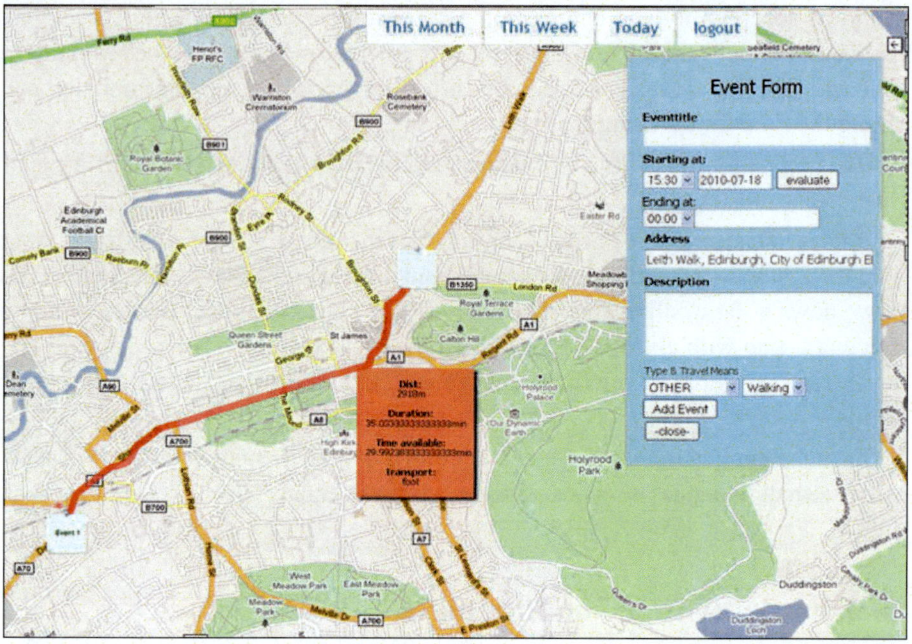

Fig. 20.7. The estimated travel time for walking from Event 1 to the next planned Event exceeds the time available.

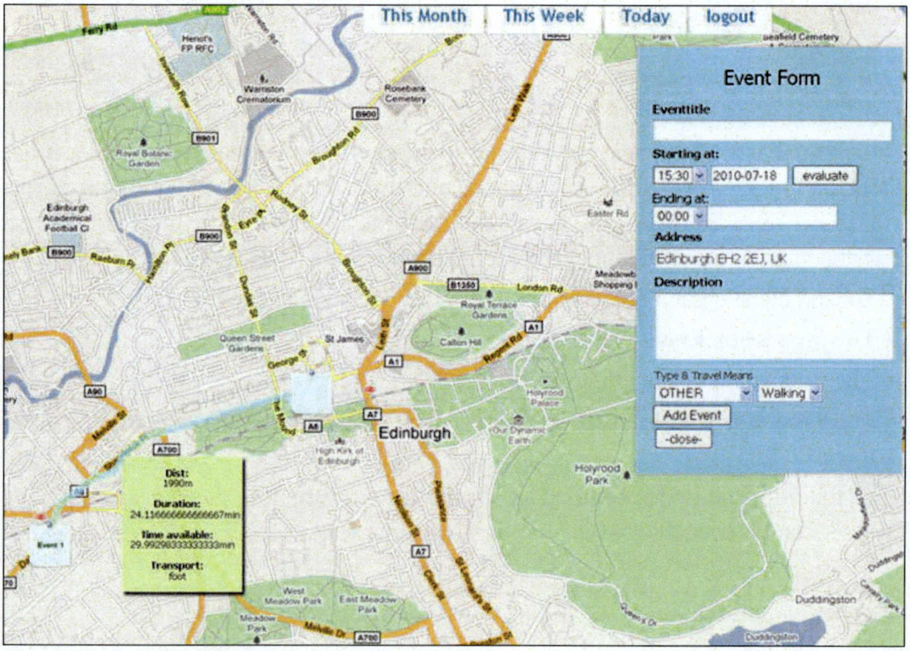

Fig. 20.8. Changing the location of one of the events is a possible solution to solve the problem.

The equation above is a formalisation of how the geographical planning aid is implemented. If the application detects a contradiction in the planning of events, e.g. the time to get to from event A to B takes longer than the time available, it alerts the user by overlaying a red coloured polyline to the map.

In such a case the user does have the ability to alter one or more of three parameters in order to reach Event B on time:

- Location
- Time
- Travel Mean

By having a visual feedback from the system the user is able to assess travel times between events and therefore plan them according to it.

20.5.2 Weather Information

The integration of weather information was suggested by two of the ten participants and therefore prioritised. Weather information could potentially change behaviour in terms of space, time or transportation. Rain might deter the user from using a bike for example, or even erase the event completely from the map. Although weather information could potentially be acquired by the user from other web sources, the tight integration of the information into the application transforms it into an integral part of the planning process.

Ideally the weather information would be added to each of the events according to time and venue, but for the sake of simplicity the web component shows weather details only for the location the map is centred on. The weather data is acquired by using an API[14]. The implementation showed that weather information could easily be tied into such an application on the basis of geographical coordinates and temporal details.

20.5.3 Routing & Location Viewing

The ability to get instant routing information on your mobile device was assessed to be the most useful feature in the application. One of the main advantages of using the android platform for mobile development is its design which "...encourages the concept of reuse..."(Meier, 2010). The platform allows the developer to use other applications data or functionalities. Therefore the Google Maps-App[15], which is very likely to be preinstalled on android devices, was integrated to handle navigation and showing event locations on Google Street View.

[14] http://www.worldweatheronline.com
[15] http://www.google.com/mobile/maps/

Fig. 20.9. Directions are fetched and displayed by the Google Maps Application on the device

The application automatically recognises which event the user is supposed to be heading towards and gets the directions to it from the current location.

20.5.4 Punctuality Alert

This functionality is the second feature (after the Geographic Planning Aid) which attempts to make use of the geo-temporal data provided. The aim is to notify the user ahead of the next scheduled event, in order to ensure punctuality. The notification time is set according to the user location, so that the alert would be triggered earlier if the user is located further away than it would be the case if near by.

The required information for implementing such an algorithm are:

- Current Location of the user
- Starting Time of the next Event
- Travel Time from the current position to the Event position (implies knowledge of the travel mean)

The equation below formalises the procedure necessary to acquire the time for scheduling the alert. In order to provide the user with some time to prepare for the departure additional 10 minutes are subtracted. For this prototypic implementation it is assumed to be sufficient, but ideally this amount will be chosen by the users themselves.

$$AT = ST_{NE} - TT_{CE} - 10$$

AT = Alerttime
ST_{NE} = Event Starttime
TT_{CE} = Traveltime from Current Position to the Event

It shows that the process is done in four steps:

1. Fetch the necessary event details from the database
2. Acquire the travel time through the API used to gain route information
3. Apply the above quoted equation
4. Set the Alert

The location listener, shown in green, is integrated to track the user movement and restarts the algorithm from step 2 in case a location change of more than 50 metres is detected. Having the algorithm rerun again, Step four will then update the alert time according to the new location and travel time.

The application in this research is restricted to walking, cycling and driving times, since public transport information is difficult to acquire and monitor. Ideally a genuine public transport web API would be provided, which services could query according to geographic location and time.

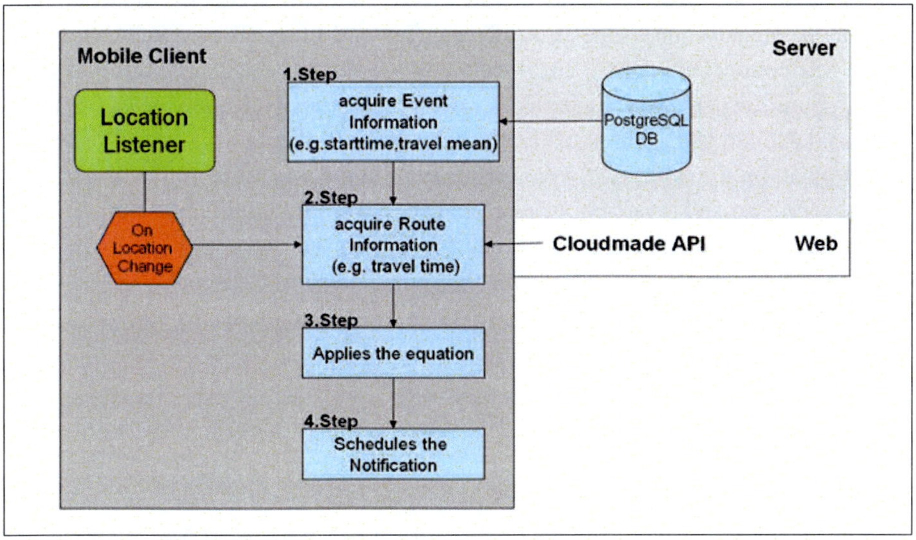

Fig. 20.10. Visualisation of the evaluation process applied for the Punctuality alert

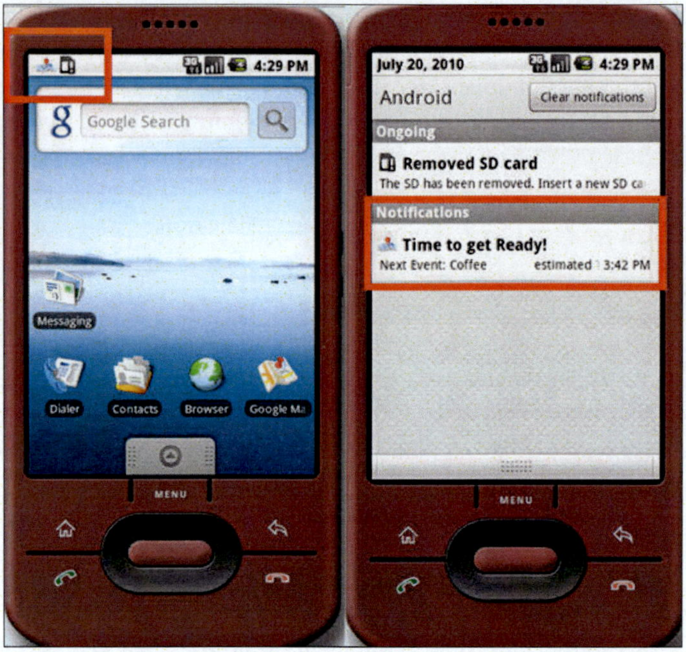

Fig. 20.11. The user is notified in the notification bar (upper left frame) when it is time to depart.

20.6 User Feedback

User feedback was crucial to evaluate the functionalities and benefits of the application, unfortunately due to the lack of resources and time, no comprehensive user study was conducted. Thus a usability study with five participants, three of them smart phone owners, was undertaken. The participants were between 20 and 30 years old and students. Three of the participants used their own phones for testing the application, the other two were equipped with a compatible mobile device and had never possessed smart-phones before. They were asked to choose a particular day, which they thought to be adequate for testing such an application. All users confirmed that walking or cycling was their main mean of transport.

The following are summaries containing the core points of the interviews conducted with the participants after their field testing.

User A

User A positively mentioned the fact that "*...you can have a diary of all the places you have to go*", which allegedly gave him a different perspective and better geographic understanding of his schedule.

An issue raised by this participant was that of the event geometry. This was identified when the user planned a drive from University to the user's hometown. The application currently only supports the storage of events in form of single-points, this means the starting time of an event represents the time the user should be at the mapped location.

Instead the user mapped the event at the location of his final destination, but set the start time to be the time intended to leave from University. This made the application assume that the user has to be there at the time he wants to head off and notified him far too early.

This confusion highlighted the fact that events might start and end at different geographic locations. Therefore a double-point representation of those would be more appropriate.

User B

User B reported an occasion on which the device did not notify him about when to leave until he reached the events location. The situation highlighted a second unaddressed matter of the application, the *approaching user*.

The Punctuality Alert updates its notification time according to the position of the user. Once the user is notified it will not trigger another notification for the same event again. The issue originated in the fact, that the user started walking towards the event before receiving a notification, which led to a constant update and postponement of the notification time. Departing before the notification lets the user potentially reach the location 10 minutes earlier to the appointment. Once the user reached the location the application stops updating the alert time and will then trigger the notification although the user is already there.

User C

The User utilised the application for a trip to an unfamiliar city. Since it was the first time the user visited the place and as there were several appointments to meet the day appeared to be an ideal setting for testing the application. Time was spent to map out the appointments the night before, in which weather and geographical information allegedly gave a good basis for planning the day. The geographic planning aid was utilised by the user to estimate travel times from one point to the other.

When the user arrived in the city at a bus station several hours time were left until the next bus was departing from the same location. Having the application made it possible for the user to conveniently explore the surroundings of the place. The application allegedly gave the user the ability to see how long it will take to the next appointment and therefore allowed the assessment of how far the user could move away from it. "*The application gave me the convenience of moving freely through the city, having in mind that it will tell me when to head back and being able to easily navigate to my next appointment*"

On the other hand the user criticised that the notification is very short and might be overheard in a noisy environment such as a street. It was also mentioned that in some occasions the notification was not triggered at all, which made the user very suspicious about the reliability of the feature.

User D

User D was similar to user E in that they were not smart phone owners. The user was therefore provided with an android phone. Nevertheless the user was able to immediately understand and interact with the mobile application which proved the simplicity of the interface design.

The weather information was perceived positively since the user planned an outdoors activity. Criticism was issued due to the fact that the web interface does not exhibit an address search functionality which, according to the user, is crucial for locating some of the event positions. It was stated that "*the punctuality alert is a very useful feature, but is in its current stage just not reliable enough*".

User E

User E was the second participant who had no significant experience with smart phones but similar to User D did not come across difficulties to interact with the application. The participant found the web and mobile interface to be easy understandable and liked the fact that both were not over loaded with information and buttons.

Similar to prior users an address search was missed in the web interface, as well as event input functionalities for the mobile application.

The user was intrigued by the map as a basis for arranging events and likened the fact that the application could potentially save time and stress by taking away the need of looking up directions prior to the head-off. What the user found most beneficial was that the application lets you focus on other things prior to an appointment since it does the work of monitoring the time for you.

It was stated that "*...the application has great potential as it saves time and takes away a lot of worries you usually have before an appointment.*"

20.7 Feedback Summary

Most users agreed that a web interface is a useful planning tool, but insisted that the mobile application should provide planning abilities, since events could be changed or cancelled during the day. The web interface was perceived by most of the users as clear and easy to use, while some added that it lacked the ability to provide address searches, since they were not able to find particular locations simply by looking at the map. All users did mention problems with the reliability of the alert, which in the author's point of view was mostly due to the lack of gps-reception indoors. Another common point of criticism was, that the permanent usage of gps drained the battery and resulted in a total shut down of the phone by the end of the day.

20.8 Conclusion

The research project implemented a prototype geo-temporal task planning application, which provided the users with new abilities to manage their tasks and errands. It was successful in addressing a *When?*-question, using simple route time estimations.

The survey demonstrated that the main benefit of the punctuality alert is not improved punctuality but the convenience the user gains by not being forced to continually verify the time. Simultaneously there was a major concern about reliability of the functionality. The notification did not prove to be stable, as the majority of the users reported missed or misplaced notifications. The issue highlighted some of the limitations which were present in the technical implementation of the idea, such as location data accuracy and data acquisition (e.g. public transport data). The study proved that precise location information is not easy to get hold of. Ideal GPS reception is not always given and the accuracy of it might vary from a few metres to a several ten meters, which could lead to inaccurate alert times. Other matters included the estimation times of the routes and the usage of several transport modes to reach an event. Overall the punctuality alert is a feature in need of deeper exploration to enhance efficiency and reliability, which potentially involves the incorporation of predictive modelling (e.g. Karimi and Xiong, 2003; Ashbrook and Starner, 2003; Thi Hong *et al.*, 2009) as mentioned in the introduction, or the integration of contextual traffic information.

An important point highlighted by the interviews is that the application was perceived most useful in unfamiliar environments. But especially when abroad users often do not have an internet connection available. Therefore it is necessary to think about what parts of the services can be provided offline.

Whilst the application proved to be of great value to the users, questions of technical and theoretic nature remain. Issues like (1) How can we model "tasks" in a more appropriate way, (2) and what are their spatio-temporal properties? Related to this is the cartographic presentation of events. (3) How can temporal factors be visualised on a map and (4) would size for example be an appropriate means to help the user distinguish between closer events or events further away in time?

In future research we hope to gain answers to these question.

References

Ashbr international symposium on Advances in geographic information systems. New Orleans, LA, USA, pp. 126–133

Klante P, Krösche J, Boll S (2004): Accessights – A Multimodal Location-aware Mobile Tourist Information System, Proceedings of the 9th International Conference on Computers Helping People with Special Needs, Paris, France, pp. 287–294.

Koeppel I. (2000): What are location services? – from a GIS Perspective, ESRI white chapter.

Lamming M. and Flynn M.(1994): "Forget-me-not" Intimate Computing in Support of Human Memory. Proceedings of FRIEND21, '94 International Symposium on Next Generation Human Interface, Meguro Gajoen, Japan.

Lonthoff J. and Ortner E.(2007): Mobile Location-Based Gaming as Driver for Location-Based Services (LBS) – Exemplified by Mobile Hunters. In Informatica V. 31, pages 183–190.

Marmasse N. and Schmandt C.(2000): Location-aware information delivery with commotion. Proceedings of the 2nd international symposium on Handheld and Ubiquitous Computing, Bristol, UK, pp.157–171.

Meier R. (2010): "Professional Android 2 Application Development", Wrox Programmer to Programmer, Wiley Publishing, Inc.

Miluzzo E., Lane N. D., Eisenman S. B., and Campbell A. T.(2007): CenceMe – Injecting Sensing Presence into Social Networking Applications. Proceedings of the 2nd European conference on Smart sensing and context, Kendal, England pp. 1–28.

Miluzzo E., Lane N. D., Fodor K., Peterson R., Lu H., Musolesi M., Eisenman S. B., Zheng X., and Campbell A. T.(2008): Sensing meets mobile social networks: the design, implementation and evaluation of the CenceMe application. Proceedings of the 6th ACM conference on Embedded network sensor systems 2007, Raleigh, NC, USA, pp. 337–350.

Nanni M.and Pedreschi D. (2006). "Time-focused clustering of trajectories of moving objects." J. Intell. Inf. Syst. 27(3), pp. 267–289.

Nivala, A.-M., T. Sarjakoski, Laakso K., Itäranta J. And Kettunen P. (2009): User Requirements for Location-Based Services to Support Hiking Activities. Proceedings of the 5th International Conference on Location Based Services and TeleCartography (II), Salzburg, Austria, pp. 167–184.

Osman Z., Maguire M. and Tarkiainen M. (2003): Older Users' Requirements for Location Based Services and Mobile Phones Human-Computer. Proceedings of the 5th International Symposium of Human-Computer Interaction with Mobile Devices and Services, Udine, Italy, pp. 352–357.

Sohn T., Li K. A., Lee G., Smith I., Scott J., and Griswold W. G.(2005): Place-Its: A Study of Location-Based Reminders on Mobile Phones. Proceedings of the 7th International Conference of Ubiquitous Computing, Tokyo, Japan, pp. 232–250.

Thi Hong N. V., Keun Ho R., Park N., (2009): “A method for predicting future location of mobile user for location-based services system.” Comput. Ind. Eng. 57(1), pp. 91–105.

Uhlirz M. (2007): A Market and User View on LBS. Proceedings of the 5th International Conference on Location Based Services and TeleCartography (II), Salzburg, Austria pp. 47–58,

Chapter 21

Accessibility to Essential Services and Facilities by a Spatially Dispersed Aging Population in Suburban Melbourne, Australia

Gang-Jun Liu and Benno Engels

RMIT University, Melbourne, Australia
{gang-jun.liu, benno.engels}@rmit.edu.au

Abstract

Advances in microelectronics and the increasing use of digital mobile technologies associated with location based service (LBS) provision and consumption, have started to profoundly change the way we make decisions about when, how and where we do things. Microelectronics based modes of communication have extended our capacity to escape the 'spatial fix' that has constrained our ability to access and consume services outside the home environment in the past. In the mean time, spatially dispersed ageing population is becoming a challenging issue for public policy makers in many developed countries. Australia has opted to deal with this unprecedented ageing issue via an 'ageing in place' strategy, encouraging the elderly or seniors to live in their own housing for as long as possible, with only minimal support from public authorities. This strategy has been widely supported by the community. However, it poses a number of serious challenges for both policy makers and seniors alike. These include the issue of personal mobility and the ability to access those things that are deemed essential to ones well-being and quality of life from outside the home. The elderly or seniors are often identified as being subject to social exclusion because of difficulties associated with travelling outside their homes to access services and facilities, especially towards the outskirts of major urban centres, where services and facilities are generally only readily accessible by car. As drivers age and are required to surrender their driving licences, accessibility to a range of services and facilities either in their immediate neighbourhood or surrounding region will become a major issue especially in those locales where public transport infrastructure is inadequate.

G. Gartner, F. Ortag (Eds.), *Advances in Location-Based Services*, Lecture Notes in Geoinformation and Cartography, DOI 10.1007/ 978-3-642-24198-7_21,

To better understand the potential benefits that LBS based technologies have for the elderly members of our society, this paper presents an empirical assessment and visualisation of micro level manifestations of location disadvantage within residential suburbs of Melbourne using GIS-based spatial analytical procedures and fine spatial resolution data sets. The paper presents the spatial patterns of accessibility in terms of the shortest network distance travelled from residential locations to their nearest essential services and facilities, such as medical centre, pharmacy, bank, library, church, shopping centre, bus stop, train station, social club, etc. This spatially dispersed aging population lives in or on the outskirts of metropolitan Melbourne, and some of whom do not have ready access to a car or public transport service. This case study reveals that the existence of social exclusion amongst non-car driving seniors is being reinforced by a regional public transport system that can not adequately service the entire municipality, as too much reliance continues to be placed on the ownership and usage of cars.

The paper points out that there is a range of specific applications of LBS technology that can benefit the wellbeing of seniors, especially non-driving seniors; the next wave of retirees will be more technologically savvy and shall have fewer difficulties using the latest microelectronic devices that can disseminate LBS information; designers of LBS applications will need to take account of this development. The approaches taken and the findings made in this study should be useful to public policy makers, government and non-government service providers and developers of LBS applications in many countries who will all be confronted by this same policy problem within the next two decades.

Keywords: aged population, Melbourne, accessibility, public transportation, social exclusion, GIS, network analysis, spatial variation, LBS.

21.1 Introduction

Advances in microelectronics and the increasing use of digital mobile technologies associated with location based service (LBS) provision and consumption, have started to profoundly change the way we make decisions about when, how and where we do things. Aided by hand-held, wireless and mobile microelectronic devices, our ability to communicate with others and find information has been phenomenal. Microelectronics based modes of communication have extended our capacity to escape the 'spatial fix' that has constrained in the past our ability to access and consume services outside the home environment.

Australians like many others who live with modern technologies are now more mobile and able to communicate with far greater ease than ever before. The widespread adoption of mobile telephones, portable computers, wireless communication networks and the Internet, has enabled us to engage others without even leaving our homes or places of employment. New LBS technologies are helping to simplify the locational decision-making processes for many. Being able to take advantage of these technological developments are however conditional upon one's ability to afford and utilize the portable hardware or hand-held devices which transmit LBS based information, as well as being sufficiently mobile enough to reach the desired destination. The elderly who are experiencing difficulties physically moving about outside the home environment will initially see little direct benefit in acquiring such devices. However if the supplied LBS information is tailored to meet their specific needs then it is quite conceivable that the decision to travel around ones neighbourhood or activity region could become a less daunting task.

To date, very little research has been done on the potential use of LBS technology for seniors and how its use could enhance the quality of life of such individuals. This focuses on a significant proportion of the population as the number of retirees in most western societies is going to dramatically increase in the coming decades (WHO, 2011). This situation is made all the more imminent with more elderly people choosing to stay at home rather than move into aged care residential facilities. This is commonly known as 'ageing in place'. Due to houses being dispersed in metropolitan areas, the task of locating and accessing services and facilities become crucial for its elderly occupants. This situation calls for a unique potential application of LBS technology.

In our technology driven societies, it is easy to forget that there are people who for one reason or another, can not take advantage of these kinds of developments. The elderly or seniors can experience social exclusion from mainstream society. Seniors are often insulated from the latest developments in both computer and communication technology because of their age. These latest technological gadgets are usually cost prohibitive for many seniors who are generally on fixed incomes or government funded pensions. Not all seniors will have access to a car and the poor public transport system can compound the location disadvantage of non-driving seniors. Many 'old and frail seniors' surrender their driving licences when their physical capacities to drive a car safely begin to diminish in their late 70s, early 80s. For such elderly, the LBS technology might be able to ease some of the difficulties they experience when travelling outside the home environment. To be able to positively age and keep living in one's home it will depend on how easy it is to access a whole range of essential services and facilities. This would include doing the weekly grocery shopping, withdrawing the fortnightly pension payment from the bank, visiting the doctor, collecting medication from the chemist, posting a letter, borrowing a library book, attending a church service, maintaining regular social contact or seeking support from persons outside the immediate and extended family. While some of these services – banking and communicating – can be accessed over the Internet, there is still a need for most seniors to travel outside the home in order to access a range of location-based services and facilities. This is where LBS can be of some assistance to seniors irrespective of whether they are still driving a car or not. However, to ascertain how valuable this might be it will be necessary to establish how many seniors might be in need of such assistance, how they can be identified spatially and what forms of LBS technologies it takes to help overcome the locational and transport disadvantage of seniors.

To better understand the potential benefits that LBS based technologies have for the elderly, this paper reports on a GIS based investigation into identifying and measuring the existence of transport related locational disadvantage amongst a cohort of non-driving seniors, who currently live in a middle distant suburban municipality of Melbourne, Australia. More specifically, this study sought to identify those locales within the case study region where the need for specially adapted LBS based technologies might be of greatest benefit, as it concentrated exclusively upon non-driving seniors. This cohort of elderly residents are the most locationally disadvantaged within the case study because they are heavily reliant upon the existing public transport system and for them to move around the municipality is much more difficult compared to those seniors who still drive or have access to a car. To have access to LBS information about services and facilities will make the ordeal of travelling within the municipal region a potentially easier task than it is presently. The paper is divided into a number of sub-sections including a brief review of the relevant literature on ageing and locational disadvantage; an overview of the case study region; the types of data collected and used to identify location based transport disadvantage amongst non-driving seniors in the Monash municipality; the methods and findings; plus a discussion of how LBS based technologies could help address the social exclusion and transport disadvantage being experienced by a cohort of seniors currently living in this case study.

21.2 Location Disadvantages

Location based advantages and disadvantages 'emerges when other benefits or penalties compound the advantages or disadvantages of particular groups by virtue of where they live' (Smith, 1977: 112) due to the uneven distribution of public and private provisions of services and facilities across urban space (Coates, B., et al., 1977; Herbert, D., and Smith, D., 1979; Kirby, 1982; Knox, 1995). Social exclusion occurs in circumstances that prevent individuals, households and even entire communities from fully participating in and better integrating with mainstream society (Vinson, 2009). People can be disadvantaged if they are living in communities that have a shortage of jobs, services or facilities (Littlewood, 1999; Percy-Smith, 2000; Favreau and Spear, 2001; Mayes and Salais, 2001; Pantazis, et al., 2006). Social exclusion and location disadvantage can be intertwined because inadequate transport and access to services and facilities might contribute to location disadvantage (Hine and Mitchell, 2003). An ability to freely travel beyond the home is essential to maintaining ones quality of life (Alsnih and Hensher, 2003; Banister and Bowling, 2004; Golob and Hensher, 2007; Schmocker, et al., 2008).

As the size of major cities continue to grow and sprawl, residents are compelled to become reliant upon cars (Dodson and Sipe, 2008), especially in the expanding outer and fringe areas where housing is relatively cheap but services like public transport are quite limited (Buchanan, et al., 2005; Currie, 2004; Dodson et al., 2004; Hurni, 2006). Not being able to drive can diminish ones quality of life and lead to feeling socially excluded particularly when the typical post-war Australian suburb was designed around the car and a capacity of residents to drive, and upgrading public transport can ameliorate select aspects of locality based forms of social exclusion, particularly for non-car owning households (DETR, 2000; Mollenkopf, et al., 2005). The most frequently identified households found to be living in transport disadvantaged locales and suffering from transport related social exclusion were non-car owning groups including the unemployed, working poor, the disabled, newly arrived migrants, children, single parent families and the elderly (Church et al., 2001; Hine and Mitchell, 2001; Hine and Grieco, 2003; SEU, 2003).

Seniors experience social exclusion because they are restricted in their capacity to travel outside the home. The need to travel can increase as illness requires seniors to visit more regularly a doctor, chemist and even a hospital and the car is often identified as the preferred mode of travel (Morris and Rosenbloom, 1997; Berry, 2007). Not all seniors will have access to a car and poor public transport services can compound the location disadvantage for non-driving seniors. The cost of maintaining a car that is used less frequently as one ages will prove to be cost prohibitive to many over time and eventually lead to its sale. When their physical capacities to drive safely begin to decrease in their mid 70s, many seniors surrender driver licences in their late 70s and early 80s (Burns, 1999; Rosenbloom, 2001; Foley, et al., 2002; Browning and Sims, 2007), putting all members of effected elderly households at risk of social exclusion unless alternate modes of travel are available.

This becomes an even more acute problem as the number of persons who already have or are about to retire in most western societies is going to dramatically surge in the coming decade. The proportion of person aged 65 years and older for metropolitan Melbourne, for example, stood at 12.8 pre cent of the total population in 2006 and it is projected to reach 26 pre cent by 2030 (ABS, 2006). This will become a significant social issue for policy makers in the next two decades as the number of non-driving seniors will significantly expand.

In addition, there are many more seniors who have chosen to remain living in their homes for as long as possible, rather than in retirement villages or other forms of aged care accommodation that may have better access to services. As a result, seniors are now more widely dispersed than ever before in Australia's sprawling cities, encountering difficulties when they have to travel in order to access services and facilities (Browning and Sims, 2007; Currie, 2009; Hensher, 2007), and making it more difficult to ensure that they have adequate access to both private and publicly supplied services and facilities.

To understand location based disadvantage and associated social exclusion for non-driving seniors it is necessary to have an understanding of the relative location of the main services and facilities that they use. To be able to positively age and keep living in ones home will be conditional upon whether it is easy to access a whole range of necessary services and facilities. This would include doing the weekly grocery shopping, withdraw the fortnightly pension payment from a bank, visit the doctor, collect medicine from a chemist, post a letter, borrow a library book, attend a church service, maintain regular social contact or seek support from persons outside the immediate and extended family (Cantor, 1979; Lawton, 1980; Golant, 1984). While some of these services (e.g. banking and communicating) can be accessed over the internet, there is still a need for most seniors to travel outside the home in order to access necessary services and facilities (e.g. to visit a doctor, a chemist or priest).

21.3 Case Study Area

Melbourne is the capital and largest city of the state of Victoria. It is the second most populous city in Australia after Sydney with almost 4 million inhabitants. The metropolitan area is large by conventional standards, covering 2152.8 km^2, but with low population densities. It is divided into 31 local government areas, including the municipality of Monash *(Figure 21.1)*. Until the early 1950s a majority of intra-urban travel in Melbourne was done by train, tram or bus, and suburban development was clung to the outwardly radiating train and tram lines (Wilkinson, 1984). With car ownership levels rose in the late 1950s (Davison, 2004), Melbourne sprawled extensively through the 1960s and 1970s, and house building occurred further away from the fixed tram and train routes, exceeding 30 km from the CBD in some directions, but with little extension of the existing tram and rail network (Mees, 2000). The journey to work length for many Melbournians increased (King, 1979; Maher and O'Connor, 1978), as did the trip to other amenities including shops, health and welfare services (Beed, 1981). By the early 1980s continued urban sprawl became recognized as a problem as the price of petrol rose alongside the cost of providing new public infrastructure to outlying areas (King, 1981). The proportion of seniors aged 65 years and older for metropolitan Melbourne stood at 12.8% of the total population in 2006, and is projected to reach 26% by 2030 (ABS, 2006).

Cheal (2003) found that 82 percent of Melbourne's population lived outside the 'transit rich areas' or the inner suburbs surrounding the CBD, and that middle and outer suburbs serviced by trains and buses experienced varying levels of 'transit poverty', with the highest levels being recorded in the outer fringe areas; some were more than 40 kilometers from the CBD. Some recent investigations on transport related social exclusion in Melbourne (Currie and Senbergs, 2007; Currie et al., 2007; Currie, et al., 2009; Currie and Delbosc, 2010) revealed that there was a clear mismatch between public transport supply and potential need in fringe and outer suburban areas of Melbourne whereas the level of discrepancy was much lower in most middle and inner areas, 'forced car ownership' in the outer and fringe areas is largely due to a shortage of public transport despite the financial stress car ownership imposes upon low income households, and non-car owning households in these outer and fringe areas had to live near activity centres that contained shops and other services.

The Monash municipality consists of 12 discrete suburbs located some 20 km south-east of the city centre and covering an area of 81.4 km^2. It is a quintessential example of the type of residential development that occurred in Melbourne at wars end (Gobbi, 2004). At the end of WWII, this area was largely rural consisting of orchards and market gardens with half a dozen small townships that had sprung up near railway stations located on two electrified railway lines that dissected the area and connected it to the centre of urban Melbourne. This area grew rapidly from the early 1950s because it had been designated as a major growth corridor that would absorb large quantities of Melbourne's post war industrial and residential expansion. Until car ownership became more widespread in the mid 1960s, travel within this area of Melbourne was either by foot or bicycle, as buses were expensive and operated on restricted routes. The purchasing of a car in these then outer areas of Melbourne was not just a matter of necessity but it was also influenced by a rapidly emerging suburban culture based around the car (Davison, 2004).

By the mid 1970s, population growth had begun to slow and by the early 1980s, 3 post offices had closed with several schools following suite in the mid 1990s as the need for these facilities declined because the population profile of the area had now started to age. At the 2006 census, 16 per cent of the total municipal population was found to be over the age of 64 years, whereas the metropolitan average was only 12.8 per cent. In fact, the municipality of Monash has numerically the single largest number of persons aged 65 years and older, within the south-eastern region of metropolitan Melbourne. Like any area that was progressively settled over an extended period of time, there is an uneven spatial distribution of its older occupants over the study area. Glen Waverly and Mount Waverley had the largest concentrations with smaller but significant numbers also to be found in Oakleigh – Oakleigh, Oakleigh east and Oakleigh south – as well as parts of Wheelers Hill. For the purposes of the current study the proportion of this population without a car is a critical concern.

There are a large number of public and privately supplied services and facilities for seniors in the municipality. Altogether, there are 257 different kinds of medical services, including doctor surgeries, dentists, pharmacies and a major public hospital. There are 15 shopping centres including 4 large shopping malls that account for some of the 84 hairdressers and barbers, 31 banks, 21 newsagents, 22 post offices, 30 butchers and 13 opportunity shops. There are many types of social and cultural facilities available to seniors, including 189 social clubs and activity groups, 26 culturally-specific social clubs and 37 service clubs, 5 libraries, 62 churches, temples, worship centres and meeting halls that can cater for the religious needs of seniors, as well as 4 cinema complexes, 23 pubs, as well as 16 approved gambling venues.

Consistent with findings made elsewhere in Australia and around the world, a disproportionate number of young seniors living in Monash are still driving cars, but this begins to decline by their late 70s and early 80s particularly for women. An examination of the 2006 census revealed that 95.1 per cent of all non-driving seniors living in Monash were not users of the internet (ABS, 2006). In other words, the vast majority of non-driving seniors living in Monash still need to travel outside the home either by walking or using the bus to do their grocery shopping, visit a doctor or chemist, attend a church service or play bingo at their preferred community social club. These features make it an ideal choice as a case study of a middle distant Australian municipality.

21.4 Data Sets, Methods, and Results

21.4.1 Data Sets

Four key data sets were collected and used in this study. First, a survey was undertaken in late 2007 and early 2008 of 16 senior groups that were all active in the municipality. These groups were randomly selected from a community service directory and identical questionnaires were administered to the respondents at the venue where they meet for their designated weekly activities. The questionnaire required respondents to provide information about themselves, what local services and facilities they used, how far they are willing to travel in order to reach certain services and facilities, plus the mode of transport they regularly used. In total, 187 seniors completed the questionnaire, all of whom were residing in the municipality at the time of the study.

Second, the 2006 population census was consulted to establish, at the census collector district (CD) level, how many persons were aged 65 years and older, plus how many of them do not drive a car. There are a total of 230 CDs in the Monash municipality, and each CD consists of about 200 households and provides the population sample from which socially excluded seniors experiencing transport disadvantage will be drawn from.

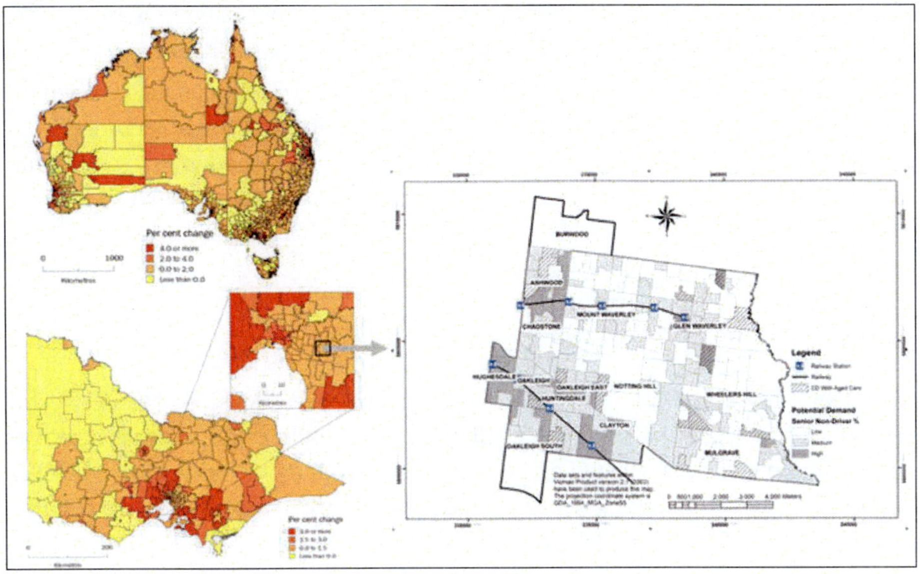

Fig. 21.1. The case study area: the Monash municipality in the Metropolitan Melbourne area in Victoria, Australia.

Third, through site visits as well as checking on-line data bases and various directories, a detailed land use audit of the Monash municipality was undertaken in 2007 and early 2008 to identify the location of 10 different types of private and public services and facilities found within the municipality that seniors might have a need to travel too, such as medical services, emergency services, municipal facilities, social and cultural facilities, retail and other shopping, and entertainment. It was simply beyond the scope of this study to investigate all different types of services and facilities found in the Monash municipality so only 10 types of services and facilities that were regarded as essential to positive ageing by the cohort of seniors who completed the questionnaire were included in the subsequent analysis, including a number of commercial services – shopping centre, bank and post office, medical services – doctor, chemist and hospital, plus social or recreational facilities – library, social club, religious organisation and park. The nomination of these services and facilities is reflected in how often they are used at different time intervals by the survey respondents (see *Table 21.2)*. Community centres and social clubs are used most frequently which is not surprising as all the respondents undertook the survey questionnaire at such a venue. The next most frequently used are shopping centres, doctors, banks, chemists and post offices. These are usually located in close proximity to each other but their usage pattern does vary. Shopping is frequently done on a weekly basis whereas a visit to the doctor and a chemist is done more on a monthly basis. Churches are visited weekly whereas libraries and parks prove to be popular at several time intervals during an average month. The least frequently visited facility is the regional public hospital which offers out-patient services to the general public including the treatment of seniors for ongoing illnesses.

Fourth, the location of each of the 28 bus routes and 277 stops that fall within the municipality and run by different private bus companies was ascertained from data supplied by the Victorian Department of Transport, and other relevant spatial datasets were adopted from the 1:25k VICMAP database, such as the addresses and road network.

21.4.2 Methods and Results

With these key datasets, a GIS-based approach has been developed to derive for each of the 230 CDs values of the following three indicators: senior non-driver concentration, bus access, service access, and an overall index representing the relative spatial accessibility to the set of selected services by senior non-drivers residing in the different CDs using public transportation.

The concentration of senior non-driver is defined as % of CD-level senior population (> 65 years old) without a driver's license and is derived directly from ABS 2006 census data, or as the proportion of CD-level senior non-drivers to the total CD-level seniors (> 65 years old), with all values fall between 0 and 1: larger values indicate higher concentration of senior non-drivers (who could have accessibility problems) and smaller values indicate lower concentration of senior non-drivers which means lower demand for accessing public transport and the selected services. *Figure 21.2* shows the spatial pattern of senior non-driver concentration over the study area. Small but significant clusters of non-driving seniors in Monash municipality are found in Hughesdale, Oakleigh, Clayton, Ashwood, Chadstone, Mount Waverley and Glen Waverley. These senior non-drivers would have to largely rely upon public transport to undertake their regular out-of-home travel, the most convenient being the bus.

Based on distance travel thresholds set by Victorian government planning authorities such as one should not be more than 400 m from a bus stop, 800 m from a train station and 1 km from other basic services including a shopping centre (Victorian Department of Sustainability and Environment, 2006: 6.12–13), and as a surrogate to public transport access, bus access is defined as % of CD-level population within 400 m of nearest bus stop, or as the proportion of CD-level population that is within 400 m of nearest bus stop to the total CD-level population, and is approximated by the % of residential address points within 400 m of nearest bus stop based on the road network distance measured within a GIS. All bus access values fall between 0 and 1: higher values indicate more accessibility to public transportation, and smaller values indicate less accessibility to public transportation. *Figure 21.3* shows the spatial pattern of bus access over the study area.

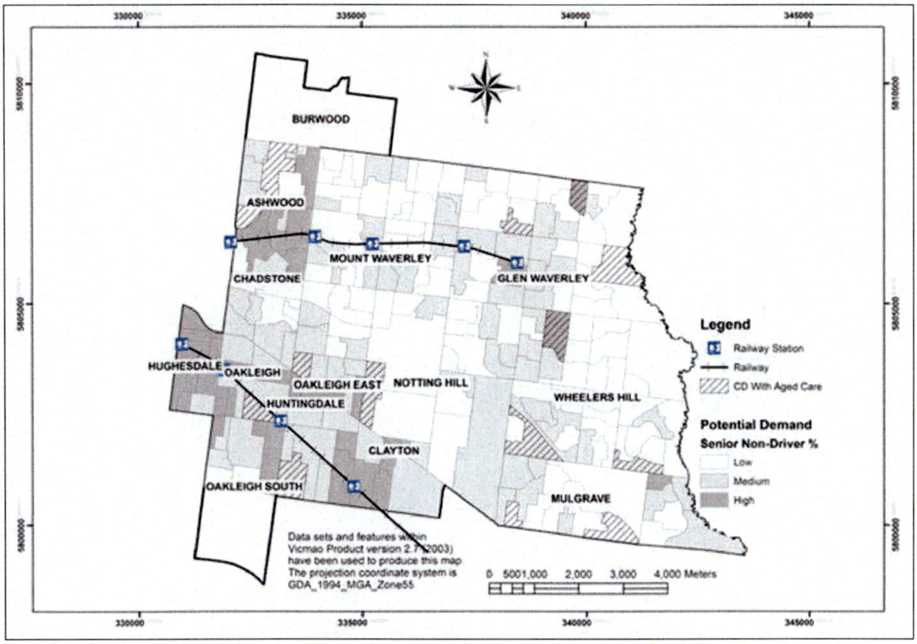

Fig. 21.2. The spatial pattern of senior non-driver concentration over the study area, showing three categories based on natural breaks

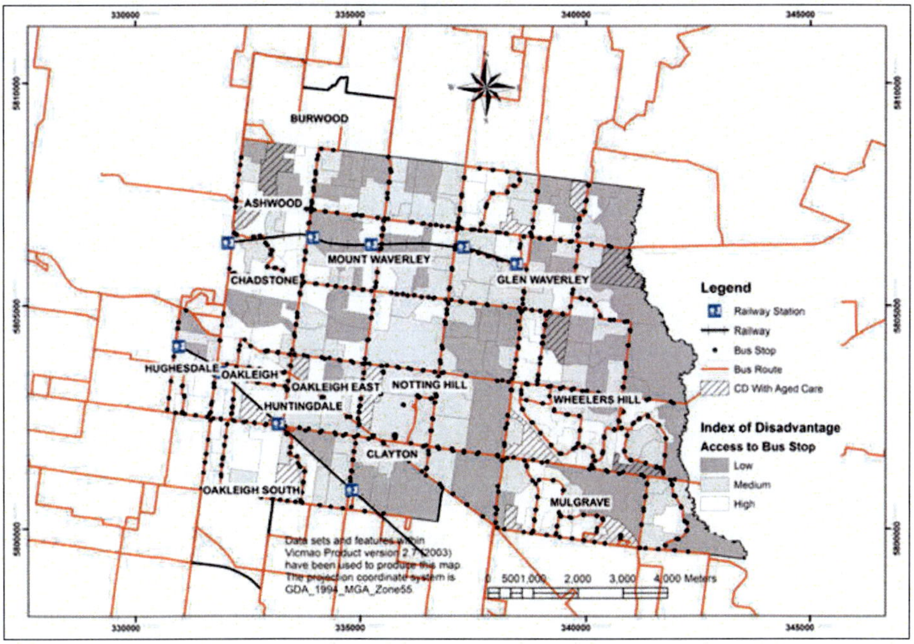

Fig. 21.3. The spatial pattern of bus access over the study area, showing three categories based on natural breaks

For each of the 10 selected types of services and facilities (i.e. doctor, chemist, public hospital, shopping centre, bank, post office, library, church, social club and park), road network distances in metres between the population centre of each CD to its nearest service facility is calculated as d_{ij}, which is then standardized into D_{ij} using the following equation:

$$D_{ij} = \frac{d_{ij} - d_{\min,j}}{d_{\max,j} - d_{\min,j}}$$

Where d_{ij} is the road network distances between the population centre of CD*i* to its nearest type *j* service facility, while $d_{min,j}$ and $d_{max,j}$ are the shortest and longest road network distances between all CD population centre to their respective nearest type *j* service facility, as listed in *Table 21.1*

Table 21.1.

Service type j	$d_{min,j}$	$d_{max,j}$
Social Club	121.27	4069.05
Shopping	304.95	3507.14
Doctor	26.94	2404.29
Chemist	228.51	3316.31
Bank	250.26	4747.95
Post Office	189.97	5119.15
Church	48.19	2528.58
Park	0.00	1707.69
Library	350.18	5942.16
Hospital	308.10	11301.56

This analysis assumes that people will choose to access their nearest service or facility using the existing road and footpath network– whether travelling by car, bus, bike, foot or a combination – in order to reach their chosen destination, although for those seniors who have access to a car they may travel beyond their immediate neighbourhood (Smith, 1988; Fober and Grotz, 2006).

The service access A_i is defined as follows

$$A_i = \sum_{j=1}^{10} w_j D_{ij}$$

Where $\Sigma w_j = 1$ and w_j is the weight value for type *j* service determined statistically from the survey data. A_i is distance-based. Its normalised values fall between 0 and 1: smaller values indicate closeness or more convenient locations, larger values indicate remoteness or less convenient locations. The procedure used for determining w_j is as follows:

- For each service type, the number of person indicating specific use frequency (daily, weekly, fortnightly, monthly, 6-monthly, 12-monthly, or other – never or missing) p_{jk} were summarized in *Table 21.2.*

- Each cell value in *Table 21.2* p_{jk} was then divided by its respective column sum $\Sigma_k p_{jk}$ to derive the relative significance of each service type for specific frequency r_{jk} which are summarized in *Table 21.3*

$$r_{jk} = \frac{p_{jk}}{\sum_k p_{jk}}$$

- Each column sum in *Table 21.2* $\Sigma_k p_{jk}$ was divided by their total $\Sigma_k \Sigma_j p_{jk}$ (which is 1256 in this case) to derive the relative significance value for each specific use frequency R_k which sum to 1

$$R_k = \frac{\sum_k p_{jk}}{\sum_j \sum_k p_{jk}}$$

- The weight value for type j service w_j is calculated by summing up the products of the relative significance of each service type for specific frequency r_{jk} and the relative significance value for each specific use frequency R_k

$$w_j = \sum_k r_{jk} R_k$$

Figure 21.4 shows the spatial pattern of service access over the study area.

Spatially, those seniors who continue to live within close proximity to the older shopping areas that had sprung up around the railway stations usually proved to possess location advantage, with easy access to shopping centres, doctor surgeries,

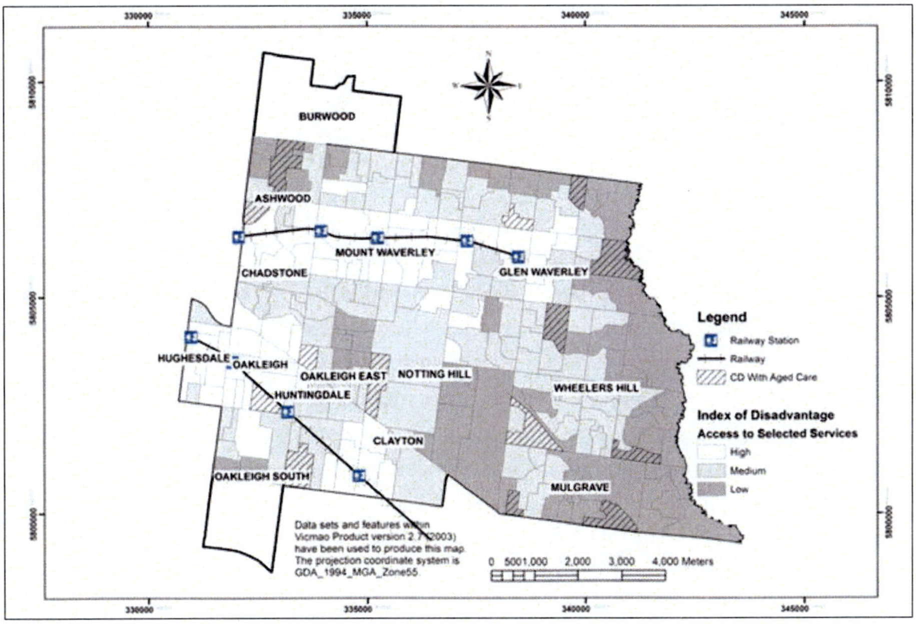

Fig. 21.4. The spatial pattern of service access over the study area, showing three categories based on natural breaks

Table 21.2. Usage patterns of services and facilities by a sample of seniors in the City of Monash compiled from authors survey (2008)

Service/Facility p_{jk}	Daily	Weekly	Fortnightly	Monthly	Half Yearly	Yearly	Other (Never, Missing)
Community Centre/Club	6	82	7	34	19	19	20
Shopping Centre	19	120	17	7	0	0	24
Doctor	0	6	20	80	50	7	24
Chemist	0	13	43	87	7	0	37
Bank	1	44	65	36	1	1	39
Post Office	2	39	37	49	13	5	42
Church	6	59	7	5	5	9	96
Park	10	19	17	18	11	8	104
Library	1	13	23	28	7	7	108
Hospital	0	5	1	6	21	34	119
$\sum_{kpjk}$	45	400	237	350	134	90	613
R_k	0.036	0.318	0.189	0.279	0.107	0.072	

Table 21.3. The relative significance of each service type for specific use frequency

Service/Facility rjk	daily%	weekly%	fortnightly%	monthly%	Half yearly%	yearly%	wj
Community Centre/ Club	0.133	0.205	0.030	0.097	0.142	0.211	0.1330
Shopping Centre	0.422	0.300	0.072	0.020	0.000	0.000	0.1298
Doctor	0.000	0.015	0.084	0.229	0.373	0.078	0.1298
Chemist	0.000	0.033	0.181	0.249	0.052	0.000	0.1194
Bank	0.022	0.110	0.274	0.103	0.007	0.011	0.1178
Post Office	0.044	0.098	0.156	0.140	0.097	0.056	0.1154
Church	0.133	0.148	0.030	0.014	0.037	0.100	0.0725
Park	0.222	0.048	0.072	0.051	0.082	0.089	0.0661
Library	0.022	0.033	0.097	0.080	0.052	0.078	0.0629
Hospital	0.000	0.013	0.004	0.017	0.157	0.378	0.0533
	1.000	1.000	1.000	1.000	1.000	1.000	1.0000

chemists, banks, post offices and other services. Seniors who live in car dependent suburbs (which includes parts of Wheelers Hill, Mulgrave, Glen Waverley and Notting Hill) have location disadvantages. Although retailing facilities expanded with the building of new car based shopping malls – The Glen, Waverley Gardens and Brandon Park – as population numbers increased during the 1960s and early 1970s, a pre-existing landscape of retail-service nodes continues to create transport advantages for seniors in some parts of Monash. Certain public facilities – hospital, churches and libraries – were located to serve large catchment areas. In some cases these were sited close to other facilities and have reinforced the existing concentrations of services around the railway stations. Some facilities were deliberately dispersed across the municipality for easy and regular use such as parks, community centres and social clubs. Very few areas of the municipality are disadvantaged with respect to them except parts of Mulgrave, Notting Hill and South Oakleigh.

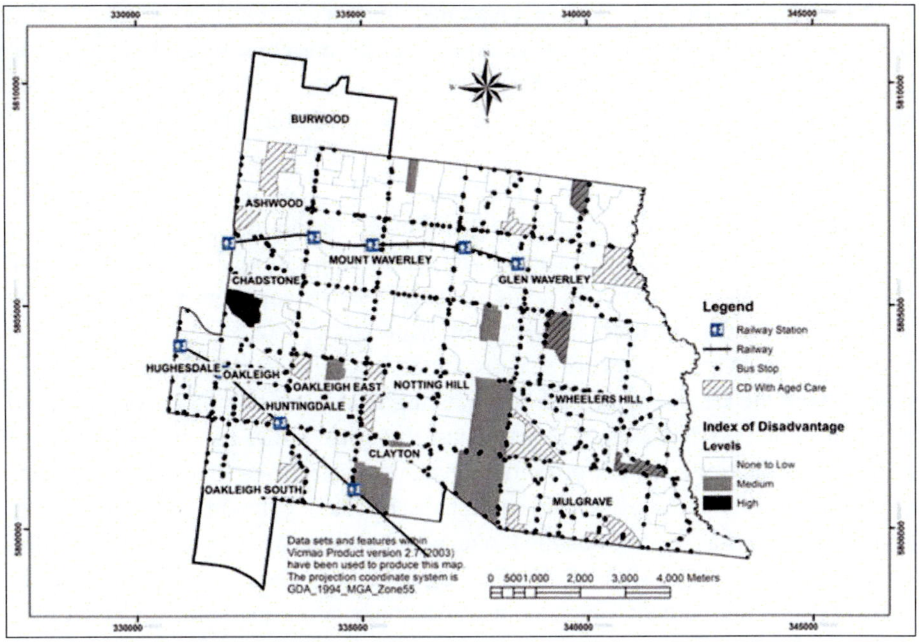

Fig. 21.5. The spatial pattern of overall location disadvantage over the study area, showing three categories based on natural breaks

The merger of all three indicators of disadvantage – location of non-driving seniors, poor access to the bus routes and distance from basic services and facilities – using GIS software allowed for the identification of those locales where location based transport disadvantage exists amongst non-driving seniors. To represent the relative spatial accessibility to the set of selected services by senior non-drivers residing in the different CDs using public transportation, the three indicators presented above were integrated in such a manner that their contributions are properly reflected. This has been achieved by multiplying the product of the service access A_i and the senior non-driver concentration S_i by the reciprocal of the bus access a_i as follows:

$$Index_i = \frac{A_i S_i}{a_i}$$

The larger the index value the more disadvantaged the CD is, whereas the smaller the index value the less disadvantaged the CD becomes. *Figure 21.5* shows the spatial pattern of overall location disadvantage over the study area.

21.5 Discussions and Conclusions

From the perspective of interaction between location-based service providers and consumers, four types of provider-consumer interactions can be identified:

(1) Service providers are fixed in locations, and service consumers have had to travel to the destination of the service provider before consumption could occur, e.g. consumers eating at drive-in fast food restaurants, purchasing groceries from shops, visiting a GP, attending a gym, etc. (2) Service users are fixed in locations, and service providers are mobile and come to service users' home locations, such as the Home and Community Care Program (HACCP) for the elderly and disabled who are wholly or partially confined to their homes; the home delivery of groceries, newspaper, milk, parcels, etc.; and service delivery performed at the users home by trade persons like plumbers, home painters, electricians, house builders, home cleaners, gardeners, repairers of electronic white goods, etc. (3) Service providers and service users are fixed in different locations, but an interaction / exchange is still possible via the telephone or Internet. Existing examples include telephone based medical advice (ie. Nurse on Call service in Victoria), e-Bay selling and buying of goods (e.g. electronic books or computer games), electronic banking and online payment of utility of bills, and paid cable TV services (e.g. Foxtel). (4) Service user and service provider are both on the move, where service delivery and consumption occur away from the user's home or the provider's business premises. A well established service that already operates in this manner is the road-side assist services that are common across Australia and elsewhere in the world for paid members. Emergency services like police, ambulance and fire could also be added here.

LBS can now simplify the search process and locate the closest service provider for type (1) interactions by providing the service user with up-to-date information about recently opened outlets and the ability to advise on traffic conditions en route. It remains a challenge for LBS in the conventional type (2) interaction to devise ways for service providers to locate service users other than advertising on electronic notice boards of their existence and services. For the expanding type (3) and type (4) interactions, it is also a challenge for LBS to devise intelligent 'applications' that may be of benefit to the service users and the service providers.

This study has sought to identify whether locality based transport disadvantage could be found to exist amongst a cohort of non-driving seniors living in a middle distant suburban region of metropolitan Melbourne. It has established that there were small pockets of housing occupied by seniors who lived in parts of the Monash municipality that were beyond established travel distance thresholds from both public transport and a range of locality-specific essential services. This finding was only possible by combining together a unique set of data types that were all locality specific in nature including census data, location of services and bus stops, plus the subjective travel preferences of a group of surveyed seniors living within the case study region.

The collection of survey data made it possible to derive empirical statistical weights for the selected types of service/facilities. The merging of these different data types has never been attempted before in the study area and it was only possible through the use of GIS-based analytical spatial procedures, e.g. the derivation of bus access and the calculation of network distances between population centres and sites of service provision, which permitted the analysis to generate visualizations of spatial patterns of transport based locational disadvantage and advantage amongst non-driving seniors who need to access services across the case study region on a weekly or fortnightly basis over the Monash municipality. Certain essential services were found to be spatially concentrated and therefore at a significant distance from the homes of some seniors including those who do not have ready access to a car and could therefore be classified as being locationally disadvantaged.

Most Australian cities have and will continue to suffer from urban sprawl and an over reliance upon the car, and many middle distant suburban areas of most Australian cities including Melbourne already have large ageing populations. Many of these seniors shall be surrendering their driving licence in increasing numbers over the next decade and, as a result, these non-driving seniors shall place an unprecedented level of demand upon the existing public transport system. Research conducted in the Monash municipality has revealed that the existing bus network is comprehensive but there are still pockets of the municipality that are under-serviced. Unless there is an upgrading of the existing public transport system in the Monash region and elsewhere across middle distant Melbourne there is then a very real prospect of much greater levels of location based transport disadvantage being experienced amongst Melbourne's non-driving seniors.

This is clearly a timely warning to both public policy makers and transport planners alike to begin planning for this next wave of non-driving seniors. This is also a prime opportunity for LBS technologies to help address this emerging problem. LBS technologies can ease the distress and discomfort that many driving and non-driving seniors experience when they are required to travel outside of the home environment.

Although a large proportion of seniors prefer to remain living in the family home for as long as possible, some decide to downsize and move to a smaller house or apartment usually within either the same neighbourhood or to a nearby suburb (see Salt, 2004). There is greater reliance made of services in the local neighbourhood as we age and our activity space outside the home shrinks. If this neighbourhood is unknown to inward-bound seniors then LBS application can fore warn them about the distances they will need to travel in order to reach a desired number of essential services. Non-driving seniors are anticipated to grow in numbers over the next decade and they will in turn place greater demand pressures upon the public transport network.

Waiting for lengthy time periods at bus stops can be frustrating for seniors especially if there are no benches for them to sit on while they have to wait or they may have incontinence issues but are too far away from the nearest public toilet. Cancelled bus services can only compound these problems. In these cases, application of LBS technology for the provision of real time information and updates regarding the arrival and departure of public transport services have proved to be very useful. Waiting to be attended to and losing precious time is a frustrating experience for anybody but sitting in a long queue can become an even more serious problem for some seniors who are either incontinent or suffer from some other kind of physical disability. Providing real time updates of service availability can make the journey by bus to a regional shopping centre where one has a number of appointments less daunting and frustrating for non-driving seniors (and others who do not have access to a car).

Many senior car drivers especially those over the age of 70 become anxious and confused if they have to drive around large car parking areas looking for an empty parking space (Morris, 2007). Application of LBS technology (e.g. the installation of automated sensory devices in each car parking space throughout its large car parking area that indicates if a space were occupied or empty) for real timely updates about the availability and location of car parking spaces would make the task of finding a car parking spot so much easier to senior drivers.

LBS technology can also be used to help seniors, whether they are travelling by car or public transport, to make a more informed decision about where they want to do their grocery shopping within a region. Being on a fixed income (i.e. the aged pension) encourages many seniors to save money by purchasing grocery items either in bulk or when they are on discount (see Engels, et al., 2009). Some essential services like grocery shopping are spatially concentrated in different locations across the urban space. Moreover, some large supermarket chains are located only in specific centre or shopping mall. If seniors have information about where the cheapest grocery items could be purchased on a particular day then a different travel plan could be adopted in order to take advantage of this opportunity. The real time provision of LBS information regarding the price of grocery items available from different chain supermarket stores located in different locations could be a significant benefit for seniors whether they travel by car or public transport. It is possible that the various large chain supermarkets operating in Melbourne could supply a LBS application accessible by mobile phones that inform grocery shoppers including seniors where they could find the latest discounts or clearance sales on a daily basis.

The use of LBS technology could also prove to be beneficial for government and non-government run service providers, who might have a need to target seniors. Once under-serviced pockets of transport and facility disadvantage have been identified, a range of LBS applications could be implemented by appropriate state government authorities to adapt their public service provisions (e.g. public transportation routes) according to the changing spatial pattern of non-driving seniors over the coming decade. Local municipal authorities could also use findings updated from more recent census data to improve their community service planning initiatives (e.g.

community bus service) for seniors. Commercially run businesses might find such locality-specific data of value for tailoring their products and services to the needs of a changing senior population. Various home maintenance services – gardening, rubbish removal, house cleaning, painting – could better target areas where seniors are spatially concentrated through a LBS application that made use of census data disaggregated at the locality level.

In conclusion, there are a range of specific applications of LBS technology that can benefit the wellbeing of seniors especially non-driving seniors. Such users may not be a priority in the design of current LBS applications as technology designers are seemingly targeting younger users who enjoy having the latest i-Phone or i-Pad. For now, this has probably been an appropriate marketing strategy, but things will and must change. The next wave of retirees who shall swell the ranks of existing seniors will be more technologically savvy and shall have fewer difficulties using the latest microelectronic devices that can disseminate LBS information. Designers of LBS applications will need to take account of this development for the simple reason that there will be more seniors in the next few decades than every before. In turn, there will be an enlarged market for LBS applications that cater for the needs of seniors, as this will also be a new market to exploit. From the perspective of seniors and the research undertaken here, it will be a beneficial development because as people age, they will progressively reduce the amount of travel they undertake outside of the home environment. LBS technology has the potential to maximise the benefits that seniors – whether they are car drivers or non-drivers – can potentially derive from the fewer trips that will need to be undertaken to access essential services and facilities. The approaches taken and the findings made in this study should be useful to public policy makers, government and non-government service providers and developers of LBS applications in many countries who will all be confronted by this same policy problem within the next two decades.

Acknowledgements

The authors wish to acknowledge funding support from the Office of Seniors Victoria, data support from ABS and Victorian DSE, and Ms. Francis Stiglec who has assisted with the collection and collation of senior survey data for this project.

References

ABS., 2006 National Census of Australia, Melbourne Statistical Division, Australian Bureau of Statistics, Canberra.

Alsnih, R., and Hensher, D., 2003 The mobility and accessibility expectations of seniors in an ageing population, Transportation Research: Part A, 37, 903–916.

Buchanan, N., et al., 2005 Transport Disadvantage and Social Status: A Gold Coast Pilot Research Project, Research Monograph No. 8, Urban Research Program, Griffith University, Brisbane.

Banister, D., and Bowling, A., 2004 Quality of life for elderly: the transport dimension, Transport Policy, 11:105–115.

Beed, C., 1981 Melbourne's Development and Planning, Clewara Press, Melbourne.

Berry, M., 2007 Ageing in Space: Transport, Access and Urban Form, In Borowski, A., Encel, S., and Ozanne, E., Eds., Longevity and Social Change in Australia, University of New South Wales Press, Kensington.

Browning, C., and Sims, J., 2007 Ageing without Driving: Keeping Older People Connected: In Currie, G., Stanley, J., and Stanley, J., 2007 Eds., No Way to Go: Transport and Social Disadvantage in Australian Communities, Monash University ePress, Clayton.

Burns, P. 1999 Navigation and the mobility of older drivers, The Journal of Gerontology, 54B (1) 49–55.

Cantor, M. 1979 Neighbors and Friends, Research on Aging, 1 (4), 434–463.

Cheal, C., 2003 Transit Rich or Transit Poor: Is Public Transport Policy in Melbourne Exacerbating Social Disadvantage? Unpublished Honours thesis, Faculty of Architecture, Building and Planning, University of Melbourne, Melbourne.

Church, A., et al., 2001 Transport and social exclusion in London, Journal of Transport Policy, 7 (3), 195–205.

Coates, B., et al., 1977 Geography and Inequality, Oxford University Press, Oxford.

Currie, G., 2004. Gap analysis of public transport needs: measuring spatial distribution of public transport needs and identifying gaps in the quality of public transport provision. Transportation Research Record 1895, 137–146.

Currie, G., and Senbergs, Z., 2007 Identifying Spatial Gaps in Public Transport Provision for Socially Disadvantaged Australians: The Melbourne 'Needs Gap' Study, paper delivered at the 30th Australasian Transport Research Forum, Melbourne.

Currie, G., et al., 2007 Eds., No Way to Go: Transport and Social Disadvantage in Australian Communities, Monash University ePress, Clayton.

Currie, G. et al. 2009 Investigating links between transport disadvantage, social exclusion and well-being in MelbournePreliminary results Transport Policy 16:97–105

Currie, G., 2009. Australian Urban Transport and Social Disadvantage. Australian Economic Review Special Forum Edition on Urban Transport, The Australian Economic Review, 42 (2): 201–208.

Currie, G. 2009 Quantifying spatial gaps in public transport supply based on social needs, Journal of Transport Geography 18(1):31–41.

Currie G and Delbosc A., 2010 Exploring Public Transport Usage Trends in an Ageing Population' Transportation Vol 37 pp 151–164.

Davison, G., 2004 Car Wars: How the Car Won Our Hearts and Conquered Our Cities, Allen & Unwin, Sydney.

DETR., 2000 Social Exclusion and the Provision and Availability of Public Transport, DETR, London.

Dodson, J., et al., 2004 Transport Disadvantage and Social Status: A Review of literature and methods, Research Monograph 5, Urban Policy Program, Griffith University, Brisbane.

Dodson, J., and Sipe, N., 2008 Shocking the Suburbs: Oil vulnerability in the Australian city, University of New South Wales Press, Sydney.

Engels, B., Nissim, R., and Landvogt, K. 2009 Under Pressure: Costs of Living, Financial Hardship and Emergency Relief in Victoria, Emergency Relief Victoria and the Victorian Council of Social Services, Melbourne.

Favreau, L., and Spear, R., 2001 Tackling Social Exclusion in Europe, Ashgate Pub., London.

Fober, S., and Grotz, R., 2006 Everybday Mobility of Elderly people in Different Urban Settings: The Example of the City of Bonn, Germany, Urban Studies, 43 (1), 99–118.

Foley, D., et al., 2002 Driving life expectancy of persons aged 70 years and older in the United States, Amercian Journal of Public Health, 92 (8), 1284–1289.

Gobbi, H. 2004 Taking its Place: A History of Oakleigh, The Oakleigh and District Historical Society, Oakleigh.

Golant, S., 1984 A Place to Grow Old: The Meaning of Environment in Old Age, Columbia University Press, New York.

Golob, T., and Hensher, D., 2007 The trip chaining activity of Sydney residents: a cross-section assessment by age group with a focus on seniors, Journal of Transport Geography, 15, 298–312.

Hensher, D., 2007 Social exclusion: Informed reality thinking on accessibility and mobility in an ageing population, In: Currie, G., Stanley, J., and Stanley, J., Eds., No way to Go: Transport and Social Disadvantage in Australian Communities, Monash University ePress, Clayton.

Herbert, D., and Smith, D., 1979 Eds., Social Problems and the City: Geographical Perspectives, Oxford University Press, Oxford.

Hine, J., and Grieco, M., 2003 Scatters and clusters in time and space; implications for delivering integrated and inclusive transport, Transport Policy, 10, 299–306.

Hine, J., and Mitchell, F., 2001 Better for Everyone? Travel experiences and transport exclusion, Urban Studies, 38 (2) 319–322.

Hine, J., and Mitchell, F., 2003 Transport Disadvantage and Social Exclusion: Exclusionary Mechanisms in Transport in Urban Scotland, Ashgate Publishing Limited, Aldershot.

Hunri, A., 2006 Transport and social exclusion in Western Sydney, Western Sydney Community Forum and University of Western Sydney, Parramatta.

King, R., 1979 Accessibility and Behaviour in Twenty Areas of Melbourne, Research Report, Australian Road Research Board, Melbourne.

King, R., 1981 Interest Rates, Energy and House Prices: Some Aspects of the Melbourne Housing Market, 1966–80, University of Melbourne, Melbourne.

Kirby, A., 1982 The Politics of Location: An Introduction, Methuen, London.

Knox, P., 1995 Urban Social Geography: An Introduction, 3rd edition, Longman, Harlow.

Lawton, M., 1980 Environment and Aging, Wadsworth, San Francisco.

Littlewood, P., 1999 Social Exclusion in Europe: Problems and Paradigms, Ashgate Publishing, London.

Maher, C. and O'Conner, K., 1978 Journey to Work Patterns: Melbourne 1961–1971, Working Paper, No. 6, Department of Geography, Monash University, Melbourne.

Mayers, D., and Salais, E., 2001 Eds., Social Exclusion and European Policy, Edward Elgar Publishing, London.

Mees, P., 2000 A Very Public Solution: Transport in the Dispersed City, Melbourne University Press, Melbourne.

Mollenkopf, H., Marcellini, F., Ruoppila, I., Szeman, Z., and Tacken, M., 2005 Eds., Enhancing Mobility in Later Life: Personal Coping, Environmental Resources and Technical Support, IOS Press, Amsterdam.

Morris, J., and Rosenbloom, S. 1997 Transportation of the Elderly in Melbourne, Driving off into the Sunset: Some Myths about Ageing, Transport Research Centre, RMIT University, Melbourne.

Morris, J. 2007 Maintaining Mobility: The Transition from Driver to Non-Driver, Policy Framework Report, Victorian Department of Infrastructure, Melbourne.

Pantazis, C., Gordon, D., and Levitas, R., 2006 Eds., Poverty and Social Exclusion in Britain: The Millennium Survey, Policy Press, London.

Percy-Smith, J., 2000 Policy Responses to Social Exclusion, Open University Press, Buckingham.

Rosenbloom, S., 2001 Sustainability and automobility among the elderly: An international assessment, Transportation, 28, 375–408.

Salt, B. 2004 The Big Shift: Who we are and where we are headed, 3rd edition, Hardie Grant Books, Melbourne.

Smith, D., 1977 Human Geography: A Welfare Approach, Edward Arnold, London.

Smith, G., 1988 The Spatial Shopping Behaviour of the Urban Elderly: A Review of the Literature, Geoforum, 19 (2), 189–200.

SEU, 2003 Making the Connections: Final Report on Transport and Social Exclusion, Social Exclusion Unit, London.

Victorian Department of Sustainability and Environment, 2006 Melbourne Atlas, Department of Sustainability and Environment, Melbourne.

Vinson, T., 2009 Social Inclusion: A Discussion Paper, Commonwealth Department of Education, Employment and Workplace Relations, Australian Government, Canberra.

WHO, http://www.who.int/ageing/en/ accessed 9/9/2011 4:58:27 PM.

Wilkinson, N., 1984 Conflict over transport in Melbourne in J. Halligan and C. Paris Eds. Australian Urban Politics: Critical Perspectives, Longman Cheshire, Melbourne, pp. 204–221.